Gaining Ground

LIFE OF THE PAST
James O. Farlow, editor

INDIANA
University Press
Bloomington and Indianapolis

Gaining Ground

The Origin and Evolution of Tetrapods

Jennifer A. Clack

This book is a publication of

Indiana University Press
601 North Morton Street
Bloomington, IN 47404-3797 USA

http://iupress.indiana.edu

Telephone orders 800-842-6796
Fax orders 812-855-7931
Orders by e-mail iuporder@indiana.edu

© 2002 by Jennifer A. Clack

All rights reserved

No part of this book may be reproduced or utilized in any form or by any means, electronic or mechanical, including photocopying and recording, or by any information storage and retrieval system, without permission in writing from the publisher. The Association of American University Presses' Resolution on Permissions constitutes the only exception to this prohibition.

The paper used in this publication meets the minimum requirements of American National Standard for Information Sciences—Permanence of Paper for Printed Library Materials, ANSI Z39.48-1984.

Manufactured in the United States of America

Library of Congress Cataloging-in-Publication Data

Clack, Jennifer A., date
  Gaining ground: the origin and evolution of tetrapods / Jennifer A. Clack.
    p. cm. — (Life of the past)
  Includes bibliographical references and index.
  ISBN 0-253-34054-3 (cloth : alk. paper)
  1. Lungfishes, Fossil. 2. Amphibians, Fossil.
3. Leg—Evolution. 4. Paleontology—Devonian.
5. Paleontology—Carboniferous. I. Title.
II. Series.
  QE852.D5 C57 2002
    566—dc21
                        2001004783

1 2 3 4 5   07 06 05 04 03 02

CONTENTS

Acknowledgments vii

Abbreviations ix

1 • Introduction   1

2 • Skulls and Skeletons in Transition   20

3 • Relationships and Relatives: The Lobe-Fin Family   46

4 • Setting the Scene: The Devonian World   78

5 • The First Feet: Tetrapods of the Famennian   105

6 • From Fins to Feet: Transformation and Transition   139

7 • Emerging into the Carboniferous: The First Phase   191

8 • East Kirkton and the Roots of the Modern Family Tree   212

9 • The Late Carboniferous: Expanding Horizons   234

10 • Gaining Ground: The Evolution of Terrestriality   278

References   333

Index   353

Acknowledgments

This work essentially represents a summary of my research career in early tetrapod paleontology. Throughout this period, I have worked with, been guided and helped by, learned from, and been supported by a large number of people and institutions. I now have the opportunity to thank them.

I begin with my father, Ernest Agnew, and my husband, Rob Clack, both of whom have been instrumental in encouraging me on this road and have helped reduce the rather long odds against my arriving where I am. It is also a delight to acknowledge the encouragement of an understanding mother-in-law, Molly Clarke.

I belong to the "Panchen" school of early tetrapod paleontology, and thanks must go largely to Dr. Alec Panchen for having taken me on as a "mature" research student after several years' work in a provincial museum. My boss at that time, Mrs. Anna Meredith, deserves my eternal thanks for having encouraged me to take the plunge and return to academia, as well as providing the opportunity that led to that outcome.

During my subsequent career, several colleagues have been key sources of advice, discussion, intellectual challenge, and stimulation, as well as providers of specimens and general paleontological and social fun: Drs. Mike Coates, Andrew Milner, Angela Milner, and Tim Smithson, all Panchen school graduates; my own research students, but particularly Per Ahlberg, whose research overlaps mine in a most satisfactory and fruitful way. At the University Museum of Zoology, Cambridge (UMZC), I am indebted to Dr. Ken Joysey, its former director, for crucial help in mounting the first Greenland expedition; to its current director, Professor Michael Akam; the curator of vertebrates, Dr. Adrian Friday; and the head of the Department of Zoology, Professor Malcolm Burrows, for continuing support both for my research and my personal progress. Mr. Ray Symonds, the collections manager of UMZC, has been a source of practical support, good-humored in even the most trying circumstances, and I thank him and his staff. Several people have assisted practically with the project over the years, including Miss Rosie Rush, Dr. Nick Fraser, and currently Dr. Henning Blom, our postdoctoral assistant.

My expeditions to Greenland could not have even begun had not Rob "mithered" me until I gave in, and nothing would have progressed without the help of Dr. Peter Friend of the Department of Earth Sciences, Cambridge, and the unwitting help of his former student John Nicholson. The expeditions were supported logistically by the Greenland Geological Survey of Denmark, led by Dr. Neils Henricksen (Oscar) and facilitated by Dr. Svend Erick Bendix Almgreen of the Geological Museum Copenhagen (MGUH). I thank both of them for the part they played in our expeditions' successes. Sally Neininger and Becky Hitchin were two of my gallant field assistants in 1998. More recent collaboration with MGUH through the good offices of Dr. Minik Rosing, its director, and Professor Dave Harper, professor of palaeontology, have allowed my team to continue working on the Devonian tetrapods we find so rewarding and fascinating.

In the United Kingdom, much of my material has been provided by the collector Mr. Stan Wood, and I record here the debt British vertebrate paleontology owes to him.

Further abroad, many colleagues have helped me by generously discussing their own research and allowing access to collections, and providing support in more subtle ways. I particularly wish to thank Professor Eric Lombard and Dr. John Bolt from Chicago, for ongoing collaboration and lots of gin and tonics; Professors Bob Carroll and Robert Reisz in Canada; and Drs. Anne Warren and Susan Turner in Australia. Other museums and their staff members have given me generous access to their collections and specimens: in the United Kingdom, Drs. Mike Taylor and Bobby Paton of the National Museums of Scotland; Dr. Neil Clarke of the Hunterian Museum, Glasgow; Mr. Steve McClean of the Hancock Museum, Tyne and Wear; Mrs. Sandra Chapman of the Natural History Museum, London; in the United States, Dr. Dave Berman of the Carnegie Museum, Pittsburgh; Dr. Ted Daeschler of the National Academy of Sciences, Philadelphia; Dr. Gene Gaffney of the American Museum of Natural History, New York; Dr. Farish Jenkins of the Museum of Comparative Zoology, Harvard, Cambridge, Mass.; and in Eastern Europe, Dr. Erwins Luksevics of the Natural History Museum in Riga and Dr. Oleg Lebedev of the Paleontological Institute in Moscow. Many other colleagues and friends directly or indirectly helped in my work on early tetrapods, and I thank them all.

The other side of my research interest, the evolution of hearing, has been greatly influenced by two people, Dr. Art Popper, who encouraged me into the world of auditory neurophysiology, and Dr. Christopher Platt, my close friend and colleague. I thank them both for welcoming a paleontologist into their world.

My research has benefited from funding from the following bodies, which I gratefully acknowledge: Natural Environment Research Council, UK; National Geographic Society; Newnham College and the Department of Zoology, University of Cambridge; UMZC; Isaac Newton Trust Fund, Cambridge; and Copenhagen Biosystematics Centre.

Specific thanks in the production of this book go to Drs. Chris Berry for help with Devonian plants, Mike Coates for help with *Hox* genes, Andrew Milner for helpful comments after reading the book in manuscript, and my editors, Jim Farlow and Bob Sloan.

Finally, I would like to draw particular attention to the part played by my preparator for more than a decade, Mrs. Sarah Finney, without whose patient, exquisite, and delicate work on the fossil material this work could hardly have happened. She also produced many of the photographs that appear in this book. Early tetrapod paleontology owes her a great debt of gratitude.

Abbreviations

AMNH	American Museum of Natural History, New York, New York, USA
ARM	Dr. Andrew Milner
CM	Carnegie Museum, Pittsburgh, Pennsylvania, USA
GLAHM	Hunterian Museum, Glasgow, Scotland, UK
MCZ	Museum of Comparative Zoology, Harvard, Cambridge, Massachusetts, USA
MGUH	Museum Geologicum Universitatis Hafniensis, (Geological Museum, University of Copenhagen), Denmark: f.n. = field number; VP = Vertebrate Palaeontology collections
NEWHM	Hancock Museum, Tyne and Wear, England, UK
NMS	National Museums of Scotland, Edinburgh, Scotland, UK
RNGC	Mr. Robert Clack
SLN	Miss Sally Neininger
SMF	Mrs. Sarah Finney
UCLA	University College of Los Angeles, Los Angeles, California, USA
UMZC	University Museum of Zoology, Cambridge, England, UK; GN = gnathostome; T = tetrapod
WDIR	Dr. Ian Rolfe
YPM	Yale Peabody Museum, New Haven, Connecticut, USA

Gaining Ground

One
Introduction

The Origin and Evolution of Tetrapods

About 370 million years ago, something strange and significant happened on Earth. That time, part of an interval of Earth's history called the Devonian Period by scientists such as geologists and paleontologists, is known popularly as the Age of Fishes. After about 200 million years of earlier evolution, the vertebrates—animals with backbones—had produced an explosion of fishlike animals that lived in the lakes, rivers, lagoons, and estuaries of the time. The strange thing that happened during the later parts of the Devonian period is that some of these fishlike animals evolved limbs with digits—fingers and toes. Over the ensuing 350 million years or so, these so-called tetrapods gradually evolved from their aquatic ancestry into walking terrestrial vertebrates, and these have dominated the land since their own explosive radiation allowed them to colonize and exploit the land and its opportunities. The tetrapods, with their limbs, fingers, and toes, include humans, so this distant Devonian event is profoundly significant for humans as well as for the planet.

Today, the modern descendants of these early pioneers are divided into two major groups (Fig. 1.1). The modern amphibians include frogs (anurans—jumping, tailless amphibians), salamanders (urodeles—tailed amphibians), and caecilians (apodans—elongate, limbless amphibians). The

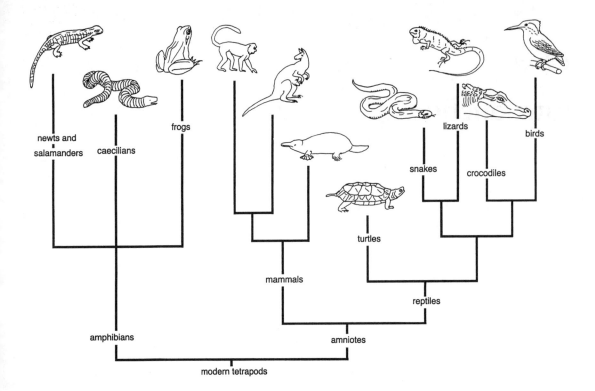

Figure 1.1. Family tree of the living tetrapod groups.

modern amniotes include mammals (which usually have fur and which produce milk, humans among them), turtles, crocodiles and their relatives the birds, and lizards and their relatives the snakes. (Note that the term *reptile* can be used to include all amniotes except mammals, provided birds are included in this group.) Tetrapods include any animal with four legs or whose ancestor had four legs, in practice. (The situation is a bit more difficult with fossils, as will be shown.) These two major radiations of vertebrates also had many relatives, many of which are now extinct—notably the dinosaurs, whose closest living relatives are the birds (see below).

Because becoming terrestrial was apparently a slow process, I have called this book *Gaining Ground* to suggest that it was achieved with some difficulty. At the same time, it has led to enormous innovations in evolutionary terms, and I hope to suggest not only something of the breadth of possibilities that it opened up, but also the contingent nature of the transition. In other words, much of this change occurred because of happenstance—being at the right place at the right time—rather than the result of a directed process.

The origin, early evolution, and relationships of tetrapods form the focus for the interaction of several disciplines. Paleontology (the study of fossils) and the related studies of paleoecology, taphonomy (how the creatures died and became fossilized), and paleobiogeography (where the creatures lived and how they were distributed in time and space), as well as modern zoology, anatomy, and developmental and molecular genetics all contribute various aspects. Most people are aware that at some stage creatures crawled out of the water and came onto land and thus can relate to the contents of this book. I hope that this book will both show how much more can be said than that, and how much more there is to know.

The study of the origin of tetrapods has gone through many phases in its history, but none has been more exciting than that of the present day. Over very recent years, more fossil material from this crucial period has been unearthed than at any time in the past, and these discoveries have helped to reshape ideas about when, where, how, and even possibly why the transition occurred at all. In the not-too-distant past, there was almost no fossil material, and ideas were based largely on informed guesswork. Speculation was intense, and as is often the case, in inverse proportion to the amount of data. To be truthful, there is still not much real data, so that speculation is still active, and whatever is concluded today may be overturned by the discovery of a new fossil tomorrow. That in some sense is to be hoped for, because only in that way can guesses be falsified and tested as scientific hypotheses.

This book tells the story of the evolution of tetrapods from their fish ancestry and puts the sequence of events into its ecological context. The story is founded on an understanding of the evolutionary relationships between tetrapods and their fishy relatives—their phylogeny—and traces the family tree of tetrapods from its roots to the point at which the major groups of modern tetrapods branch off from its original trunk. The tetrapod family tree is in fact more like a bush, with several main branches, some of which have died out during the course of evolution and some of which have become large and important from small beginnings.

This book looks at the changes that occurred in the transition from creatures with fins and scales to those with limbs and digits in an attempt to understand how, as well as when, the changes occurred, and to do this, it is necessary to understand something of the anatomy of the animals involved. Chapters 2 and 3 are devoted to these parts of the story. Chapters 4, 5, and 6 set out what is currently known of the earliest tetrapods and their lifestyles. By careful analysis of what is known of them from fossils, and by comparison with modern animals that live at the transition between water and land, it may be possible to understand a little of how the early tetrapods worked as animals. After the tetrapods had become established, they radiated into a range of forms requiring modification of the original tetrapod pattern. Chapters 7, 8, and 9 carry the story forward from the origin of tetrapods to their ultimate conquest of terrestrial living. The final chapter draws together some of the threads that have been taken up in the preceding chapters and shows how they impact the study and understanding of tetrapods today.

I hope that this book brings the excitement of this field of study to a wider public, shows something of how paleontology progresses and what it can and cannot do, and of course, most importantly, shows people a little more of how they fit into the broader picture of evolution.

Tools of the Trade: The Geological Framework, Fossilization, and Family Trees

The Geological Framework

To put the evolution of terrestrial tetrapods in its context, it is necessary to have an understanding of Earth's history in general outline. This is not the place to discuss dating methods or techniques of stratigraphical correlation, and these can be found in readily available geological textbooks such as those by Briggs and Crowther (1990) or Raup and Stanley (1978). However, it is necessary to explain the approximate dates and approximate lengths of time over which the story takes place so that it can be put in the context of other major events in the story of evolution.

The geological column is the name that scientists give to the succession of times, dates, and names into which Earth's history is divided. There are several ways of expressing this. It can be expressed in a way that accords each time interval a space proportional to its length, usually as a vertical column, always with the oldest at the bottom, or as a sort of clock face. The problem with this method is that only a small proportion of Earth's known history is represented by an abundant fossil record. The planet is estimated to be about 4500 million years old, and the first signs of life (fossil bacteria) are dated at about 3500 million years. Complex multicellular animals first appear commonly in the fossil record only about 550 million years ago, so that to use the clock face method has practical problems in that most of it would effectively be empty. Another way is simply to set out the list of dates and names in their relative order, again with the oldest at the bottom, and this is the way shown in Figure 1.2. The numbers show the dates of the boundaries between the divisions and the lengths of time for which they lasted.

The idea that the Earth is as old as this is a relatively recent one, dating back only to the early 19th century, and its appreciation has changed the perspective from which we view our place in its history. The concept has been called "deep time." As an example, one of the important factors that study of deep time reveals is the complexity of climate change through Earth's history, culminating in the appreciation of the possibility of human-induced global warming. This would not be possible without study of the Earth's climate over the past few million years. Talking of perspectives, when considering the period in Earth's history covered by this book, climate changes far more radical than recent ones are obvious. If global warming continues as predicted, eventually the climate may become something like it was about 15 million years ago in the Miocene period, but it will be a long way from that which current study suggests prevailed when early tetrapods were alive.

The interval for which there are abundant fossils in the rocks is called the Phanerozoic, meaning "visible life," and it represents a time of about 600 million years. The Phanerozoic is divided into three eras, originally named according to what proportion of its biota resembled that of the modern world. These divisions are named the Paleozoic ("ancient life"), Mesozoic ("middle life"), and Cenozoic ("recent life"). The ages are divided into periods and the periods into stages. To a large extent, the boundaries of the divisions are based on the fossils of animals and plants that lived at that time, although the names they receive do not necessarily reflect this. Names of the stages, for example, are often based on where the representative strata were first found or on where they are most clearly seen. These large time periods, rock sequences, and their names, as well as the basic faunal complement of each, were worked out during the 19th century and have not changed very much since then. What has happened over subsequent decades is a process of refinement, increasing resolution of time intervals, and precision of dating and correlation between sequences in different parts of the world.

The story of the origin and early evolution of tetrapods and the timeframe of this book takes place almost entirely within the later part of the Paleozoic, during the Devonian, Carboniferous, and Permian periods

Figure 1.2. (opposite page) Time scale showing the time of origin of major groups and other events in Earth history. The shaded area indicates the period covered in this book.

	Millions of years ago		Event
Recent	1		● humans, 100,000 years ago
C E N O Z O I C	65		major diversification of mammals **Major extinction event**
M E S O Z O I C	144	CRETACEOUS	● last dinosaurs
	213	JURASSIC	● earliest lizard ● earliest bird, *Archaeopteryx* ● earliest frogs and salamanders
	248	TRIASSIC	earliest crocodile ancestors earliest dinosaurs earliest turtles earliest mammals
			Major extinction event
P A L E O Z O I C	286	PERMIAN	major diversification of amniotes
	320	Upper CARBONIFEROUS	● earliest true amniotes major diversification of tetrapods *Crassigyrinus, Whatcheeria*
	360	Lower	East Kirkton *Lethiscus*, adelogyrinids, *Casineria* ● Tournaisian tetrapod
	408	DEVONIAN	*Acanthostega, Ichthyostega* earliest known tetrapods earliest tetrapod-like fish, Escuminac Bay, *Eusthenopteron*
	438	SILURIAN	● early land plants
	505	ORDOVICIAN	
	590	CAMBRIAN	● earliest vertebrate ● Major radiation of multicellular life-forms

Introduction • 5

(Fig. 1.2). More details of the stages into which these periods are divided are given in later chapters. The story really begins about 370 million years ago, although some chapters set the scene by describing the history of plants and animals that were already present as the tetrapods started their evolutionary journey. It takes the story through about 122 million years to a time about 248 million years ago as the Permian period comes to a close. For comparison, the first dinosaur is dated at around 225 million years, and the last died out 65 million years ago, a comparable period of time. The earliest tetrapods that feature in the story are nearly twice as old as the oldest dinosaur. Humans can trace their lineage back to a split from the common ancestor of apes and humans about 5 million years ago; *Homo sapiens* as a species is currently reckoned to be about 100,000 years old.

Fossils and Fossilization

The only means of finding out about animals and plants that lived so long ago is from fossils. These are the preserved remains or traces of these ancient organisms, and to understand the story more fully, it is necessary to look at how and under what circumstances these remains are preserved and discovered.

Types of fossils can be categorized in a variety of ways: according to what is preserved, or according to how it is preserved. What is preserved is usually the harder parts of an animal or plant; for example, bones or shells make good fossils. This type of fossil is often called a body fossil, to distinguish it from another category, that of trace fossil. Trace fossils are preserved impressions of features that an animal (usually) has made—for example, footprints or burrows. Body fossils show the anatomy of the plant or animal it preserves, and from this, it is possible to work out something of its evolutionary relationships and functional morphology. Trace fossils can sometimes be even more telling in that they can provide information about the behavior of the animal and clues to its lifestyle that body fossils cannot give. Both kinds of fossil are known in the story of the origin of tetrapods.

Body fossils are preserved in many different ways. Usually the process involves water, and fossils of aquatic animals are much more common than those of terrestrial animals. Fossils of terrestrial animals are usually found only when the creatures' bodies have been accidentally washed into bodies of water. Generally, the animal sinks to the bottom of the lake or sea and the soft, fleshy parts usually decay quite rapidly. The remains may be scavenged by other animals and disintegrate so that bones become isolated, but gradually they are covered with sediment that over the millennia hardens to preserve the bones. Most favorable to preservation are deep, still waters where the sediment particle size is small. Sediments can then take up small details of the bones or shells by filling in even the smallest crevices, and then not being disturbed again. If decay happens more slowly—for example, in water low in oxygen, where predators are few and bacterial action is slow—the carcasses may be preserved in a more complete form.

The remains may be preserved subsequently in a variety of ways. They may be more or less unaltered. Shells made of calcium carbonate, for example, may retain the same chemical structure they had in life if they are preserved in limestone. Bones, which are the main concern here, are formed of calcium phosphate, which may also be unchanged chemically in many instances. However, bones are not solid but have pores, or spaces, in them for blood vessels, nerves, and fluid or even air, to allow them to grow, and

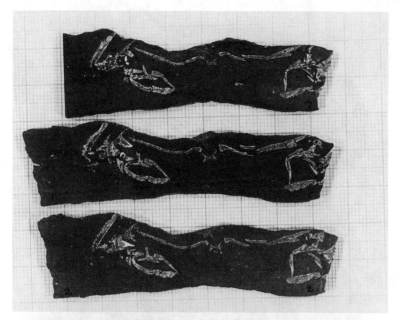

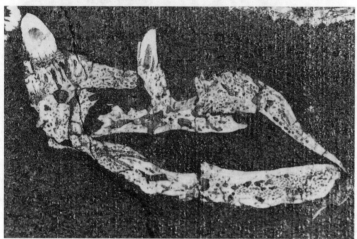

Figure 1.3. Photograph of sections through a skull of Acanthostega *(MGUH f.n. 1604), placed on a centimeter graph for scale. The palate and both lower jaws can be seen in the sections. The enlargement below shows dentary and coronoid teeth. Photograph from UMZC archives.*

to make them lighter and stronger than solid bone would be. During fossilization, these spaces are often suffused with solutions of chemicals that later precipitate out and harden, so that the fine internal structure of the bone is preserved. This is why fossil bone is often much heavier than recently dead bone. Figure 1.3 shows sections through part of the skull of *Acanthostega,* a Devonian tetrapod, to show how well the internal structure of the bone can sometimes be preserved. The upper picture shows the faces of three consecutive slices through a skull of *Acanthostega* (the skull roof is missing in the sections), and the lower picture shows a close-up of a section through one of the lower jaws. This sort of detail can be used to explore aspects such as cell size, growth rates, annual or seasonal cycles, and microarchitecture.

Figure 1.4. Two photographs of "Grace" (MGUH f.n. 1300, Acanthostega). Anterior is to the right. Photographs by S.M.F. Scale = 10 mm.

Sometimes even the hard parts of the bone are dissolved away, leaving a natural mold of the original. This may be preserved as a space in the rock, or it may be replaced by another mineral to form a natural cast. Other hard parts of organisms that can be preserved are the chitin of arthropods, the silica spicules of sponges, the woody (cellulose) parts of plants, and the teeth of vertebrates, the enamel layer of which is often completely unaltered chemically even in ancient specimens.

Occasionally, the remains have been buried and preserved so rapidly or in such unusual circumstances that decay has not had time to occur fully. In some very exceptional cases, even the softer parts of the organism may be fossilized. Some fish fossils, for example, retain details of the gill structure, and in such cases, muscle tissue shows such fine detail that individual cells can be identified. More commonly, soft tissue may be colonized by bacteria in the decay process, and they may take up the shape of body outlines, fur or feathers, or gut contents.

For ancient fossils such as those from the Paleozoic, the risk is high that during the period from their preservation to their discovery, the fossils will have been altered in many different ways: by dissolution by acids that remove calcium and allow bones to bend into unnatural shapes without breaking; by rock movements, faulting, or heat from igneous activity that can distort and fracture them; or by exposure of the sediments to erosion in the air, which may remove or degrade them. Figure 1.4 shows a skull of *Acanthostega* from above and from the side to show examples of the sort of distortion that can take place (compare this with the reconstruction based on examination of several specimens, shown in Chapter 5, especially Fig. 5.20). Fossils may have disappeared when their sediments were subducted under other continental plates by tectonic movements. In short, fossils from earlier periods in Earth's history are usually rarer and less well preserved than those of more recent periods.

Sometimes fossils can be preserved without the aid of water-borne chemicals or sediments. Amber is a resin produced by certain trees that occasionally traps insects or even small vertebrates such as frogs or lizards. When the amber is eventually hardened and preserved (again, usually in a water-laid sediment), access is even sometimes possible to the molecular structure of the animals. Unfortunately for the story of early tetrapods, amber-producing trees had not evolved in the Paleozoic. Sometimes animals die in arid conditions, such as deserts, and their hard parts become mummified (that is, completely air-dried) before being buried in sand. When the sand hardens to sandstone, the soft tissue is sometimes preserved as well as the bones. Several kinds of dinosaur skin are known from this type of preservation, but so far, no early tetrapods have been found with soft tissue preserved in this way. Further details of how animal remains can be fossilized can be found in general textbooks.

When a fossil is found, it very often remains in its surrounding rock, called matrix, and this has to be removed before study of the animal can begin. This special skill, called preparation, requires dedication and patience because it may take many years to complete a large or difficult specimen. Preparation may be done using mechanical means, such as a fine needle used under a binocular microscope, or by chemical means, such as dissolving away matrix by use of dilute acid. Figure 1.5 shows a preparator using mechanical tools to remove the matrix from a fossil from Greenland. In this case, a dental mallet is being used to chip tiny particles of rock off the fossil. The handpiece converts the drill action of this dentist's apparatus into a reciprocating one, and both the "throw" and rate of the action can be adjusted. Some matrix that adheres strongly to the fossil may require

Figure 1.5. Sarah Finney, the preparator who has done most of the work on the Devonian tetrapod Acanthostega *and many of the early Carboniferous forms described in this book. The photograph is taken in the Palaeontological Laboratory in the Department of Zoology, University of Cambridge, Cambridge, England, UK. Photograph from UMZC archives.*

Figure 1.6. Photograph of "Grace" showing the palatal surface prepared. Note the details of the teeth on the palate and the gill bars. Anterior is to the left. Scale = 10 mm.

finer work with a hand-held mounted needle to remove the last particles grain by grain. For large tracts of bulk matrix, a pneumatic pen, like a tiny jackhammer, is the tool of choice. Figure 1.6 shows the interior of the same skull of *Acanthostega* as in Figure 1.4 in which the palate and internal surface of the lower jaw have been exposed after bisection of the specimen with a diamond wire saw. (The wire in this case had a diameter of only 0.3 mm, so very little of the original specimen was lost in the process.)

In the case of natural molds, specimens may be prepared by the use of a form of silicone or latex rubber to make casts from the original. Sometimes bone is badly preserved, in which case it is occasionally more useful to dissolve away the bone artificially and to use a rubber peel from the resulting mold to perform the study. Some of the Carboniferous fossils described in Chapter 9, such as those from Linton, Ohio, are treated in this way. The surface detail obtained by this method is sometimes so fine that images produced by a scanning electron microscope can give valuable information.

Good articulated (i.e., those still joined together) fossil skeletons of Devonian and Carboniferous tetrapods, those that are described in this book, are very rare. Only a handful of animals are represented by more or less complete skeletons; the rest are represented by individual bones such as lower jaws, partial skull roofs, or limb fragments. Usually the matrix in which they are preserved is hard, and the specimens have taken many hours of effort to extract. Occasionally, this general rule is broken: some early tetrapod fossils from the Baltic states are preserved in sandy material so soft that it can be washed off. The bones from these localities are three-dimensional and pale in color, so that they look almost like modern bone. The downside of this is that because the consolidation of the matrix is so poor, so is the consolidation of the fossil, and this creates a problem with conservation. It is difficult to extract them for the opposite reasons: they frequently fall apart without great care and the use of consolidating plastics.

Understanding Phylogeny

One of the first clues to the idea that life on Earth had evolved by a gradual process of change over time was the discovery that animals and plants could be placed in groups showing a hierarchical order. Small groups of forms showing detailed similarities could be placed with other small groups into larger ones showing more generalized similarities. The arrangement resembles a family tree, and it was this that first suggested the idea of "descent with modification"—the realization that the similarities that exist between more and more inclusive groups of animals result from ancestor–descendant relationships. Although the hierarchical classification of animals and plants had been recognized since the 18th century, and the idea that animals had changed throughout time had been suggested several times, what this meant remained mysterious and debatable until the mid-19th century. At that time, Darwin and Wallace finally realized that it could be explained by genealogical relationships and put forward compelling evidence for the idea. They also independently suggested a mechanism—now known as natural selection—that could have brought about the changes. Everything that has been discovered since in modern genetics and developmental biology has only served to strengthen the evidence for this kind of relatedness among animals and plants. Many recent books discuss the patterns and processes involved in descent with modification, such as that by Jones (1999), which provides an update to Darwin's original observations and expands the evidence he presented to include such topics as genetics, development, and plate tectonics.

The groups are given names that express this relationship. A species is the smallest group, usually meaning a group of organisms that can and do reproduce with one another. (This is not possible for extinct animals, of course, so paleontologists try to use other criteria. It is sometimes very difficult, especially if only fragments of an animal or plant are known.) Each species is given its own unique name. Species are grouped into genera, (singular, genus) and share a generic name. Genera can be grouped into families, families into orders, and orders into classes (Fig. 1.7). In practice, in recent years, categories higher than genus are less used than formerly because they are in many ways incompatible with new methods of classification (see paragraphs below about cladistics).

An animal or plant species will have two names—a binomial. The first is the generic name, and the second is the specific name. For example, one of the animals that features in this book is *Panderichthys rhombolepis*. *Panderichthys* is its generic name and *rhombolepis* is its specific name. Other species of the genus *Panderichthys* exist, such as *stolbovi* and *bystrowi*. Usually the first scientist to describe a new species will name it, unless it turns out to be a member of a genus that is already known, in which case the discoverer will only have to think up a new specific name. Each combination is unique to a species, and the name is designed to reflect something about the animal, such as who found it, where it came from, or some interesting feature of its anatomy. Scientists who name animals often have great fun doing so, although the international rules preclude facetious or vulgar names. Recently, there have been suggestions to abandon this system because it was not designed to fit with recently developed methods

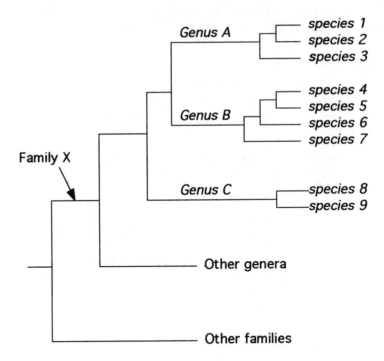

Figure 1.7. Chart showing Linnean hierarchy of species grouped into genera and genera into families.

of working out relationships between groups. However, such suggestions are not universally accepted, and it remains to be seen whether they will stand the test of time and usefulness (see, for example, Benton 2000; de Queiroz and Gauthier 1992; and other references listed in Chapter 3).

To understand the course of evolution, the evolutionary relationships between different groups must be worked out. Phylogeny is the name given to the evolutionary relationships and history of the animal in question, and the practice of working this out is called phylogenetics. The system of groupings in which an animal is placed is called its classification, and one of the main goals of evolutionary studies is to make classification reflect phylogeny.

Two major problems are encountered here. The first is that of discovering the best way of working out the phylogeny, and the second is that the phylogeny that is finally decided to be the best by scientists is often at odds with the classification used in everyday language.

In recent times, one method for working out phylogeny that has become widely accepted is called cladistics. Named from the Greek word for "branch" (a clade), cladistics has some fairly strict rules about how to judge relationships between organisms and uses its own language to express them. Often, commonly used words have a subtly different meaning in cladistics from those in everyday use, which may be confusing for the lay reader. One of the main rules is that only features (characters) that the groups (taxa) share uniquely (shared, derived characters) should be used to assess the relationship between them. This is in contrast to using characters that are more generally present in the larger grouping to which the organism might belong (shared primitive characters).

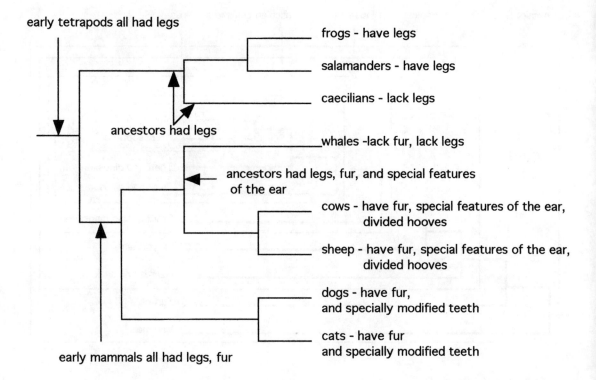

To give an example of this, in a classification of dogs and cats, possession of fur would not be a character used to unite them into a clade, because fur is found in all kinds of other animals. Fur is a character unique to a much larger group, the mammals. Dogs and cats instead share features of the dentition only seen in members of the mammalian order Carnivora. Recently, it has emerged that whales show some specialized characters of the inner ear that are otherwise found only in ungulates—cows, sheep, hippopotamuses, and their relatives. It is thought the most likely explanation is that such characters arose in a common ancestor of all these groups and have been inherited by all the descendants (Fig. 1.8). Another example might be the three modern amphibian orders: frogs, salamanders, and caecilians. The fact that frogs and salamanders have legs and caecilians do not would not be used to suggest that frogs and salamanders were more closely related to each other than either one is to caecilians (Fig. 1.8). Legs are the common property of tetrapods (the main subject of this book) and have been lost by the ancestors of caecilians. (Indeed, some fossil caecilians are now known that have legs.)

Problems arise with this method for two reasons. One is the means by which primitive characters are distinguished from derived ones, and the other is how genuinely shared derived characters (homologies) are distinguished from similarities derived independently (analogies). An example of an analogous character might be the lack of limbs in snakes and caecilians. Such characters may show something about lifestyle, but they show nothing about relationships.

These problems can get complicated, especially when large numbers of extinct taxa are involved. There are usually many characters in the animals

Figure 1.8. Chart showing use of derived characters in phylogeny: even though frogs and salamanders all have legs, these characters are not used to unite them; instead, a suite of others, such as specialized receptors in the inner ear and glands in the skin unique to all modern amphibians, is used; although whales lack legs, they are more closely related to ungulates such as sheep and cows than to carnivores, as shown by shared characters of the ear found only in that clade.

Introduction • 13

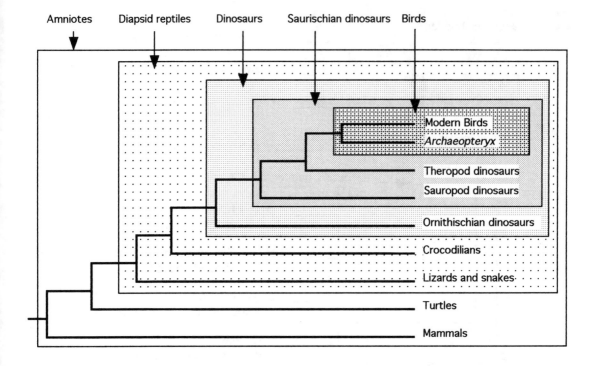

Figure 1.9. Chart showing the phylogeny of birds as a nested group within the dinosaurs and the relationship of dinosaurs to other amniotes.

for which no data exist, and many characters show conflicts (incongruence) in their distribution. In other words, the distribution of primitive and derived characters do not always overlap cleanly and simply in the way one might expect. The answer these days is to use a computer. The taxa are listed along one axis and characters along the other, and a matrix is filled in showing the state of each character found in each taxon. Then the computer can find the branching arrangement (tree) of the taxa that involves the least incongruence and implies the fewest evolutionary changes (the most parsimonious). The tree or trees that are discovered represent the best available hypotheses of relationship drawn from the data used, but are very much provisional. They will be subjected to much testing and are likely to be overturned by the addition of new data.

When a tree like this is drawn up, the task is then to name some of the branches. Because the animals are believed to be united into a clade by shared derived characters, all the members of the clade should be united by the same name to express this relationship, rather than arbitrarily split off some members from their closest relatives. This has led to some problems. For example, most paleontologists now believe that birds are most closely related to a group of small theropod dinosaurs (theropods) (Fig. 1.9). In other words, birds *are* dinosaurs, evolutionarily speaking. Therefore, they should not be promoted to a higher taxonomic status than dinosaurs and distinguished by being called a new class of vertebrates. In old classifications, this was what happened. The new method takes the evolutionary relationships of an animal group to be more important in its classification than its physical appearance.

A similar problem directly concerns the animals in this book. Because I will be dealing with a group of animals (tetrapods) that evolved from

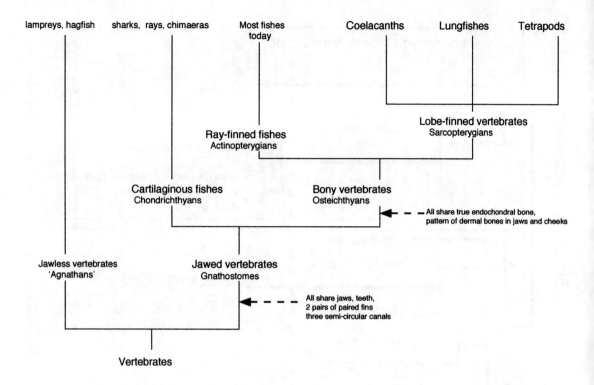

Figure 1.10. Broad phylogeny of vertebrates, showing how lobe-finned and ray-finned vertebrates are all part of the bony vertebrates known as osteichthyans, and how these nest into the larger picture of vertebrates. Some of the shared characters uniting the larger groups are indicated.

within a group of fishes (the lobe-finned fishes, or sarcopterygians), technically, tetrapods are members of the lobe-finned fishes (Figs. 1.10, 1.11). Perhaps counterintuitively, a lungfish (a lobe-finned fish) is more closely related to a cow (a tetrapod) than it is to a salmon (a ray-finned fish); classifications should reflect this. The results of this will be explored in later chapters.

Yet another problem of phylogenetic classification arises because evolution is a continuous process. This means that it is sometimes difficult to know where to draw a line on the classification and give the branch a new name. With tetrapods, the problem has not arisen until very recently. Tetrapods share a number of unique derived characters, most notably limbs with digits, by which they have previously been distinguished. However, this is no longer such an easy distinction, and problems arise in deciding exactly where tetrapods begin (see Chapter 3). There is a useful chapter on cladistics and its methodology in Briggs and Crowther (1990).

The Tetrapods and Their Place in the Family Tree

To understand how tetrapods fit into the pattern of evolution, the first requirement is to understand a little of their relatives, living and extinct. To see which features are unique to tetrapods, distinguishing them from all other vertebrates, it is necessary to recognize those they share with others. These common features betray their ancestry and relationships and are the basic material upon which the forces of evolution have acted to produce all the new features that tetrapods display.

Tetrapods belong to a once large and numerous group of vertebrates known as the sarcopterygians, or Sarcopterygii. This name literally means

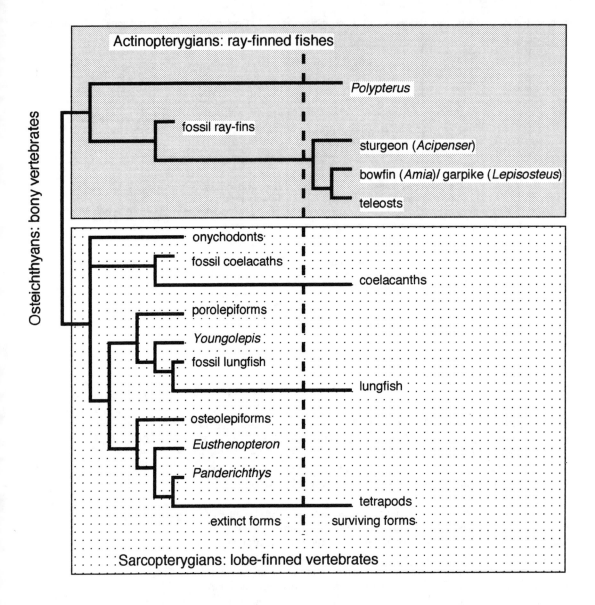

Figure 1.11. Phylogeny of bony vertebrates showing how the modern groups fit with some of their extinct relatives. To the left are extinct forms, and to the right are surviving members. This cladogram shows onychodonts in an uncertain relationship to coelacanths and the remaining lobe-fins. Compare this with Figures 3.1 and 3.25.

"fleshy-finned fishes," though the group is better known as the "lobe-fins." To put these animals into context among their relatives, the wider family tree of the vertebrates as a whole must be considered (Fig. 1.10).

Most fishes, including most modern and extinct ones, carry two pairs of paired fins (except where they have been lost through specialization), a fore pair—the pectorals—and a hind pair—the pelvics—which are carried on bony supports called girdles. These paired fins contrast with the midline fins such as the dorsal, anal, and tail fins, which are single. More details of this construction are given in Chapter 2. Paired appendages such as these are in fact characteristic of a larger, more inclusive group of vertebrates, those whose defining characters also include possession of jaws and teeth. These are the gnathostomes ("jaw holes").

The gnathostomes share a whole suite of complex features not found anywhere else among the animal kingdom; these features seem to be innovations related to their adoption of an active predatory lifestyle. Intimately connected with the evolution of jaws is the possession of a series of jointed gill bars used in the ventilation of a series of paired gill slits piercing the throat or pharynx. The gill filaments used for collecting oxygen from the water lie within pouches, each with its gill bar and set of muscles, nerves, and blood vessels to operate the system.

The gnathostomes include fishes such as sharks, which form their skeletons from cartilage and are called chondrichthyans (cartilaginous fishes), and those such as carp, which form their skeletons from bone and which are known as the osteichthyans (bony vertebrates), or more formally, Osteichthyes. Figure 1.10 gives more details of the osteichthyan family tree.

Bony vertebrates comprise the majority of vertebrate species alive today, and they are united by the fact that their skeletons are composed at least in part of a special kind of bone. This is known as endochondral bone because it has been formed by replacement of a cartilaginous precursor. The bones of human arms and legs, spinal column (backbone), and most of the skull is formed from this kind of bone. It contrasts with a bone formed in the skin without a cartilaginous precursor, called dermal bone. Even the most primitive known vertebrates have some form of this latter kind of bone, although in true osteichythans it is arranged into distinctive patterns that can be recognized throughout the group. Humans retain some of these, such as the tooth-bearing jawbones, the maxilla, premaxilla, and dentary.

Another feature that it seems that early bony vertebrates shared was the ability to form an extra pair of pouches behind the standard set of gill pouches in the throat. This pair was used as an air chamber, and in many early bony vertebrates, they acted essentially as lungs. Some of these animals retained the lungs and used them regularly for breathing air. This needs to be borne in mind when considering the evolution of tetrapods. Air breathing is and was common among bony vertebrates and is not unique to land-living ones.

The animals that belong in the bony vertebrates fall into two groups: the ray-finned fishes (the actinopterygians or Actinopterygii) and the lobe-fins. The difference between them is for the most part quite straightforward and easily recognized. In the ray-finned fishes, the paired fins articulate with the shoulder girdle through a series of parallel bones called radials. These in turn support the bony fin rays, or lepidotrichia. At least in early fossil forms, the fins had a broad base and were not very maneuverable, though this is no longer the case in most modern forms. This group encompasses almost all modern fishes.

The lobe-fins, by contrast, attach their paired fins by a single radial to the shoulder girdle and form the main axis of the fin by stringing the rest of the radials in a chain growing outward from the body. The radials bear muscles running between the bones, thus making the lobe of the fin both segmented and muscular and giving it a narrow, flexible base that allows it to rotate as well as be raised and lowered. It is this formula that humans and other tetrapods share most obviously with their lobe-finned relatives—and one you can contemplate each time you look at your own arm or leg (Fig. 1.12).

Although humans do not usually think of themselves as fishes, they nonetheless share several fundamental characters that unite them inextri-

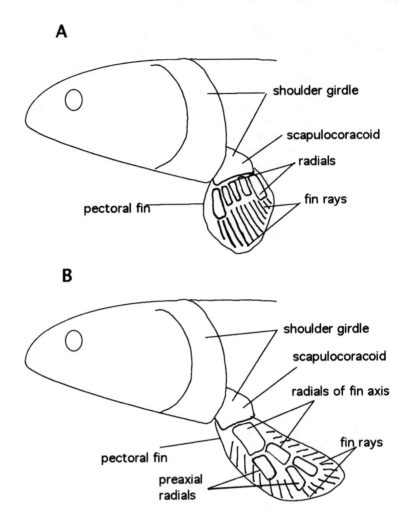

Figure 1.12. Diagram of contrasting fin-support structure for (A) ray-finned and (B) lobe-finned fishes. The pattern in B is that which is found in tetrapods.

cably with their relatives among the fishes. If one of the aims of classifying animals is to reflect their relationships and phylogeny, then inevitably humans and other tetrapods fall within the same grouping as other members showing these characters and sharing the same common ancestor. Thus, humans are mammals and are defined as belonging within mammals, despite what might be seen as a suite of distinguishing characters. In the same way, mammals are tetrapods because they have four legs (or their ancestors did) and a set of other characters that unite them with other tetrapods (amphibians plus amniotes). So humans are also tetrapods. So far, so good. But the next step often causes a problem. In phylogenetic classification, tetrapods are sarcopterygians (lobe-fins). They belong with other forms, the lungfishes and coelacanths and a range of extinct forms, within the lobe-fins. Because "sarcopterygian" is usually translated as "lobe-finned fish," it is sometimes a problem to appreciate that tetrapods really belong here. "Sarcopterygian" is better translated as "fleshy-limbed vertebrate" so this can be more easily understood. Tetrapods did not evolve *from* sarcopterygians; they *are* sarcopterygians, just as one would not say that humans evolved from mammals; they *are* mammals. Similarly, sarcopterygians, along with their sister group the actinopterygians (ray-fins), both belong within the osteichthyans or bony vertebrates, and thus humans are also osteichthyans, even though this word is sometimes translated as "bony fish" (Figs. 1.10, 1.11).

Any speculation about how evolutionary changes ("processes") took place must be set in the context of a testable phylogenetic hypothesis. If not, any ideas about such matters will be set at the level of storytelling and have little or no scientific validity. They may, of course, precipitate a new look at fossil material that may eventually end up with more characters to add to the data matrix and thus a revised phylogeny. This kind of reciprocal enlightenment certainly goes on. However, each time, the phylogeny into which the ideas are nested must be explicit and testable by other people. Discussions of the detailed phylogenies followed here will be found in the appropriate chapters, as will some examples of how different phylogenetic ideas can influence how fossil material is interpreted.

Two
Skulls and Skeletons in Transition

This chapter is an introduction to the skeletal anatomy of animals that exemplify the fish–tetrapod transition. The first part examines how the skulls and skeletons of lobe-finned fishes and tetrapods were built and introduces the terminology used for the bones. Unfortunately, many of the terms will be unfamiliar to the nonspecialist reader, but at least some of them need to be assimilated because in most cases, there simply are no other words available to describe them. Throughout the book, reference will also be made to the embryonic origins of certain tissues, so the second part of the chapter sets out the basics of how a vertebrate embryo forms, where various tissues come from, and what structures each becomes. A section on recent work in developmental genetics, especially on *Hox* genes, is necessary for understanding certain aspects of tetrapod evolution, where work on living animals has had an impact on paleontological studies.

Once the basic anatomy of a fish, such as the Devonian lobe-finned fish *Eusthenopteron,* and a tetrapod such as the Devonian *Acanthostega* or the Carboniferous temnospondyl *Dendrerpeton,* is understood, the differences that came about at the fish–tetrapod transition can be approached. These are outlined in the last part of this chapter. However, the features that fish and tetrapods share must be dealt with before those that separate them can be appreciated. Many of these features can be found in modern vertebrates as well as in very early ones.

Understanding Skeletal Structure

Skull

Dermal Skull Roof

Figure 2.1 shows the skull of the Devonian lobe-finned fish *Eusthenopteron,* which will be used as an exemplar to show fish skull anatomy. Early bony fishes, early tetrapods, and some modern ray-finned fishes share some common features of skull construction, which are considered to be primitive. In these animals, the skull is built of an outer covering of armor called the dermal skull roof, whereas the braincase is a separate, relatively small box containing the brain proper, the ear capsules, and the roots of the nerves serving the other sense organs. Hinged to the back of the dermal skull roof is the lower jaw; the palate forms the roof of the mouth. In most modern fishes and tetrapods, much of the dermal skeleton found in Paleozoic members has been lost, which is why their skulls look rather different from those of these more ancient creatures.

The dermal skull roof can be divided into several regions. At the top and back lies a flat, more or less rectangular lid called the skull table. It is made up of paired bones, sometimes given different names in fish and tetrapods. Reaching out from the front of this and running between the eye sockets, or orbits, runs a series of paired bones forming the interorbital region. In *Eusthenopteron,* the main bones in this region are the parietals surrounding the parietal foramen, sometimes called the pineal foramen. The parietals attach at the front to the snout, which has the external nostrils on each side of it. In *Eusthenopteron,* this is made up of a mosaic of small bones. Running back from the snout beneath the orbits and the skull table is the cheek region. The main cheek bones are the lacrimal, jugal, squamosal, and quadratojugal. At the back of the cheek, which is usually drawn backward well beyond the hind edge of the skull table, is the region where the lower jaw hinges to the skull roof, and this whole back portion of the cheek is sometimes called the suspensorium. Each of these regions consists of several mostly flat, platelike dermal bones joined or sutured together. These form in the skin as the animal grows. The skull roof protects the braincase, which is suspended beneath the skull table. It encloses the eyes and nasal capsules and anchors some of the muscles responsible for closing the jaws. Of critical importance, it also bears the marginal arcade of teeth on the premaxilla and maxilla.

Behind the main part of the skull in fishes, and running beneath the lower jaw, is a series of dermal bones called the operculogular series. Two or more large dermal bones form a flaplike extension to the back of the cheek, protecting the gill (or opercular) chamber and forming a seal for use in the ventilatory cycle. This flap is extended downward as a flexible covering to the throat through an interlinked chain of dermal plates, ending with the gular plates under the chin.

Between the opercular bone and the skull table is a notch where, in *Eusthenopteron,* as in some modern ray-finned fishes, the opening of the spiracle is found. The spiracle is the remnant of the gill pouch once associated with the hyoid arch (see below).

Running over the skull bones at certain points are the lateral line canals. This sensory system is explained in more detail in Chapter 6.

Braincase

The braincase (Figs. 2.1, 2.2) consists of two main parts. The back part is formed from the ear capsules and the occipital region, with which

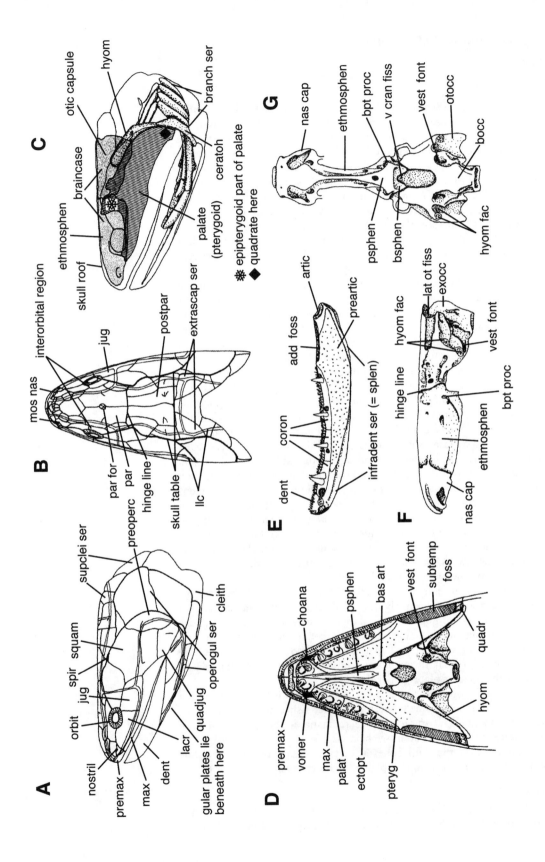

the vertebral column articulates. These form from separate units in the embryo, seen in all modern jawed vertebrates and implied by the construction seen in fossil forms (Fig. 2.2). These are the otic capsules, which house the vestibular system and semicircular canals for monitoring posture, balance, and movement, and the sensory cells for hearing, when these are present; there may or may not be a roof to the tube housing the brain, joining the two otic capsules in the midline. Bones of the otic capsules in the adult are called the pro-otics and opisthotics. Underlying this complex is the occipital arch complex, consisting of the paired parachordal cartilages, and growing up from them is a series of arches or columns that may or may not join up with the otic capsules. In the adult, these structures form the basioccipital, exoccipital, and supraoccipital bones. Otic capsule and occipital arch are separated by a gap called the lateral otic fissure. The complete unit of otic capsules plus occipital regions is called the otoccipital region. It bears facets at which the hyomandibular articulates (see below) and a large unossified gap called the vestibular fontanelle separates the otic and occipital regions.

The front part of the braincase consists of a midline partition separating the eyes and carrying the olfactory nerves to the front of the snout, where the nasal capsules, also part of the braincase, are situated. The front part consists of the ethmosphenoid (ethmosphen) (in fishes) or sphenethmoid (in tetrapods) and a more posteriorly lying portion called the basisphenoid. This bears articulations by which the braincase and palate link up called the basipteryoid process. The front and back parts of the braincase are separated by a gap called the ventral cranial fissure. Unlike the skull roof, the braincase is made of bone that forms from the ossification of a cartilage precursor. This is called endochondral bone, and the braincase is part of the endoskeleton.

Palate

A third component of the skull is the most difficult to explain and visualize, perhaps because it has several different jobs, and its structure is a reflection of its complex history. This is the palatal region (Fig. 2.1). At the front, the palate forms the roof of the mouth and supports the nasal capsules in the snout region. Along its outer edges, it fastens to the inner margins of the cheek. Moving back along the skull, just behind and internal to the eyes, a socket in the palate houses the basipteryoid process, a peg from the braincase, supporting the braincase at a crucial point. Behind this basal articulation, the palatal complex sends a process, the epipterygoid, upward to contact either the otic region of the braincase (marked with a star in Fig. 2.1), or the underneath of the skull roof. In this region, the palate curves in cross section from an approximately vertical component near the midline of the skull to a much less steeply angled or even nearly horizontal plane at the edges. The vertical part stretches backward, in contact with the back edge of the suspensorium, whereas the more horizontal one leaves its contact with the cheek to form a gap between the two regions, through which the jaw muscles pass downward. This is called the subtemporal fossa. At the very back, it is part of the palatal complex that provides the hinge where the lower jaw attached, called the quadrate.

The palatal complex is a mixture of dermal and endochondral components. The major contributor is the dermal pterygoid. This forms the main plate of the palate and the part that clasps the quadrate, called the quadrate ramus. Bordering the pterygoid and linking the palate to the cheek on each side, a row of bones bears an inner row of teeth—the vomer, palatine, and

Figure 2.1. (opposite page) Skull of Eusthenopteron *to show structures.*
(A) Lateral view of skull roof and lower jaw.
(B) Dorsal view.
(C) Lateral view with skull roof bones shown transparent and palate, braincase, and gill arches visible beneath.
(D) Ventral (undersurface) view with braincase in position.
(E) Medial (internal) view of lower jaw.
(F) Lateral view of braincase.
(G) Ventral view of braincase.
Based on Jarvik (1980).

Abbreviations:
add foss = adductor fossa;
artic = articular;
bas art = basal articulation;
bocc = basioccipital;
bpt proc = basipteryoid process;
bsphen = basisphenoid;
branch ser = branchial arches;
ceratoh = ceratohyal;
coron = coronoids;
dent = dentary;
ectopt = ectopterygoid;
ethmosphen = ethmosphenoid;
exocc = exoccipital;
fen vest = fenestra vestibuli;
hyom = hyomandibula;
hyom fac = hyomandibular articulates;
infradent ser = infradentary series;
jug = jugal;
lacr = lacrimal;
lat comm = lateral commissure;
llc = lateral line canals;
lat ot fiss = lateral otic fissure;
max = maxilla;
mos nas = mosaic of small bones;
nas cap = nasal capsules;
operogul ser = operculogular series;
otocc = otoccipital region;
palat = palatine;
psphen = parasphenoid;
par for = parietal foramen;
par = parietals;
posttemp foss = posttemporal fossae;
preart = prearticular;
premax = premaxilla;
pteryg = pterygoid;
quadr = quadrate;
quadjug = quadratojugal;
spir = spiracle;
splen = splenial series;
squam = squamosal;
subtemp foss = subtemporal fossa;
v cran fiss = ventral cranial fissure;
vest font = vestibular fontanelle.

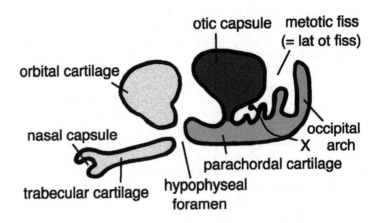

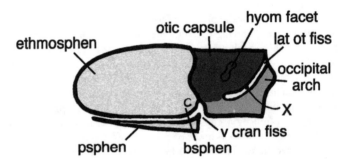

Figure 2.2. Diagram of embryonic braincase components, above, and the regions of the braincase that they become in the adult, below. X marks the exit point and course of Cranial Nerve X, the vagus. Abbreviations as in Figure 2.1.

ectopterygoid, from front to back. The front end of the palate in some species embodies a small endochondral component, but the main endochondral elements are the epipterygoid forming both the vertical (otic) process and the socket for the basal articulation, and the quadrate.

A midline dermal element called the parasphenoid underlies the braincase and forms part of the palate. Like the rest of the palate, it may bear teeth or denticles. At the junction between the premaxilla, maxilla, vomer, and palatine is a hole in the palate known as the internal nostril or choana.

Lower Jaw

The lower jaw is essentially a tube formed from dermal bones, of which some bear the teeth (Fig. 2.1). An outer arcade of teeth on the dentary opposes those of the premaxilla and maxilla and an inner row on a series of coronoids opposes those of the vomer, palatine, and ectopterygoid. Beneath the dentary on the outside lies a series of bones known as the infradentary series in fish, and the splenial series in tetrapods. On the inside the main bone is the prearticular, and it is often covered with denticles to match those on the palate. As with the upper part of the jaw hinge, the lower part is also an endochondral bone, called the articular, sandwiched between inner and outer faces of the dermal jaw bones. In front of the articular is a hollow between the inner and outer faces, called the adductor fossa, into which the jaw muscles inserted. Within the tube in some cases lies a bar of endochondral Meckelian bone.

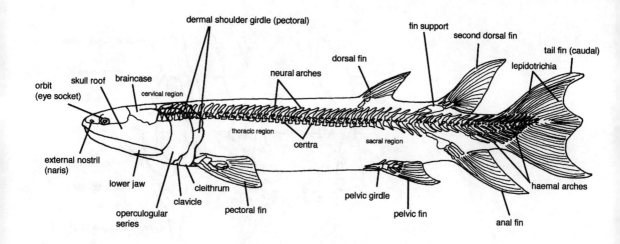

Figure 2.3. Skeleton of Eusthenopteron *showing main features of the skeleton and body.*

Hyobranchial Skeleton and Gill Openings

The hyobranchial skeleton (Fig. 2.1) is another term for the gill-supporting bones and their associated structures. In fishes, there is usually a series of five jointed gill bars, or branchial arches, supporting the gill filaments. The first couple of the series hinge onto the braincase and the rest attach to the more anterior ones. They join at the base, under the "chin" of the fish, to a midline bone. This structure provides a flexible, elastic basket that allows the gills to move in and out and the gill chamber to expand and contract as the fish breathes. In front of the standard set of gill arches, there lies a larger and highly modified set called the hyoid arch. The uppermost element of this set is called the hyomandibula, and the lower part is the ceratohyal. Each gill arch is associated with an opening from the throat region, known as the pharynx, to the outside, through which water passes as it is processed by the fish during breathing. The hyomandibula has a crucial and literally pivotal job to do in the architecture of the skull of most fishes. It was associated in most early fishes with a modified gill opening called the spiracle. In tetrapods, the form of the bone and its function changes and it is known as the stapes. These functional and anatomical differences are explained in more detail below.

The gill and hyoid arches are formed, as with the braincase, from endochondral bone, although in this case, the embryological origin of the cartilage is different from that of most other endoskeletal elements. Its formation is initiated by a unique vertebrate embryonic tissue called neural crest (see below). This tissue is also responsible for bringing about the formation of the quadrate, articular, epipterygoid, and other endochondral parts of the palate that may ossify. For this and other reasons, scientists consider these jaw elements to be members of the same series as the gill arches.

Postcranial Skeleton

The postcranial skeleton in all fishes and tetrapods consists, like the skull, both of endochondral and dermal elements. It can be divided into different regions, shown in the skeleton of *Eusthenopteron* in Figure 2.3.

Skulls and Skeletons in Transition • 25

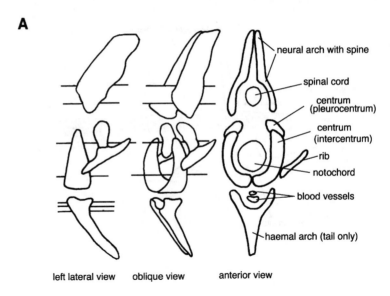

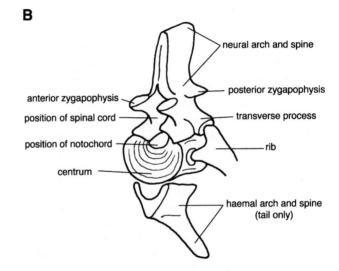

Figure 2.4. Construction of vertebrae in Eusthenopteron *and a tetrapod.*
(A) Vertebrae of Eusthenopteron *in three views.*
(B) Oblique view of a tetrapod vertebrae such as an early amniote.

Axial Skeleton

The axial skeleton consists of the vertebrae, together with, in the case of fishes, the midline fins. A vertebra consists of a neural arch that sits over the spinal or nerve cord and rests on a centrum, which forms around the embryonic supporting bar called the notochord (Fig. 2.4) (see below). Centra may be simply formed from a single element or be a compound of single and paired elements. Neural arches may be joined one to another or not. Ribs may be present, articulating with a neural arch, a centrum, or both. In the tail, hemal arches may be present under the centra, allowing passage of blood vessels serving the posterior part of the body. There may be differentiation of different regions of the vertebral column, such as the cervical region at the front, thoracic in the middle, and sacral near the pelvis, although in fishes, the differences are minimal and the terms are not really applicable. All these elements are endochondral in origin.

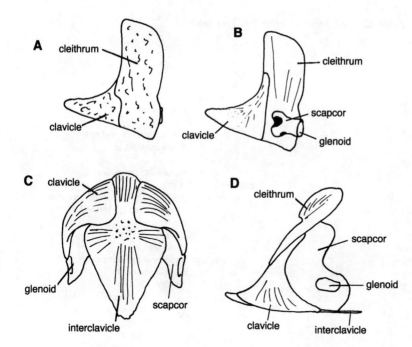

Figure 2.5. Shoulder girdles of Eusthenopteron *and a generalized early tetrapod.*
(A) Left lateral view of cleithrum and clavicle of Eusthenopteron.
(B) Medial view of right cleithrum and clavicle of Eusthenopteron *showing the scapulocoracoid (scapcor) on the internal face.*
(C) Ventral view of the shoulder girdle of an early tetrapod; note the large interclavicle.
(D) Left lateral view of the shoulder girdle of an early tetrapod; note the reduced cleithrum and enlarged scapulocoracoid.

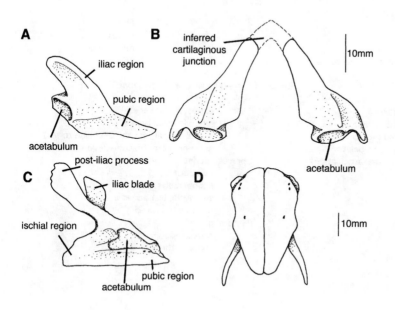

Figure 2.6. Pelvic girdles in Eusthenopteron *and a tetrapod (*Acanthostega*).*
(A) Right lateral view of Eusthenopteron.
(B) Ventral view of Eusthenopteron.
(C) Right lateral view of Acanthostega.
(D) Ventral view of Acanthostega.

Skulls and Skeletons in Transition • 27

Table 2.1 Anatomical Terms for Tetrapods

Skull
- Dermal skull roof
  - Snout
    - nasal
    - lacrimal
    - frontal
    - prefrontal
  - Cheek/suspensorium
    - jugal
    - quadratojugal
    - squamosal
    - postorbital
  - Skull table
    - postfrontal
    - parietal
    - postparietal
    - intertemporal
    - supratemporal
    - tabular
  - Marginal toothbearing bones
    - premaxilla
    - maxilla
- Palate
  - Dermal bones
    - pterygoid
    - vomer
    - ectopterygoid
    - palatine
  - Endochondral bones
    - epipterygoid
    - quadrate
- Braincase
  - Otic capsule
    - proötic
    - opisthotic
  - Occipital arch
    - (supraoccipital in a few forms)
    - exoccipital
    - basioccipital
  - Ethmoid region
    - sphenethmoid
    - basisphenoid
    - basipterygoid process
      - (links to palate
      - via basal articulation)
- Lower jaw
  - External face
    - dentary
    - splenial
    - postsplenial
    - angular
    - surangular
  - Mesial face
    - parasymphysial plate
    - coronoids (usually three)
    - prearticular

Postcranial skeleton
- Axial skeleton
  - vertebrae
    - neural arch
    - centrum
    - haemal arch (in tail only)
    - zygapophyses
  - ribs
  - supraneural radials (primitively)
- Appendicular skeleton
  - Pectoral girdle and limb: dermal components
    - anocleithrum (primitively)
    - cleithrum
    - clavicle
    - interclavicle
  - Pectoral endochondral components
    - scapulocoracoid
    - forelimb
      - humerus
      - radius
      - ulna
      - carpus
      - metacarpus
      - manus
  - Pelvic girdle and limb
    - ilium
    - ischium
    - pubis
    - hindlimb
      - femur
      - tibia
      - fibula
      - tarsus
      - metatarsus
      - pes

Bones of fish skull lost in early tetrapods
- Skull
  - extrascapular series
  - multiple nasal bones
  - lateral rostral (except Devonian forms)
  - anterior tectal (except Devonian forms)
- Opercular series
  - opercular
  - subopercular
  - submandibular series
  - gular plates
- Pectoral girdle
  - posttemporal
  - supracleithrum

Appendicular Skeleton

The appendicular skeleton comprises the limbs and girdles, at the front the shoulder or pectoral set, and at the back, the hip or pelvic set. Figure 2.5 shows views of the shoulder and Figure 2.6 the hip girdles of *Eusthenopteron* and a tetrapod.

The pectoral girdle is a mixture of elements, consisting of dermal bones such as the cleithrum, clavicle, and interclavicle and the endochondral scapulocoracoid. It is this latter bone that provides the articulatory socket, or glenoid, for the pectoral appendages, and it fits within and behind the sheathing series of dermal bones. The pectoral girdle is not attached to the vertebral column except indirectly by soft tissue or through other linking bones. The pelvic girdle is entirely endoskeletal in origin and carries the socket, or acetabulum, for the pelvic appendage. Fishes also have an additional part to the skeleton, the body scales and the bony fin rays or lepidotrichia, which support the fin web. These are all dermal in origin. More details of the structure of the appendages and of the girdles will be dealt with in the section on differences between fishes and tetrapods. A summary of anatomical terms for tetrapods, along with their location and type, is given in Table 2.1.

All the features mentioned in this section are common not only to all lobe-finned fishes and tetrapods, but also to ray-finned fishes. These are the shared inheritance of all bony vertebrates, the osteichthyans.

Developmental and Embryonic Origins

Tissues and Skeletal Structures

In order to understand the physical and developmental relationships among the various types of skeletal tissues and the features that they form, an explanation of their embryonic origins is necessary. Some of the characteristics that demonstrate the phylogenetic relationships between animals are best seen in the embryos of the animals, and the developmental history of some skeletal elements shows something of their commonality of descent. More details of this early stage in animal development will be found in a standard textbook on vertebrate biology, such as that by Kardong (1998), or in a textbook on developmental biology and embryology, such as that by Wolpert et al. (1998).

Each animal starts out as a single cell that, when fertilized, grows into a ball of cells (Fig. 2.7A–D). After it attains a certain size, surface-to-volume ratios come into effect, and to get nutrients into and waste products out of the ball of cells, it has to change shape. It does this by invaginating. In this process, a small hole appears in the ball of cells, and the cells surrounding the hole are swallowed into it, at first creating a dimple, followed by a deeper and deeper hole, called a blastopore (Fig. 2.7E). Cells continue to pour into this hole until eventually, the ball of cells is no longer solid, but hollow (Fig. 2.7F). Even by this early stage, most of the cells have already had their future fates decided, and the ball already has a defined top and bottom, front and back. One important landmark in the ball of cells is the upper region of the margin of the hole, the dorsal lip of the blastopore. As cells pass over this lip to the inside, they are marked out for particular duties in the growing embryo, and the whole of this part of the animal's development is called gastrulation. It is probably the most crucial moment of the animal's life, during which much of its developmental fate is decided. This part of the developmental process in some form is common to all multicellular animals.

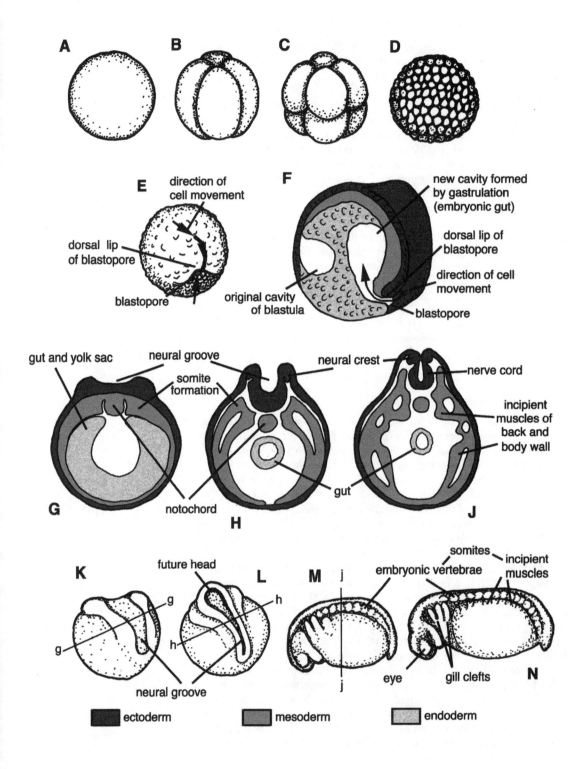

The ball of cells to begin with consists of two layers, and some animals, such as sea anemones, stop there. However, in most animals (called triploblastic), a third region of cells arises in the space between the inner and outer layers. The three cell "layers" (the middle one is more than just a layer) form different parts of the developing embryo: in vertebrates, the outer layer is called the ectoderm and forms the thin outer layer of skin called the epidermis and a number of other features; the inner layer is called the endoderm and forms the lining of the gut and other outgrowths from it such as the liver; the tissue in the middle is called mesoderm, and it forms most of the structures of the body, such as most of the bones, the muscles, and most of the internal organs such as kidneys, blood vessels, heart, and glands (Fig. 2.7G–J).

The next stage in the development of the embryo is called neurulation, and it is at this point that some of the most basic features of all vertebrates develop (Fig. 2.7G–N). From each side of the embryo, two ridges of tissue begin to gather up into parallel crests, with the blastopore at one end (it becomes the anus in vertebrates and a few other animal groups related to them, such as echinoderms). Cells from the outer, ectodermal, layer stream toward the ridges and gather up like waves. The two waves eventually meet and zip together along the embryo's length, forming a loop at one end and petering out at the other. As the two waves meet, they enclose a tube beneath. The tube is the nerve cord and the loop becomes the head, with its swollen end becoming the incipient brain. The ridges are formed within the ectodermal layer, so the odd fact is that the brain and nerve cord are part of the outer surface or ectoderm of the embryo that has become enclosed (Fig. 2.7G–J).

As the embryo grows, beneath the ectodermal nerve cord, the middle layer forms a midline structure of its own. The rodlike notochord forms here, circular in cross section and stretching the length of the embryo except for the front part of the incipient head end. To either side of the notochord, and to an extent regulated by it, the middle layer of mesoderm begins to divide into segments, starting from the head end and working backward (Fig. 2.7). The segments, called somites, become the muscle blocks that lie beside the vertebral column and control its movements. Nerves and blood vessels follow. The limbs or fins develop later from buds out of the lateral wall of the embryo. Bones of the internal limb skeleton, vertebral column (usually), some parts of the skull and braincase, the ribs, and some parts of the limb girdles are formed first in cartilage originating in the mesoderm that is then replaced by bone as the animal grows.

Just behind the mouth of the growing embryo, in the region called the pharynx, a series of pockets arises in the gut lumen, formed by the endodermal tissue. On the outside, a matching series of dimples appears, formed by the ectoderm. Eventually, the pockets meet up with the dimples and they break through the intervening tissue to form perforations. These will become the gill slits.

Although most of the cells of an embryo essentially stay put, obey instructions, and follow an orderly pattern, one population of cells, especially active in the head region, undertakes rather unexpected behavior: they begin to migrate around the body. These are called the neural crest cells (Fig. 2.7J). They are formed just where the embryonic ridges meet and touch, so they are ectodermal in origin, but they migrate into mesodermal structures to reach their destinations. Here, they initiate the development of a whole range of structures that turns out to be unique to vertebrates. The evolution of this highly active tissue probably accounts for the development of vertebrates as a diverse and successful group of animals. Neural

Figure 2.7. (opposite page) Embryonic development of an amphibian from egg to neurula.
(A) Single-cell stage.
(B) Four-cell stage.
(C) Eight-cell stage.
(D) Multicell stage before invagination.
(E) Invagination stage, showing the blastopore and direction of cell flow into it.
(F) Section through the embryo at a slightly later stage than E, showing a section through the blastopore region, the invaginating cells, and the incipient outer and inner layers of the embryo.
(G, H, and J) Sections through a later embryonic stage equivalent to those in K, L, and M, at which the neural tube begins to form, and the three germ layers are established.
(N) Neurula, a stage at which the neural tube is complete and body segments (somites) with incipient vertebrae, muscles, and nerves as well as gill slits have formed. Based on information from Gilbert (2000) and Kardong (1998).

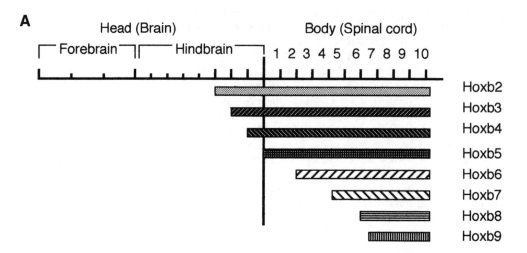

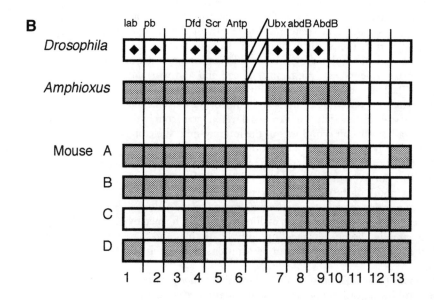

Figure 2.8. Hox genes in vertebrate development. (A) The relative sphere of influence of eight of the Hoxb series of genes in the head and anterior part of the trunk and the more anterior body segments (as well as parts of the hindbrain) that are affected by more anteriorly situated genes.

(B) The comparable segments of Hox genes in the fruit fly Drosophila, the primitive chordate Amphioxus, and the mouse (an amniote). Drosophila and Amphioxus each have one set of such genes, whereas the mouse has four, labeled A, B, C, and D. Homologous genes in Amphioxus and the mouse are shaded. Equivalent genes in the fly are shown with a diamond, and the abbreviations of the gene names are shown above. Based on information in Gilbert (2000).

crest initiates the jaws, the gill arches (including the hyomandibula), the trabecular cartilages of the braincase, the sensory capsules housing the sense organs, the sensory placodes that form the lateral lines, pigment cells, and, acting in concert with mesoderm and normal ectoderm, the teeth, scales, and dermal bones. The difference in origin between dermal and endochondral bone is important, as is the difference between bone formed by neural crest and that formed by mesodermal tissue (all dermal bone is neural crest in origin, but not all neural crest–derived bone is dermal). Their different roles and fates in evolution will become evident as the story of the origin of tetrapods unfolds.

Hox Genes

Some of the insights into changes occurring during the fish–tetrapod transition have come not from fossils or from studies of embryonic development in modern animals, but from discoveries at an even more fundamental level of organization—the genes. Studies of developmental genetics can now be integrated with those of phylogeny and paleontology to reveal a whole realm of new ideas and interpretations. Reference will be made to some of these in Chapters 6 and 10, so at this stage, it is necessary to introduce some of the terminology and explain the concepts (Fig. 2.8A, B). More details will be found in textbooks on developmental biology, such as those by Gilbert (2000) and Wolpert et al. (1998).

Among the most far-reaching discoveries of the past decade has been the existence and function of homeobox genes. These are genes that coordinate some of the most basic processes during the early development of almost all organisms. They determine such basics as the front-to-back and top-to-bottom axes of embryos and the order and position in which certain organs and structures appear. These genes occur on a particular chromosome (or chromosomes), like beads threaded on a string, and some of them are known as *Hox* genes. The genes themselves appear to be related to one another biochemically, but more surprising is that the order in which they occur corresponds to the order in which their effect is seen (that is, the order in which they are expressed) in the animal. For example, genes that occur at the front end of the chromosome (known as the 3' end) have their effect mainly at the front end of the animal, whereas those at the back (known as the 5' end) have their effects further posteriorly. Figure 2.8A shows how members of the *Hoxb* cluster act in sequence along the head and body in the same order as they appear on the chromosome.

Most animals have one or more clusters of such genes, and indeed, plants too have comparable (although not necessarily homologous) ones. Among animals (although only a few have been studied in detail, these genes have been found in a wide variety of invertebrates as well as vertebrates), the genes making up these sequences can be equated quite closely from one animal to another, and groups with similar structure and function have been given names or numbers. Vertebrate *Hox* genes are made up of around 13 groups, each group is given its own number, so that, for example, *Hox6* in a vertebrate can be compared with its equivalent in a worm or a fly (Fig. 2.8B). These are known as paralogous groups. Quite often, similarly positioned paralogous groups have comparable functions in different animals, even those not very closely related. For example, the instruction "make legs" uses a similar paralogous group in the fruit fly *Drosophila* and in tetrapods. The implication is that the genetic instructions go far back in animal evolutionary history, at least to the most recent common ancestor of insects and vertebrates, which cannot be more recent than about 600 million years ago.

Most organisms only have one such cluster of *Hox* genes, but during the evolution of vertebrates, some extraordinary events took place resulting in duplication of the *Hox* string. All modern jawed vertebrates have at least four copies of the string (e.g., mouse, Fig. 2.8B), whereas some, such as zebra fish, have up to seven. The lamprey, a primitive, jawless vertebrate, has only three sets (Sharman and Holland 1998), and one possibility is that there were some duplication events that increased the number of sets in the history of jawed vertebrates (Holland and Garcia-Fernàndez 1996).

There have been duplications as well of the individual genes. It looks as though the duplications have meant that although one set carries out its original function, the new set can go on to acquire new ones without disrupting the development of the animal. This might have allowed more rapid and more profound evolutionary events to have occurred than would otherwise have been possible. In mammals and birds, because of these additional duplications within each set, the four sets each differ very slightly from each other and they are given letters to distinguish them: *Hoxa, Hoxb, Hoxc,* and *Hoxd.* Each set has its own subtly different paralogous groups—*Hoxa13* is different in structure and function from *Hoxb13,* for example. Alongside gene or cluster duplication, sometimes genes have been deleted during the process of evolution.

Each gene makes the instructions for the production of a specific protein whose presence can be detected, by means of several different techniques, in the embryonic animal. It is the detection of these proteins that makes it possible to determine the order and position in which the genes are expressed. There are many other subtleties to the system, however, because whether the products of the *Hox* gene are expressed at any particular point depends on the interplay of several other chemicals. These may be other proteins, such as those known as transcription factors, or more simple chemicals called morphogens. Details of how these work are beyond the scope of this book but may be found in any modern text on developmental biology such as those recommended above. As will be seen, it seems to be changes to the *Hox* genes that have been largely influential in evolutionary events, such as the acquisition of limbs with digits.

How to Turn a Fish into a Tetrapod

In the next section, the basic features of a fish skeleton are compared with those of a tetrapod to see what modifications were made during the transition. The contrasting features of a fish such as *Eusthenopteron* and a tetrapod such as the Devonian tetrapod *Acanthostega* or the early tetrapods *Dendrerpeton* or *Pholiderpeton* (these animals are described in more detail in Chapter 9) can be used to illustrate each end of the spectrum encompassed by the fish–tetrapod transition.

Not surprisingly, many of the characters that show the most difference in the comparison between fishes and tetrapods are those concerned with locomotion and the contrasting needs of movement in water and on land. However, by no means all of the differences are connected with locomotion and support, nor are the most profound changes associated with the postcranial skeleton. Some of the most important differences between fishes and tetrapods lie in the skull structure and have to do with the needs of breathing and feeding. Other differences are harder to account for and may result from some unknown mechanical, behavioral, or genetic reorganization.

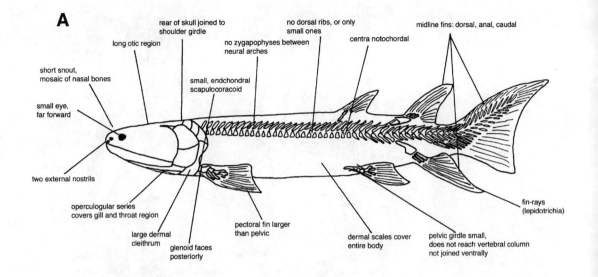

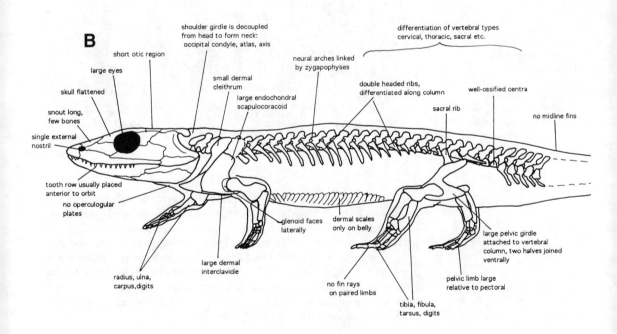

Figure 2.9. Skeleton of (A) a fish (similar to Eusthenopteron *or* Osteolepis*), compared with (B) a tetrapod such as* Dendrerpeton, *with the major differences highlighted.*

Skulls and Skeletons in Transition • 35

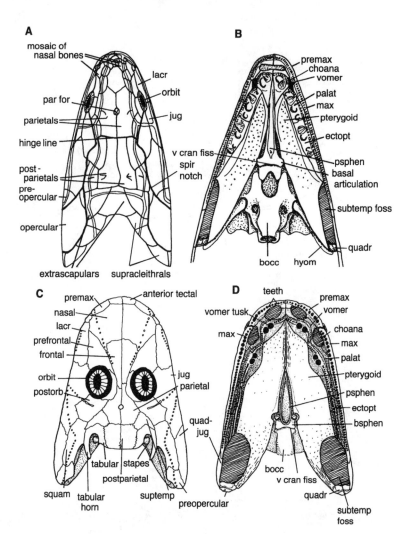

Figure 2.10. Skulls of Eusthenopteron in (A) dorsal and (B) ventral views and Acanthostega in (C) dorsal and (D) ventral views, compared to show some of the contrasts. Abbreviations as in Figure 2.1 plus postorb = postorbital; suptemp = supratemporal.

Figure 2.9 summarizes the differences between a sample fish skeleton (similar to but not actually *Eusthenopteron*) and a tetrapod such as the Carboniferous *Dendrerpeton*. Some of these changes took place in a coordinated way and are interdependent, but for others, timing and sequence remains unknown. These will be discussed in later chapters. The anatomy of *Eusthenopteron* is based on Jarvik (1980) and that of *Dendrerpeton* on Holmes et al. (1998). In some cases, comparisons are made between a *Eusthenopteron*-like fish and the early tetrapod *Acanthostega* on the basis of the work of the author, and of Coates (1996).

Skeletal Changes to the Skull and Shoulder Girdle

Some of the changes that are seen in skull construction are obvious, but some are quite subtle and involve changes in proportion. Figure 2.10 shows the skull of *Eusthenopteron* compared with that of the Devonian tetrapod *Acanthostega*. Probably the most obvious differences concern the bones covering the gill chamber and throat region in fishes, the operculogular series. This whole series was lost in tetrapods. It is tempting to think that the loss was related to abandonment of gill breathing, but for

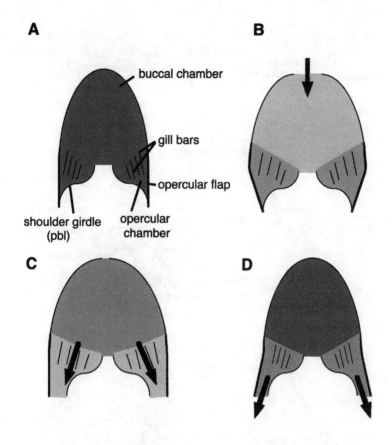

Figure 2.11. Operation of the buccal pumping mechanism. Diagrammatic horizontal sections through a fish's buccal and opercular chambers.
(A) System at rest.
(B) Mouth opens and buccal chamber expands lowering the pressure, so water enters the mouth. Pressure in the opercular chamber is also lowered and the spaces between the gill arches expand.
(C) Opercular chamber opens while the mouth begins to close, raising the water pressure and forcing water through the gill curtain.
(D) Mouth closes and buccal chamber contracts, forcing the remaining water out from the opercular chamber, and the system comes to rest again. Postbranchial lamina (pbl) of the shoulder girdle forms the seal against which the opercular flap closes.

reasons put forward in later chapters, this is not necessarily the case. However, it was an important change because it occurred in conjunction with a suite of others, although it is not clear yet whether these all occurred at the same time or whether one followed another consequentially.

The opercular series plays a key role in the operation of the pump mechanism by which a fish ventilates its gills—that is to say, in the series of movements that causes water to pass across the gills to oxygenate the blood. It is appropriate here to describe this mechanism because it is one of the functions of the skeleton that changes during the fish–tetrapod transition. Understanding this mechanism will help to explain some of these changes as they are described later. The pumping mechanism is powered by movements of the mouth (buccal) chamber and opercular system, and is known as the buccal pump. It operates in distinct phases, illustrated in Figure 2.11. In ray-finned fishes and in primitive lobe-finned fishes such as the basal tetrapodomorphs, the skull roof and cheeks were flexible and allowed the buccal chamber and opercular chamber to expand and contract slightly out of phase with each other. This permits pressure differentials to exist between the buccal and opercular chambers, causing water to move from one to the other. In tetrapods, gill breathing is first reduced and then lost altogether in adult, metamorphosed amphibians and all amniotes. To begin with, lungs were ventilated by a version of the buccal pump employing the hyoid skeleton, as in modern amphibians, but eventually, it was taken over by the body wall muscles. The hyoid skeleton was converted for other uses.

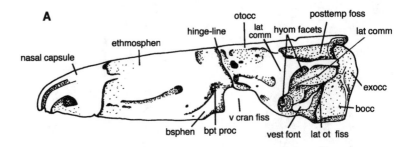
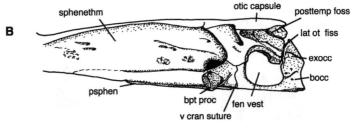
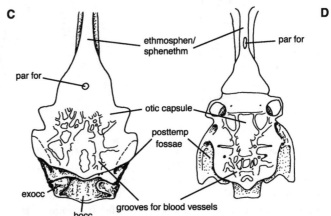

Figure 2.12. Braincases of Euthenopteron *and* Acanthostega *compared.*
(A) Eusthenopteron *in lateral view.*
(B) Acanthostega *in lateral view.*
(C) Acanthostega *in dorsal view.*
(D) Eusthenopteron *in dorsal view.* From Clack (1998c).

As well as losing the operculogular series, the tetrapod skull lacks another series of bones found in fishes. That is the series called the extrascapulars and supracleithrals, which in fishes join the dermal shoulder girdle to the skull roof at the back. The extra scapulars are the bones at the back of the skull table—usually three in *Eusthenopteron*—and the supracleithrals are the bones that lie above the cleithrum in the shoulder girdle. In *Eusthenopteron*, these bones link to one another at the back of the head (Figs. 2.1, 2.10). Loss of the operculogulars, extrascapulars, and supracleithrals results in a space being made between the suspensorium and the shoulder girdle. In effect, the head is no longer supported by the shoulder girdle, so the vertebral column and its muscles must do the job instead. Thus, tetrapods have necks.

Construction of a neck itself involves a number of changes. Muscles that hold the head up must have a solid foundation for their origin. The back of the tetrapod head became gradually adapted to provide anchorage and space for these muscles to attach. One of the threads that this book will follow is that of how changes to the back of the skull took place in tetrapod evolution. It did not happen all at once. The neck muscles require strong insertion points along the spine, and elaboration of the neck vertebrae is another such gradual change to be seen in tetrapod evolution. Some of the muscles supporting the head and joining it to the vertebral column insert into pockets at the back of the skull called the posttemporal fossae (Fig. 2.12). The neck joint, or occiput, built to allow the head to move freely on the vertebral column, was yet another feature whose evolution will be followed, and it too involved gradual changes to the back of the skull and front of the vertebral column.

Other changes to the braincase are associated with the development of a neck. Running through the vertebral column and into the braincase of a fish is the notochord. It helps the body to return to shape after the bending produced during swimming by the muscles of the body wall. In a fish's vertebral column, the notochord is sheathed by the serially arranged centra, and it runs forward right into the braincase through an opening below the foramen magnum. In most tetrapods, the notochord is replaced by bone in each centrum and no longer runs into the braincase. This process can be traced in the transition between fishes and tetrapods, because like the evolution of the neck joint, it occurred gradually. Figure 2.12 provides diagrams of the braincase of *Eusthenopteron* compared with that of *Acanthostega*.

Perhaps associated with this change are a number of others concerning the braincase. One is the elimination of the hinge found between the front and hind parts of the braincase seen in *Eusthenopteron*. Most current estimates of the phylogeny of these animals do indeed suggest that it was a loss of this feature by tetrapods rather than a development of it by osteolepiforms. In fishes, the dermal parasphenoid bone underlies only the front (ethmosphenoid) part of the braincase. In tetrapods, however, the former hinge line was sealed by the elongation of the dermal parasphenoid to underlie the back part of the braincase as well as the front. Along with sealing of this braincase hinge went the elimination of the hinge across the skull roof, as can be seen by comparing the dorsal views of *Eusthenopteron* and *Acanthostega* in Figure 2.10.

Moving forward in the skull, a number of changes are seen in the part that lay behind the eyes. As a general statement, the back part of the fish's skull is longer than the equivalent part of an early tetrapod's. This applies both to the braincase and the skull roof. If a series of early tetrapod skulls from the Devonian to the Permian were taken and assembled in chronological order, a trend of shortening of the back of the skull can be found. The reason for this appears to be that the otic capsules, which lie under the skull table, get gradually smaller in tetrapods relative to the rest of the skull. The beginnings of this trend can be seen even in very early tetrapods, although the reason for this is not clear. More details of this process and possible reasons for it are put forward in Chapter 6.

One of the more mysterious changes that took place in the braincase was possibly associated with this shortening of the otic region. The part where the hyomandibula attached to the otic capsule in most fishes, called the lateral commissure in *Eusthenopteron,* disappeared in tetrapods (Fig. 2.12). The head of the hyomandibula came to occupy the resulting hole, inserting instead into the side wall of the braincase. At this point, the bone becomes known as a stapes, and the hole into which it inserted is the fenestra vestibuli.

The hyomandibula in fishes does several jobs. From its articulation on the braincase wall, this long, flattened bone runs down the back of the palatoquadrate and supports the bones that form the roof of the mouth. The roof of the mouth has to be moveable to accommodate the expansion and contraction of the buccal cavity that occur during the fish's breathing cycles and in the movements of its feeding mechanism. These movements are controlled by the hyomandibula. This bone is also attached to the gill covers, the operculogular series, that also operate during the breathing cycle, so the hyomandibula also coordinates their movements with those of the palate. The palatoquadrate also joins to the braincase at other points: at the front and around the middle of its length at the basal articulation. However, these joints are all mobile. In fishes, then, the palatal bones are only loosely attached to the braincase, chiefly through the hyomandibula, and this type of skull is described as hyostylic.

In many later tetrapods, this kind of flexible skull construction is eliminated, and the joints become firm and sutured together. The palate has its own firm attachments to the braincase and skull roof bones, so it does not have to be hung from the hyomandibula. This may be one of the reasons why the stapes in tetrapods is so much smaller than the hyomandibula of fishes. Tetrapod skulls with this firm kind of construction are described as autostylic. Instead of disappearing, the hyomandibula took on a new role in tetrapods: that of a hearing ossicle. This change did not happen quickly or all at once, and Chapter 10 describes how it may have happened in different tetrapod groups.

This change ties up with others that occurred in the branchial system. Gill breathing was finally lost in tetrapods, and the gill arches became not only reduced in number but in size. However, the bones that support the gill arches and the muscles that move them did not disappear. Instead, they were redeployed in the formation of the tongue supports and musculature. Tetrapods, once they became terrestrially feeding organisms, could not use the water to support food while they grasped it or use its force in suction feeding, as many fishes do. Instead, that uniquely tetrapod invention, the grasping and manipulating tongue, was built from the former gill arch components. Probably this could only occur once gill breathing had been entirely lost.

At the back of the skull of a fish such as *Eusthenopteron,* the spiracle opened into the apex of the slot known as the spiracular notch or cleft. Most early tetrapods, although not all, had a rounded or V-shaped embayment at the same location in the skull, which has been variously called the otic notch, the spiracular notch, or the temporal notch. The function of this embayment has been debated by paleontologists for years, and the story behind it is tied up with the change from gill breathing to air breathing, and in the function and form of the hyomandibula/stapes. It is set out in more detail in Chapters 6 and 10.

Although the back of the skull was generally shorter in tetrapods than in comparable fishes, in early tetrapods, the snout appeared to grow longer, resulting in the eyes appearing further back. One associated consequence was that more of the tooth row came to lie in front of the eyes in tetrapods compared with its position in fish. The eyes also seem to have increased in size and became positioned closer to the top of the head in early tetrapods than in fishes. Proportional changes such as these are difficult to explain and hard to separate from each other. They may be linked with changes to both sensory and feeding requirements.

Lengthening of the snout seems to have been associated with stabilization of the bone pattern of the front part of the skull and a reduction in the number of component bones. Where fishes had a mosaic of small bones in the nasal region, tetrapods evolved a single pair of nasal bones, with remnants of this mosaic remaining only in the earliest ones. These, the lateral and median rostrals and anterior tectal bones, were lost early in tetrapod history.

Other, more subtle changes, such as those to the cheek bones, the internal and external nostrils, the spiracular notch, the dentition, and the aquatic pressure sensory system called the lateral line, which occurred during the fish–tetrapod transition, are best seen by studying particular fishes and particular tetrapods in context, as shown in later chapters.

Profound changes to the shoulder girdle occurred during the fish–tetrapod transition, associated with a shift in emphasis of the role that the complex played, and these can be seen in Figure 2.5. In fishes, the shoulder girdle is integral with the skull, continuing the profile of the skull back along the body in a smooth line, producing the hydrodynamic torpedo shape typical of many fishes. It consists mainly of dermal elements, the supracleithral bones, the cleithrum itself, the clavicle, and a small ventral midline interclavicle. This series of bones attaches to the skull at the top via the lateral extrascapular bones and forms a hoop around the body. Under the body, muscles run from the midline of the shoulder girdle to the gill arches and jaws to control their movements. Along the anterior edge of the cleithrum and clavicle a groove and an internal flange called the postbranchial lamina (Fig. 2.11) (see also Chapter 5) not only helps direct water flow out of the gill chamber but helps to form a seal against which the operculum closes during breathing movements.

A small endochondral scapulocoracoid is wedged on the inner surface of the cleithrum, bearing a small, backwardly facing articulation point for the fin. In *Eusthenopteron,* the scapulocoracoid is a small tripod of bone with its three "legs" attached to the cleithrum.

In almost complete contrast, the shoulder girdle of all but the most primitive tetrapods consists mainly of an enlarged scapulocoracoid, which may be divided into two or three separate ossifications: the scapula above and one or two coracoids below (Fig. 2.5). The scapular portion is clasped along its anterior margin by a slender dermal cleithrum, although this too is lost in most later tetrapods. The clavicle remains in most tetrapods. In early ones, it consists of a broad triangular ventral plate whose apex, positioned at the side of the body, is produced into an upwardly projecting prong. This sheathes the coracoid portion of the girdle and contacts the slender cleithrum. The dermal interclavicle may be a large shield-shaped plate of bone, or a rather more slender T-shaped one, but in early tetrapods, it is always larger than that of the corresponding bone in fishes. The scapulocoracoid in tetrapods bears a large concave articulation surface for the limb, which faces outward from the body.

The shoulder girdle in fishes is mainly concerned with hydrodynamics, but it also provides anchorage for gill- and jaw-moving muscles. Bearing the pectoral fins seems a relatively minor role. In tetrapods, bearing the limbs and providing anchorage for the muscles that move them are the most important jobs. The shoulder girdle is detached from the skull, which allows the limbs and the head greater freedom of movement than in fishes. Correspondingly, the outer dermal components are larger in fishes, whereas in tetrapods, these are reduced in favor of the inner endochondral elements.

The shoulder girdles of the earliest tetrapods are not obviously halfway in structure between those of fishes and those of later tetrapods but have some unique and some unexpected characters.

Skeletal Changes to the Pelvic Girdle, Limbs, and Axial Skeleton

Turning to the rear half of the animals, one of the biggest structural contrasts between a fish and a tetrapod lies in the size and function of the pelvic girdle. In most fishes, exemplified by *Eusthenopteron,* the pelvis is a relatively small, simple ossification anchored in the body wall without any bony connection to the vertebral column. In *Eusthenopteron,* it consisted of a single pair of boomerang-shaped elements oriented apex downward, which did not meet each other at the midline—there was no symphysial junction. The pelvis bore the articulation for the fin at about the apex of the boomerang. These can be seen in Figures 2.3, 2.6, and 2.9.

Even in the earliest known tetrapods, the pelvic girdle had become very different in structure from that of a fish. It was a large, robust element with a long ventral edge along which both halves of the girdle met to form the pelvic symphysis. The articulation point for the hindlimb, the acetabulum, was longer than the head of the femur, to allow movements of the hindlimb. Most characteristically, there were long processes dorsally, internal to which one of the ribs articulated to form a sacral joint. Muscles to move the legs and tail inserted on these processes and on the enlarged platelike region below the acetabulum. The region surrounding the acetabulum was strengthened by various buttresses. In tetrapods from the Carboniferous onward, each half of the pelvic girdle consisted of three separate ossifications: the ilium dorsally, to which the sacral rib attached, and below it the pubis anteriorly and ischium posteriorly (Fig. 2.6). These regions were not separately distinguishable in the earliest tetrapods.

The differences in girdle structure between fishes and tetrapods reflect the differences in limb construction and the differences in the emphasis placed on the different parts of the body used in locomotion. In fishes, the main muscles used for propulsion are those of the body wall, attached to the axial skeleton. The axial skeleton is flexible and the musculature complex. In tetrapods, the emphasis shifts to the appendicular skeleton—that is, to the limbs and girdles—for propulsion. The limbs and girdles are correspondingly enlarged and their musculature more differentiated, the body-wall musculature is relatively reduced, and the axial skeleton is stiffened. The neural spines bear articulations one with another, called zygapophyses (Fig. 2.4), and provide anchorage for substantial ribs. Fishes' ribs are usually thin and flexible, and true zygapophyses are absent. The vertebral centra in fishes are relatively poorly ossified and often remain hollow cylinders around the notochord; in tetrapods, they become more solidly constructed and often completely replace the notochord. In terrestrial tetrapods, this construction allows the vertebral column to function as a bridge between the girdles, keeping the body off the ground.

In fishes, where much of the propulsion comes from the tail and body, the fins are largely used for steering and braking. Usually the pelvic fins are small, and in fossil lobe-finned fishes, they were usually smaller than the pectorals. By contrast, in unspecialized tetrapods from the Paleozoic onward, the hindlimbs provide most of the thrust and are larger than the forelimbs. That was not the case in the very earliest tetrapods, for reasons to be explored later.

The feature that many people would pick to distinguish fishes from tetrapods is that which gives the name to the group—the structure of the

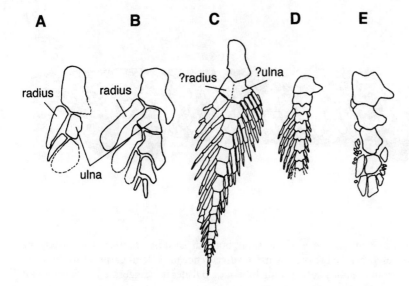

Figure 2.13. Selection of lobe-fin pectoral fin skeletons.
(A) Panderichthys.
(B) Eusthenopteron.
(C) The Australian lungfish Neoceratodus.
(D) The porolepiform Glyptolepis.
(E) The coelacanth Latimeria.

limbs and the possession of digits. As set out in the next chapter, there are some technical problems with this, but it is appropriate here to consider the similarities and differences in structure between the paired appendages of fishes and tetrapods.

The fore and hind sets of paired fins of a lobe-finned fish such as *Eusthenopteron* consist of similar bones. Figure 2.13 shows the pectoral fin skeletons in a range of lobe-finned fishes, to illustrate their common features as well as their differences. In common is first the single element, called the first axial radial, which attaches to the girdle, identifying the animal as a sarcopterygian. From the other end of this element, further elements arise and articulate with it, forming a chain called the metapterygial axis. In *Eusthenopteron* and tetrapods, there are two of these. In *Eusthenopteron,* the more posteriorly situated of these in turn gives rise to two more. The precise pattern varies among lobe-fins. Fins of the various lobe-finned fish groups differ more in structure further along the fin than they do nearer the base. Some groups, such as lungfishes, have elongated, paired fins supported by a long series of segments, each with a branching radial springing from it. In most lobe-fins, the branches occur mainly on the anterior (leading) edge of the fin and are called preaxial radials. In lungfishes, both pre- and postaxial radials are found. In osteolepiforms, the fin skeleton remains short, whereas in *Panderichthys,* the elements are broad and flattened (Fig. 2.13).

The first axial radial of the fore fin has flanges and foramina for muscle attachments and passageways for nerves and blood vessels. Bounding the outer rim of the fin is a fringe of bony elements derived from the skin (and thus dermal in nature) called fin rays, or lepidotrichia.

The similarities in construction of tetrapod limbs and lobed fins lies in the first three skeletal elements, and the pattern of a single element attaching to the body and giving rise to two more from its far end. These bones can be called by consistent names: in the fore fin or limb, the humerus, radius, and ulna; and in the hind fin or limb, the femur, tibia, and fibula. The humerus of an early tetrapod bears flanges and foramina recognizably

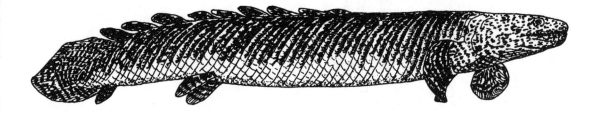

Figure 2.14. Polypterus bichir. Illustration by J.A.C.

the same as those of the same bone in *Eusthenopteron*. Thereafter, the resemblance breaks down. The divergence in skeletal pattern further away from the body seen among lobe-finned fishes is exaggerated in the contrast between fins and limbs.

In most tetrapod limbs, the two elements, ulna and radius or tibia and fibula, articulate distally with each other and with a series of blocklike bones connected by complex joint surfaces, allowing much freedom of movement. These produce the characteristic elbow and knee or wrist and ankle joints (Fig. 2.9B). They allow the ends of the limbs to be brought into contact with the ground at appropriate angles for bearing weight and transmitting thrust. At the ends of each limb, digits arise by sequential budding of a number of radial elements. The way that this happens will be explained in Chapter 6, but in effect, digits equate in position to the postaxial radials of the fish fin, which have become jointed.

Body Scales and Fin Rays

Two more kinds of structures need to be considered in the contrast between fishes and tetrapods. These are both formed from the dermal component of the bony skeleton and are, respectively, the body scales and the fin rays.

In fishes, scales are formed within the middle layer of the skin, called the dermis, which is the part of the skin containing structures such as blood vessels, touch or pressure sensors, glands, and some superficial muscle tissue. In early lobe-finned fishes, the scales were bony in texture, and in some species, they were covered in shiny, enameloid material matching the dermal bones of their skulls. The modern ray-finned fish *Polypterus* shows similar bony, enameloid-covered scales (Fig. 2.14) to those of many Paleozoic forms. In almost all modern fishes, however, the bony layer has been lost, leaving the scales only as thin proteinaceous sheets.

The scales in the early lobe-finned and ray-finned fishes were interlocked in diagonal rows that encompassed the body and met each other dorsally and ventrally at an acute angle (Fig. 2.15). Each row articulated with the one in front of and behind it, and along a line running down middle of each side of the body, each scale bore a segment of tube containing part of the lateral-line organ.

In the early tetrapods, much of this scale cover was lost and the remaining parts modified. The scales of the early tetrapods are known as scutes or gastralia, and the flexible armor that they formed covered only the

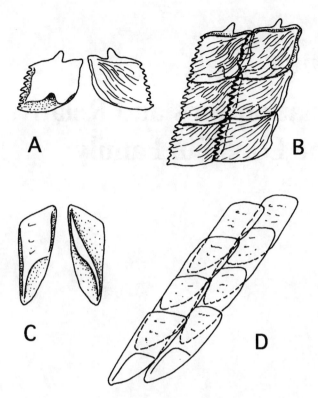

Figure 2.15. Scales and gastralia of fish and tetrapods.
(A, B) Actinopterygians scales; A shows internal (left) and external views, B shows intercalating series.
(C, D) Tetrapod gastralia; C shows external (left) and internal (right), D shows intercalating series, with overlapped areas indicated by dotted lines.

belly and parts of the sides. The gastralia were bony but bore no enameloid layer. They were often oval in shape, bearing an internal ridge and groove where each element articulated with the ones dorsal and ventral to it, but not with those to either side of it (Fig. 2.15). The rows met ventrally in a forward-pointing V, except around the shoulders and interclavicle, where the orientation was reversed and the V pointed backward. Some modern amphibians belonging to the group known as caecilians still retain vestiges of these primitive dermal gastralia. However, dermal gastralia or scutes are not equivalent to the scales of modern reptiles. Because gastralia in early tetrapods were dermal, they would have been covered by a thin layer of outer skin called epidermis. In modern reptiles, the scales are formed in the epidermis itself.

The fins of bony fishes are supported by dermal rays, which are essentially narrow, elongated scales forming a fringe suspending a thin web of skin. In early lobe-finned fishes, narrow fringes of fin web surrounded each paired fin, the tail, an anal fin, and two dorsal fins. Dorsal and anal fins were additionally supported by a series of radials, the pattern of which varied according to the family. In tetrapods, almost all the fin webbing disappeared, along with all traces of dorsal and anal fins. The only remnants are seen in the two earliest known tetrapods, in which some tail fin web was still retained.

Three
Relationships and Relatives: The Lobe-Fin Family

Introducing the Lobe-Fins

This chapter introduces the tetrapods' closest relatives, explains how tetrapods fit into the scheme of relationships with other lobe-fins, and explores how ideas about the ancestry of tetrapods have evolved with changing perspectives.

There are a few characteristics of lobe-fins that distinguish them as a group, but the most conspicuous is the eponymous lobed fin, described more fully in Chapters 1 and 2, in which the paired fins are anchored to the respective limb girdle by a single bone. The structure appears to be a true shared, derived character of the group. Another shared feature is the possession of two dorsal fins, instead of the single one found in early ray-fins, and a third is the occurrence in the tail of a second series of fin rays growing above the body portion. This has enabled lobe-fins to evolve symmetrical tails without too much modification of the existing pattern, and they have done so in a number of separate lineages. The latter two features may after all be primitive for gnathostomes, but this is not clear. In addition, most early members of the lobe-finned group show an intracranial joint or hinge line. A hinge linked the front and back parts of the skull roof just behind the eyes, reflected in a matching hinge across the underlying braincase (see Chapter 2). It is not seen in ray-finned fishes and is also lost from several of the lobe-finned groups independently, including tetrapods.

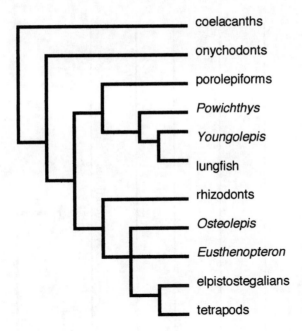

Figure 3.1. Cladogram of lobe-fins (from Cloutier and Ahlberg 1997).

Most early lobe-finned "fish" were moderately sized animals. Although a few grew to enormous size, very few small species are known, so it does not appear that there were lobe-finned equivalents of anchovies, for example. It is not at all clear why this should be the case. Early lobe-fins lived in both marine and freshwater conditions, but almost always in relatively shallow water.

Today there are very few lobe-finned "fishes" remaining, as distinct from tetrapods (which are also members of the lobe-fins). The surviving "fish" species are the coelacanth *Latimeria* and three genera of lungfish. In the Paleozoic, however, lobe-fins were dominant fishes for many millions of years, both in number of species and in size. Several lineages died out at the end of the Devonian, although some survived into the Early Permian. By the end of the Paleozoic, only the lungfishes, coelacanths, and tetrapods remained, and the ray-finned fishes took over dominance of the water.

Figure 3.1 gives one idea of the relationships of lobe-finned fish groups (Cloutier and Ahlberg 1997), including tetrapods, to one another, and that is the scheme that will be followed in this book. Only with an explicit phylogeny behind it can any ideas about what occurred in evolution have any validity. The phylogeny must take precedence in drawing up scenarios about what went on in the distant past; otherwise, although any possible story can be put forward, it will not be a scientific one. However, readers must bear in mind that not all paleontologists agree with this particular phylogeny, and it is subject to change as new finds and ideas emerge. This implies that should a different phylogeny emerge for the relationships of tetrapods, some of the conclusions drawn throughout this book might be refuted.

Figure 3.2 puts the animals on a time scale of the Devonian period. Note the subdivisions or stages into which the Devonian is divided.

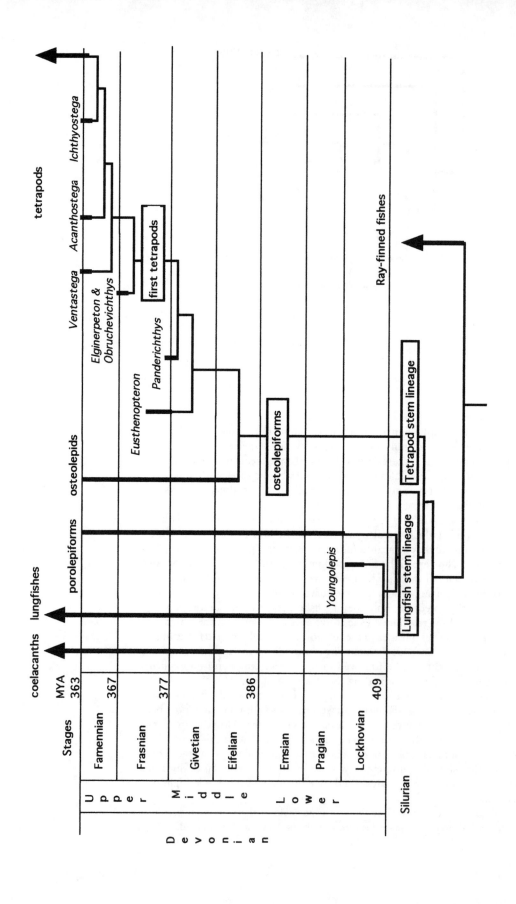

Getting to Know the Lobe-Fin Groups

Survivors

Coelacanths

Until about 60 years ago, the coelacanths (formally known as Actinistia) were considered to have become extinct during the Cretaceous period, probably at the end of it, along with the dinosaurs. In 1938, however, the first modern coelacanth was recognized (see Smith 1956), and since then, knowledge of it has increased greatly from the several specimens that have been recovered, preserved, and dissected, and from video footage taken from deep-sea submersibles (e.g., Fricke et al. 1987, 1991). These have shown that *Latimeria* does not use its fins in any kind of "walking" activity, but for slow propulsion in swimming. They have also shown how the fishes rest together in caves in mixed groups of different-aged individuals. The composition of each group appears to remain relatively stable, and it is likely that they recognize each other. Each individual has a unique pattern of spots on its flanks.

The species *Latimeria chalumnae* (Fig. 3.3) is found predominantly around the east coast of the Comoro Islands in the Indian Ocean, where it is fished up from depths of about 300 m by line fishermen. Most recently, coelacanths have been found off the Indonesian island of Sulawesi, representing a second and potentially important population of these intriguing fish, perhaps a separate species from the Comoro Islands form (Erdmann et al. 1998).

Because coelacanths were known first as fossils, ideas about them and their relationship to tetrapods were challenged by the discovery of a living form. Some ideas were proved correct, whereas others, not surprisingly, were wrong (see Forey 1998; Musick et al. 1991). One of the ideas that was tested concerned the nostrils. *Latimeria* proved to have a pair of nostrils on each side of its face, one for incurrent water and one for excurrent, and these are used for olfaction—smelling—and not for breathing. This it shares with most other fishes, both bony and cartilaginous. However, certain other fossil lobe-fins and all tetrapods, to which *Latimeria* was supposedly closely related, have only a single nostril on each side, connected to an internal one inside the mouth on each side. In many other features of its physiology too, *Latimeria* has proved unusual and more like sharks than bony fish. Most scientists take these features to show a general gnathostome condition rather than a direct relationship to sharks.

One feature suggested by fossils and confirmed by *Latimeria* is that coelacanths were ovoviviparous—they produce large eggs, maintained in the body cavity, until fully formed young are released. One fossil coelacanth was described by D. M. S. Watson in 1927 within which two tiny coelacanths are preserved, and although at the time they were considered to be embryos, another possibility was that they were stomach contents, with an adult having eaten the small fry. Discoveries of advanced embryos within female specimens of *Latimeria* have confirmed the observation of embryos in the fossils (Atz 1976).

In common with other bony fishes, coelacanths possess a derivative of the swim bladder, an outpocketing of the pharynx behind the gill pouches. In many bony fishes, this is an air sac and is used as a lung, whereas in others, it acts as a gas-secreting bladder to control buoyancy. In some fossil coelacanths, this structure has extremely thin but strong bony walls, suggesting that it could resist changes in water pressure as the animal dived. In

Figure 3.2. (opposite page) Time scale of the Devonian period with a phylogeny of lobe-finned vertebrates superimposed to show the times of origin and relationships of the main groups.

Latimeria, this bladder is instead filled with fat, which would have the same effect (Forey 1998).

Coelacanths have some peculiarities of skull structure seen in both living and fossil forms (Figs. 3.3, 3.4, 3.5). They have lost the maxillary bones of the upper jaw so that all their upper teeth are borne on the palate. Throughout their evolution, many other skull bones, present in early forms, are lost or reduced. Across the top of the skull and through the braincase runs the intracranial joint, which they share with many fossil lobe-finned fishes and which is characteristic of the whole group in its early history (see above and Chapters 2 and 6). This device has been studied in some detail and seems to provide coelacanths with added impetus to their method of suction feeding (Thomson 1967, 1969). With their deep tails, the coelacanths can produce a sudden surge of movement as they lunge at passing prey, usually small fish.

Latimeria has the characteristic fleshy lobed fins of its group, which it moves alternately or independently in a way that resembles the limb movements of a tetrapod (Fricke et al. 1987). However, at some time during their evolution, coelacanths developed a second dorsal and an anal fin that have an almost identical structure to the paired fins, each complete with half a girdle (Ahlberg 1992). The tail too looks as though it has been duplicated, mirror fashion, about a line running horizontally halfway along the body (Fig. 3.3). Perhaps some strange genetic accident duplicated the paired fin structure where midline fins should be, and also duplicated the structure of the tail. It is possible to put a rough date to the time at which this occurred. The most primitive known coelacanth, *Miguashaia* from the Canadian locality of Escuminac Bay (see below and Chapter 4), does not show this unique feature, but instead retains the primitive form of tail and has midline fins with a structure similar to those found in most early lobe-fins and ray-fins (Cloutier 1996). This animal is Late Devonian in age, dating from around 370 million years ago. The genetic change that characterizes all later coelacanths happened around this time, because in other Late Devonian coelacanths, it had already occurred.

Although only one genus of coelacanth remains, they were numerous and varied throughout the Paleozoic and Mesozoic eras. Paleozoic forms seem to have been euryhaline—that is, they could migrate between freshwater and seawater, and at least some lived in shallow water much of the time (Forey 1981, 1998). In contrast, the modern form lives exclusively in deep marine waters. Also during the later Paleozoic, variations in body form are found, especially with shifts in positions of fins, paralleled later by ray-finned fishes (Lund and Lund 1985) (Figs. 3.5, 3.6). A comprehensive review of fossil and extant coelacanths as well as a discussion of their relationships can be found in Forey (1998).

Although the living coelacanth was hailed as "Old Four Legs," in fact, it now appears that the similarities between coelacanths and tetrapods are not as close as once appeared. In the last section of this chapter, the evidence behind these ideas will be explored.

Lungfishes

In contrast to coelacanths, the lungfishes or dipnoans (formally Dipnoi) were first known from their living representatives, and only in 1871 was it realized, by Günther, that the fossil forms from the Paleozoic were related to the modern forms. (They had been described many years previously but not recognized as related to living animals.) There are three modern genera, all living exclusively in freshwater. *Neoceratodus* inhabits a few river basins around the east and south coasts of Australia, whereas

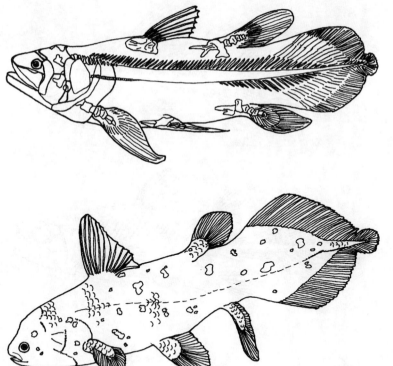

Figure 3.3. (left) Latimeria chalumnae *skeleton (top) and whole animal (bottom).*

Figure 3.4. (below) Skull roof of Rhabdoderma elegans *in lateral view (left) and dorsal view (right).*

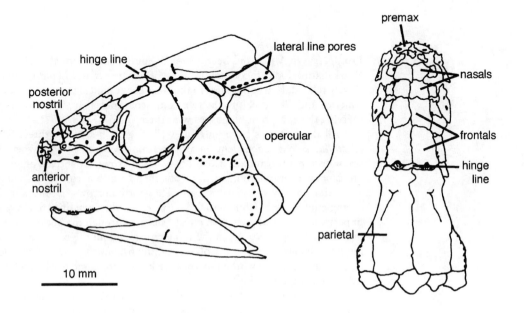

Relationships and Relatives • 51

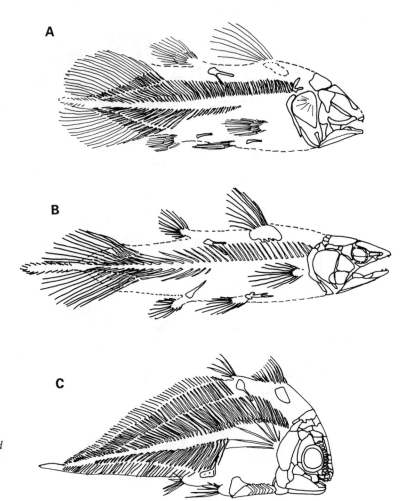

Figure 3.5. Drawings of fossil coelacanths.
(A) Macropomoides orientalis, Late Cretaceous, Lebanon.
(B) Rhabdoderma elegans, Late Carboniferous, United States and Europe.
(C) Allenypterus montanus, Early Carboniferous, United States. From Forey (1998).

Protopterus in Africa and *Lepidosiren* in South America inhabit lakes and rivers that periodically dry up. Of these, *Protopterus* and *Lepidosiren* are closely related, bearing witness to the close geographical position of South America and Africa before the two continents split apart in the Jurassic, carrying their cargo of lungfishes with them (Fig. 3.7).

All three species have lungs as well as gills, but of the three, *Neoceratodus* is least dependent on air, breathing with lungs only when the water becomes too anoxic. The other two will drown if deprived of air but are capable of estivation, which means they go into a kind of suspended animation to survive the dry season. They form burrows of mucus in the drying mud at the bottom of their lake or river. This then dries to make an impervious seal, and the fish in effect "sleeps" through the hot dry season in a cocoon (Fig. 3.8). When the rains return and fill the basin again, the mud-and-mucous cocoon dissolves, and the fish swims away after a new gulp of air. Fossil estivation burrows are known for some fossil lungfishes from the Permian period.

The skull of a lungfish, living or fossil, has a suite of unique features that makes it physically very unlike the more "conventional" skulls of

Figure 3.6. (left) Laugia groenlandica, *a fossil coelacanth from the Early Triassic, that shows modifications to the position of the pelvic fins. They are placed much further forward than in any other genus of coelacanth, and in this way parallel modifications found in teleost ray-fins.*
(1) First dorsal fin.
(2) Second dorsal fin.
(3) Pectoral fin.
(4) Pelvic fin. UMZC specimen GN 243. Photograph by S.M.F.

Figure 3.7. (below) Map showing distribution of modern lungfish genera.

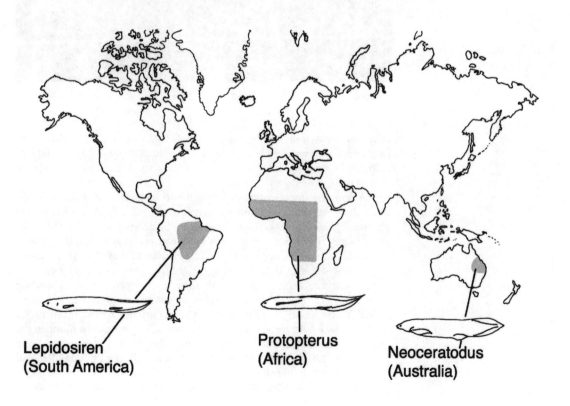

Relationships and Relatives • 53

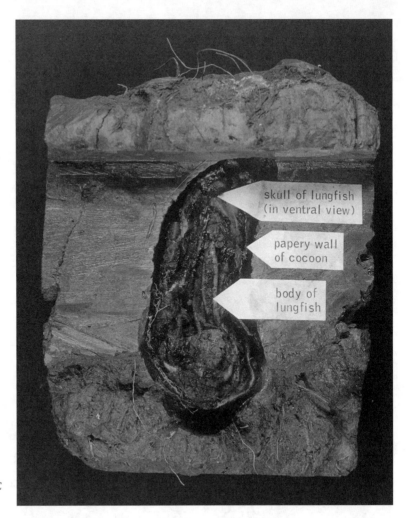

Figure 3.8. Preserved lungfish cocoon. Photograph from UMZC archives.

other lobe-fins or indeed the ray-fins (Figs. 3.9, 3.10, 3.11). Because lungfishes have featured very heavily in the struggle to understand the relationships and origin of tetrapods, it is necessary to understand this strange structure in some detail.

Among the most commonly found and most characteristic of fossil lungfish elements are their tooth plates (Figs. 3.9, 3.10, 3.11, 3.12). These are usually more or less triangular structures carrying sharp-crested, radiating ridges. There are two pairs in the lower jaw that bite against corresponding pairs in the upper jaw, the ridges meeting in a scissorlike action. Occasionally, rather than radiating ridges, the plates are equipped with denticles, either forming a uniform surface or organized into radiating rows. The pattern of ridges or denticles are characteristic of their genera or species (for a range, see Fig. 3.12). The occurrence of the tooth plates attests to the loss of the marginal teeth that would normally be found in the maxilla, premaxilla, and dentary around the mouth of most other bony vertebrates. Lungfishes lack all trace of the maxilla and premaxilla, and only the most primitive have the remnants of the dentary. The tooth plates form on the prearticular of the lower jaw and the pterygoid of the upper jaw (see Chapter 2).

 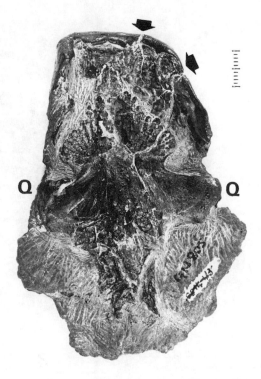

Lack of the maxilla and premaxilla, as well as other modifications to the skull roof bones and palate, have had an impact on the nostrils of lungfishes. Both nostrils in lungfishes appear in the roof of the mouth, inside a fleshy (in modern forms) or bony (in fossil forms) lip (Fig. 3.9). There is an anterior pair for water intake and a posterior pair for water outlet after the water has passed over the olfactory membranes. The identity of these nostrils and their equivalence to those of other vertebrates have been a source of debate and confusion since the early part of the 19th century, as they have featured in the debate about how lungfishes are related to tetrapods (see below).

The tooth plates appear adapted for a crushing action to use on hard prey items, and this, it has been suggested, correlates with other specializations of the lungfish skull. The skull roof of fossil forms, unlike those of most other bony vertebrates, has its bones arranged in a pattern best described as "hopscotch"—that is, a single bone alternates with a pair, in the sequence from the back to the front of the skull. Each bone tends to have a polygonal shape as a result, and it has been a problem to see how lungfish skull roof bones equate to those of other lobe-fins (Fig. 3.9). In consequence of this pattern, the lungfish skull roof lacks any trace of the lobe-fin hinge line but is rather a solidly constructed unit in most early fossil members. Modern forms have lost most of the dermal roofing bones in the same way that they have reduced or lost their scales (Figs. 3.10, 3.11).

Other features tie in with this overall consolidation of the skull. The palate is a solid unit, with the plate-bearing pteryoids sutured to the braincase, whereas the jaw joint, formed by the quadrate, lies well forward in the skull rather than at the rear, as in most lobe-fins (Figs. 3.9, 3.10,

Figure 3.9. Skull of the fossil lungfish Dipterus *in dorsal view (left) and palatal view (right). Note the "hopscotch" pattern of the skull roofing bones and the ridged tooth plates on the palate. Note also the incurved bony upper "lip" of the skull, which is toothless, lacking the premaxilla and maxilla. Arrows mark the notches for the anterior and posterior nostrils. Q marks the position of the quadrate. Bar = 10 mm. UMZC specimen GN 805. Photograph by S.M.F.*

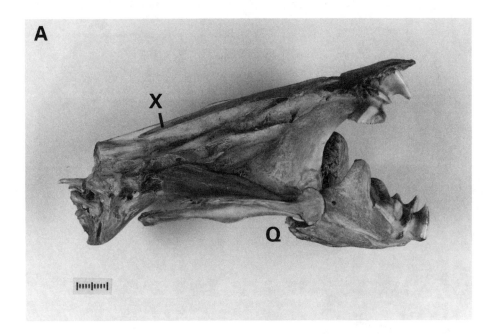

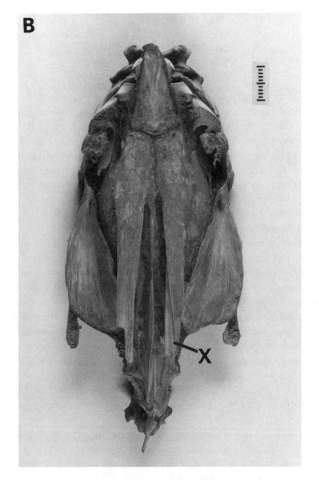

Figure 3.10. Skull of the modern lungfish Protopterus *in (A) left lateral and (B) dorsal views. Note the ridged, scissorlike tooth plates and the forward position of the quadrates (Q). Most of the dermal skull bones have been lost. X marks one of the remaining elements. Bar = 10 mm. UMZC specimen. Photograph by S.M.F.*

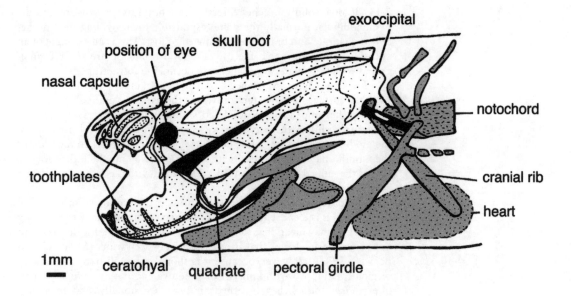

Figure 3.11. Diagram of the skull of Protopterus *showing the cranial rib. Elements not present in the skull in Figure 3.10 are shaded. From Long (1993).*

Figure 3.12. Fossil lungfish tooth plates. (A) Ctenodus, Lower Carboniferous, Scotland, UK. (B) Sagenodus, Upper Carboniferous, Northumberland, England, UK. (C) Ceratodus, Upper Triassic, Gloucestershire, England, UK. UMZC specimens. Photograph by S.M.F.

Relationships and Relatives • 57

3.11). It too is solidly attached directly to the braincase. Because of this, the hyomandibula, normally the bone responsible for providing the link between quadrate and braincase, is reduced or lost. This is an example of an autostylic skull, in which the quadrate and braincase are securely integrated with the palate. A parallel case occurs in most tetrapods.

Lungfishes have a very specialized method of inhaling air by gulping it from the surface of the water. It involves the use of a "cranial rib" unique to lungfishes and a highly modified shoulder girdle that swings round beneath the chin as the buccal cavity expands (Fig. 3.11). At least some of the modifications to the palate described above, and others to the floor of the mouth, may be associated with the unique way in which they form a seal as the gulped air is swallowed, rather than with the requirements of a crushing dentition.

The development of these adaptations can be followed in the fossil record, and it is clear that the earliest lungfishes did not have them (Long 1993). This may mean that they did not breathe air at all, for they seem to have been marine forms that lived in relatively deep water (Campbell and Barwick 1987). However, because it is thought that all bony fishes were originally equipped with lungs or their equivalent, it may have been that, like *Neoceratodus*, they breathed air only occasionally, and then in a relatively unspecialized way.

There are many features of soft tissue anatomy and physiology in which lungfishes resemble tetrapods, causing many people to believe the two groups to be very closely related (Rosen et al. 1981). However, it is true that most of these features are concerned with air breathing and were quite likely to have been absent in the earliest lungfishes. If so, at least these similarities must have evolved in parallel among lungfishes and tetrapods. The fossil evidence nonetheless still seems to point to a closer relationship than that between tetrapods and coelacanths, although more evidence from modern forms needs to be gathered before this debate can finally be settled (see below).

Early lungfishes show the lobed fin in almost an idealized form, with long lobes formed of many radials. This is seen today almost unmodified in *Neoceratodus*. As well as the series of branches that run along the anterior, preaxial, edge of the fin in other lobe-fins, lungfishes have another set postaxially. In the modern lungfishes *Protopterus* and *Lepidosiren,* the paired fins have become whiplike appendages used in steering and upon which they sometimes have been seen to support themselves and "walk" on the bottom of the lake bed. Early lungfish tails were heterocercal: they had an asymmetrical tail, with the body lobe running up the leading edge of the fin, and the main part of the fin webbing suspended below it as a flag from a pole, with a lesser fringe growing above the body lobe. This style of fin seems to have been common to a number of early fishes. During their evolution, lungfish tails changed from the early asymmetrical heterocercal form to a symmetrical form in which upper and lower lobes became equal, and the upper lobe merged with the anal and dorsal fins to form a single long fin (summarized in Long 1993) (Fig. 3.13).

Extinct Lobe-Finned Fishes

During the Paleozoic, many lobe-finned families of diverse form and lifestyle existed that have no direct living descendants. The structure and relationships of some of these forms will put tetrapods and other modern lobe-fins into closer context.

According to some modern analyses, the lobe-finned lineage produced, in addition to some less speciose groups, two major branches, one

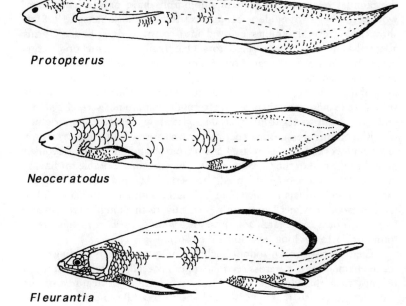

Figure 3.13. Body forms of some extinct and extant lungfishes: Protopterus *from Africa;* Neoceratodus *from Australia;* Fleurantia *from the Upper Devonian of Canada;* Dipterus *from the Middle Devonian of Scotland (from Ahlberg and Trewin 1995).*

that is more closely related to lungfishes and another that is more closely related to tetrapods (Figs. 3.1, 3.2). An alternative idea is that coelacanths are closer to tetrapods (Zhu and Schultze 1997, 2001), but more details of this debate are discussed later in the chapter.

In the scheme of relationships used here, the tetrapods are related most closely to the osteolepiforms and their relatives, in a group that has been named the tetrapodomorphs. With the lungfishes belong the porolepiforms and their relatives. The coelacanths are usually placed outside this dichotomy, having split off from the family tree earlier on, along with a small group called the onychodonts, a group whose anatomy and relationships are even more mysterious.

The tetrapodomorphs consist of several families, including some primitive forms often referred to as osteolepids, the rhizodonts (which are the least well known), the tristichopterids and some individual genera, *Panderichthys* and *Elpistostege* and *Livoniana,* all of which feature in the story of tetrapod origins.

Tetrapodomorph Lineage

The heyday of fishlike tetrapodomorphs spanned the Middle Devonian to the Early Permian. Over the last decade, many new genera of this lineage have been described from different parts of the world, including Kazakhstan and Australia. Indeed, the number of described genera has approximately doubled in the last decade. Known since the 19th century, the familial relationships of tetrapodomorphs have only recently been worked out by use of modern techniques (Ahlberg and Johanson 1998). It shows that several families evolved into relatively long-bodied forms adapted as lurking, ambush predators. One family, the tristichopterids, is particularly important because one famous example, *Eusthenopteron,* has been described in meticulous detail and has been used as a model for the kind of fish that might have given rise to tetrapods. The group as a whole shows some close resemblances to tetrapods that are not seen elsewhere, and most scientists regard these as demonstrating their common origins.

Rhizodonts are a group of poorly known forms whose anatomy is only just being worked out. They are known from the middle of the Devonian period to the end of the Carboniferous. They are most notable for having produced the largest bony-fish predators of their day, and estimates suggest that some individuals reached 7 m in length, with teeth to match. (The name *rhizodont* means "root-tooth.") Isolated teeth—long, curved, blade-like, and not a bad match for those of *Tyrannosaurus rex*—have been found up to 15 cm long (excluding the root).

Characteristics of rhizodonts include a unique pattern of folding of the tooth enamel, so that a cross section will easily establish whether a tooth belongs to a rhizodont or not. Also unique are some features of the shoulder girdle, so that one isolated shoulder girdle and fin from the Devonian of the United States called *Sauripteris* is put with reasonable certainty in this family. This specimen is important because it gives a good indication of the fin skeleton, and when first found, it was used as the model for the kind of fin structure that might have given rise to the limbs and digits of a tetrapod (e.g., Gregory 1915, 1935). More recently, however, that role has been superseded by fins belonging to other osteolepiforms.

Recent work on rhizodonts has helped to clarify their relationships a little. In Australia, a primitive rhizodont called *Goologongia* has been described from almost complete specimens (Johanson and Ahlberg 1998). New specimens of *Sauripteris* have been discovered in the United States, and work on the Carboniferous forms from Scotland has helped to shed light on their later evolution (Daeschler and Shubin 1997). They are considered to be basal members of the tetrapodomorph lineage, likely to have branched off from this clade quite early in its history. This means that the digitlike form of the fin rays in *Sauripteris* are likely to have been acquired in parallel to the digits of tetrapods (see also Chapter 6 and Fig. 6.8). They all have relatively broad, fanlike pectoral fins that may have supported the front end of the body either in water or with the head slightly emersed.

One striking feature of all rhizodonts is the elaboration of the lateral line system (for the structure and function of lateral lines, see Chapter 6). In all of them, it is exceedingly well developed both on the skull and body, and in the more derived forms has branches that ramify onto the shoulder girdle, a unique condition. A triple or possibly quintuple row of lateral line pores seem to have been present along the flanks of the fish, which may have been used as a phased array of detectors to give a very accurate picture of pressure waves in the water around them.

During their evolution, changes to the body form and increase in size can be seen in the rhizodontids, paralleling those in tristichopterids, suggesting that both groups adopted similar modes of life. The changes to the head morphology and reduction in fin size seen in tristichopterids were also echoed in the evolution of tetrapods, suggesting that such a predisposition characterized the whole tetrapodomorph lineage (Johanson and Ahlberg 1998).

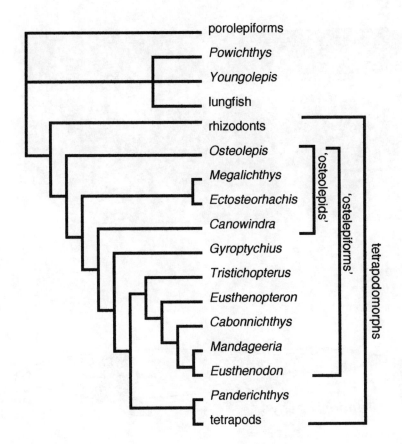

Figure 3.14. Cladogram from Ahlberg and Johanson (1998). The terms in quotation marks are paraphyletic and do not include all the descendants of the clade.

The most recent cladistic analysis of "osteolepiforms" has shown that it is paraphyletic with respect to tetrapods (Ahlberg and Johanson 1998). It includes those fishes that once were placed in the larger family known as osteolepidids. These have now been teased apart into a number of smaller more or less primitive families, with the more specialized family, called tristichopterids, including *Eusthenopteron,* lying at the crownward end of the tree near to the tetrapods (Fig. 3.14).

Osteolepis, one of the most primitive forms from the Middle Devonian, is known from deposits in Scotland and exemplifies early members of the group (Figs. 3.15, 3.16). It had the characteristic intracranial joint and a heterocercal tail. Like other early lobe-fins, *Osteolepis* had two dorsal fins. Its scales and head bones were covered in cosmine, a glossy-looking, enamel-like substance, which is seen in other early lobe-fins as well, including the widespread Carboniferous form *Megalichthys*. One of the last surviving genera, *Ectosteorhachis,* known from the Early Permian of the United States, is a rather primitive member of the osteolepiform lineage.

One Late Devonian locality in Australia, called Canowindra (the locals pronounce it "Canowndra"), has provided a spectacular number of fossil fishes in extensive bedding planes that probably represent mass mortality events. Most of the fishes are placoderms, but several lobe-fins were also present. All of them were large predators, up to a meter long. Three tetrapodomorph genera have been described, two of them only discovered in the last few years. They are *Canowindra* (the fish's name is

Figure 3.15. (above) Osteolepis macrolepidotus *from the Middle Devonian of Scotland, UK. Bar = 10 mm. UMZC specimen GN 766. Photograph by S.M.F.*

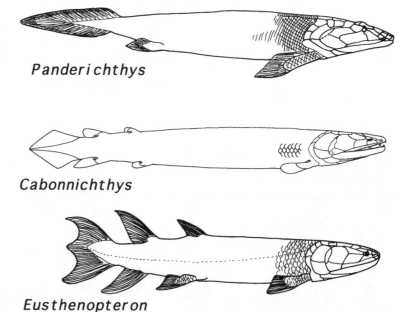

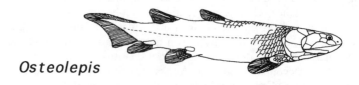

Figure 3.16. (right) Drawings of some of the so-called osteolepiforms: Osteolepis, *Middle Devonian, Scotland, UK;* Eusthenopteron, *Late Devonian, Canada;* Cabonnichthys, *Late Devonian, Australia (from Ahlberg and Johanson 1997), and* Panderichthys, *Late Devonian, Baltics and Russia.*

pronounced as written), *Cabonnichthys* (Fig. 3.16), and *Mandageria*. *Canowindra* is the most primitive of the three, whereas *Cabonnichthys* and *Mandageria* are both highly specialized members of the Tristichopteridae (Ahlberg and Johanson 1997; Johanson and Ahlberg 1997). They are both long-bodied forms with relatively small but almost symmetrical tail fins and dorsal and anal fins positioned near the rear end of the body. The shape suggests that they had evolved into ambush predators, characterized by the ability to accelerate quickly, but not designed for sustained swimming. They are both quite closely related to a similar-shaped form, *Eusthenodon*, from Greenland and Russia, showing that the tristichopterids were a wide-ranging group, not restricted to a small part of the world. The tristichopterids as a whole show an evolutionary increase in size through the Late Devonian, and the proportions of the head show matching changes in which the anterior part of the skull became progressively longer with time.

Eusthenopteron comes from a rich Frasnian locality called Escuminac Bay in Quebec, Canada, which has also yielded many other species of fish (see Chapter 4). The survey of the skeleton of a lobe-fin in Chapter 2 is based on *Eusthenopteron*, so that its anatomy is described in more detail there. However, this fish has its own unique features. One of them is the shape of its tail, which, rather than being asymmetrical like that of *Osteolepis*, is symmetrical, somewhat similar to that of a coelacanth, although different in detail (Fig. 3.16).

Eusthenopteron is sometimes figured as crawling out of the mud of a Devonian lake, apparently with the intention of finding another pool to swim in. There are two main reasons for this picture, relating to the origin of tetrapods, and they will be discussed in more detail in Chapter 4. But briefly, because *Eusthenopteron* was once cast in the role of the "ancestor" of tetrapods, tetrapodlike behavior was attributed to it. Second, the form of the fin skeleton was rather chunky, and its first three radials bear close comparison with the humerus, radius, and ulna of tetrapods. However, taking the whole morphology of the fish, with its streamlined torpedo shape, and dorsal, anal, and pelvic fins placed near the back of the body, it seems that the lifestyle of *Eusthenopteron* was much more like that of a modern pike (*Esox*), a fully aquatic lurking predator.

Panderichthys was, until quite recently, placed within a family called the panderichthyids along with a second genus, *Elpistostege*. *Panderichthys* is more closely related to tetrapods than any osteolepidid or even *Eusthenopteron*. However, although they share several similarities, the most recent work has pointed out differences between *Panderichthys* and *Elpistostege* in which *Elpistostege* is even more tetrapodlike than is *Panderichthys* (Ahlberg et al. 2000). Indeed, the first specimen of *Elpistostege*, which consisted just of a partial skull roof, was thought to be a tetrapod when it was first found (Westoll 1938). Confirmation of its true status only came when a second, more complete specimen was discovered in the 1980s (Schultze and Arsenault 1985). Now, both genera are placed as successive genera on the stem lineage close to tetrapods. The family name has become meaningless because they are not uniquely related to each other (as distinct from being branches off the tetrapod stem lineage).

Elpistostege comes from the same locality and time period as *Eusthenopteron*, Escuminac Bay, although, because it is only known from two specimens, its anatomy is not nearly so well understood. *Panderichthys* comes from a number of localities in Eastern Europe, with the most remarkable specimens from the Frasnian of Lode in Latvia. They include some more or less complete animals. Because these are still covered in their scales, the internal skeleton cannot be seen, so much of their bone structure remains unknown.

Figure 3.17. Skull of Panderichthys rhombolepis. Specimen in the Institute of Evolutionary Morphology in Moscow. Photograph by J.A.C.

Panderichthys was quite a large fish, with a skull about 300 mm long, and a total body length of over a meter (Figs. 3.16, 3.17). Its body and skull were flattened and the snout rather pointed. The eyes were placed quite close together on the top of its head and were set beneath ridges, giving the impression of eyebrows and creating a subjectively tetrapodlike appearance. Other characters of the skull were also very tetrapodlike (Vorobyeva and Schultze 1991). *Elpistostege* is still relatively poorly known, and for understanding the story of the origin of tetrapods, *Panderichthys* will provide a satisfactory guide. By comparing its skull with that of a very early tetrapod such as *Acanthostega,* those tetrapodlike features that were present already in *Panderichthys* can be contrasted with those that had yet to evolve. This gives some ideas about the order and timing of the appearance of some tetrapod characters, and these will be considered in Chapter 6.

Livoniana multidentata is the most recently recognized Frasnian tetrapodomorph (Fig. 3.18) (Ahlberg et al. 2000). It is known from only two fragments, both of the anterior end of the lower jaw. It comes from the area around the border between Latvia and Estonia, an area whose ancient name was Livonia. The most striking thing about the jaws is their multiple rows of small, rounded teeth, which give the appearance of a portion of corn on the cob. There are up to five rows of teeth, a feature not known in any other tetrapodomorph. The jaws do, however, show two or three features that are otherwise only found in animals close to true tetrapods like *Acanthostega.* When all the available information is pooled and put into a computer analysis, *Livoniana* appears sharing a node with *Elpistostege,* just below *Obruchevichthys* on the tetrapod stem lineage. It was probably not a full tetrapod in the sense of having limbs with digits, but it was probably close to the cusp of tetrapod evolution, and so is one of the few concrete pieces of evidence for the range of animals that actually existed between truly "fishlike" stem tetrapods and more fully "tetrapodlike" ones. Its peculiar dentition suggests that there was a radiation of specialized animals at the boundary, of which one went on to produce tetrapods proper.

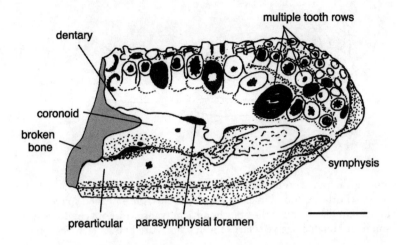

Figure 3.18. Specimen drawing of Livoniana. Based on Ahlberg et al. (2000).

Dipnomorph Lineage

The earliest member of the porolepiforms ("hollow scale forms") comes from the Early Devonian of Canada and is thus one of the earliest lobe-finned fishes for which other than scattered remains are found. This genus is called *Powichthys,* named after Prince of Wales Island, where it was found. The best-known members of the group include *Glyptolepis* and *Holoptychius,* from the Middle Devonian of Scotland and Late Devonian of East Greenland, respectively (Ahlberg 1991a). Porolepiforms are not found beyond the end of the Devonian.

There are several differences between osteolepiforms and porolepiforms, although their superficial similarity once caused them to be put together in a group known as the Rhipidistia. However, because they are not each other's closest relatives, this name has fallen into disuse. Both groups retain the intracranial joint in the braincase as well as the skull roof, and one of the easiest differences to spot is that where osteolepiforms have four bones ranged across the skull table behind the hinge line, porolepiforms have only two. Where osteolepiforms have short, compact pectoral fin skeletons, porolepiforms have long, lanceolate fins like early lungfishes. Indeed, this is but one of the characters linking porolepiforms to lungfishes. A couple of very primitive genera, *Youngolepis* and *Diabolepis* from the Early Devonian of China, corroborate this relationship by providing glimpses of what the common ancestor of these two groups was like (Chang 1995; Chang and Smith 1992).

Onychodonts

This final "fish" group is one that, along with coelacanths, cannot be slotted into the lobe-fin family tree very easily. Although excellently preserved specimens have been found in Devonian marine sediments of Australia, the onychodonts have not been fully described scientifically and remain a missing piece, possibly a key piece, in the jigsaw puzzle of lobe-finned fish relationships. Their most characteristic feature is the spiky whorl of teeth that the fish bore at the front of each lower jaw ramus (Fig. 3.19). These teeth are S-shaped and would have been hard to dislodge once engaged in their prey. The pattern of growth of these teeth seems to be

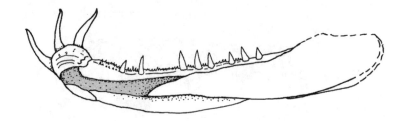

Figure 3.19. Lower jaw of an onychodont showing parasymphysial teeth, set on a moveable plate.

Figure 3.20. (opposite page) Cladogram showing the differing definitions of Tetrapoda according to the crown group and total group definitions. Note that definition of tetrapods by the key character of limbs with digits would include at least Acanthostega and everything more "crownward" (i.e., more derived than Acanthostega). Because it is not known for certain that Elpistostege had digits, its position within the tetrapods would be uncertain. Note also that this is a notional cladogram of the tetrapods by whichever definition is used; see Chapter 9 to see that some definitions would exclude temnospondyls and anthracosaurs from the crown group. Under the crown group definition, all the animals that fell outside the crown group (i.e., those directly related to modern tetrapods) would become members of the stem group, or the tetrapod stem lineage. The term tetrapodomorphs can be used to indicate this lineage. Tetrapodomorpha might also be an acceptable term for the total group.

among the most primitive found among jawed vertebrates, and similar tooth whorls are found in early porolepiforms, the so-called spiny sharks or acanthodians, and in the first chondrichthyans. Most onychodonts were small, although some very large tooth whorls are known (Janvier 1996).

What Is a Tetrapod and How Is the Group Defined?

This review of the lobe-finned fish groups is not complete without the tetrapods, because this is where, evolutionarily speaking, they (and humans) belong. Modern tetrapods include on the one hand the amphibians—frogs, newts, caecilians, and their kin—and on the other the amniotes—mammals plus the "reptile" groups, including turtles, lizards and snakes, and crocodiles and their closest living relatives, the birds. It includes creatures that, although they do not have legs (limbs with digits) themselves, are descended from some that did. So bats and whales are tetrapods, as are birds and snakes. It also includes all the fossil forms such as dinosaurs, flying or swimming reptiles, and many other more bizarre and less well-known kinds, so long as they are descended from ancestors with legs (Figs. 1.1, 1.9, 1.10, 3.20).

Most current views maintain that tetrapods are a natural group, tied together by numerous unique characters that show that the group had a single common ancestor. Among the unique features that tetrapods share is the possession of limbs with digits; the structure of these was explained in Chapter 2.

Although there is a suite of characters unique to most known tetrapods, as this book emphasizes, not all these characters evolved at once, and the sequence of their evolution is unknown in many cases. This is to be expected if tetrapods evolved by a gradual process, which is what most scientists believe happened. Furthermore, the same must be true of limbs and digits—they evolved gradually by stages. Thus, difficult questions arise: when can an appendage really be called a limb, and so at what point does a tetrapod really become a tetrapod? In fact, if evolution is gradual, then there is no precise point in the continuum at which a line can be drawn to distinguish indisputably a limb from a fin, or indeed a tetrapod from a fish. What exactly would one choose? Loss of the fin rays? The evolution of the wrist or ankle bones that interarticulate? The evolution of multiple, jointed, postaxial radials? If so, how many would be the critical number? To add to the difficulties, because there are so many unique characters known for tetrapods, limbs and digits may not be the key feature that most usefully separates them from fishes. Some people might, for example, make the case that the evolution of a neck with loss of the operculogular series makes a more biologically significant character to separate the groups. Or

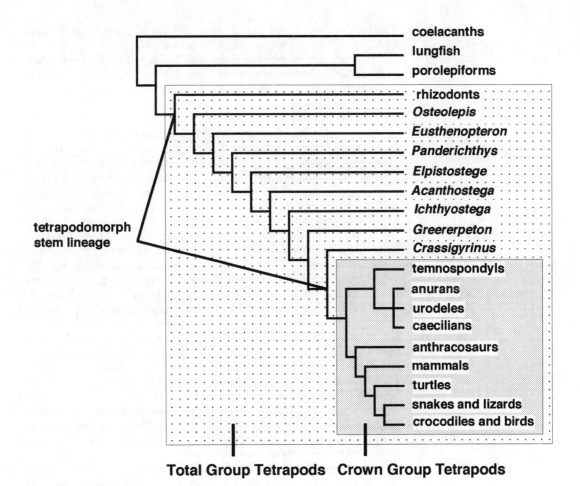

others may choose the changes to the braincase that resulted in the stapes penetrating the wall of the otic capsule.

Until recently, the distinction between fossil "fish" and fossil "tetrapods" was usually easy to make, once the position of lungfishes had been resolved. However, the discovery of more fossils representing fishlike tetrapods and tetrapodlike fishes, and the discovery of fragments of such creatures for which the limbs are not known have raised issues of some importance. In the future, the problem will most likely become even more severe as the transitional series becomes better known, and arguments over what exactly a tetrapod is could become highly contentious.

In an attempt to sidestep the problems and provide a more stable means to define the limits of key groups of animals, such as mammals, birds, or tetrapods, methods have recently been proposed that are thought to be entirely independent of any single "key" character. These make use of the phylogenetic relationships of fossil forms to recent ones and are called the "node-based" and "stem-based" methods (Fig. 3.20). Each has its problems, and as yet, there is no general agreement about which should be used, or even if either form is really more stable or more useful than the old one that used key characters (e.g., Benton 2000; de Queiroz and Gauthier 1992, 1994; Lee and Spencer 1997; Patterson 1993).

In brief, both forms look at the relationships of modern representatives and define tetrapods in terms of these forms. Amphibians and amniotes must have had a common ancestor at some stage among the early tetrapods that was related equally to both but strictly belonged to neither. (Readers should note that because the earliest tetrapods are not always or easily referred to either of these modern lineages, they are referred to as tetrapods and not as amphibians.) Although this common ancestor will probably always remain hypothetical, it would lie at the node where the lineages of amphibians and amniotes join. Any fossil form that can be shown to belong to either one of these lineages is said to lie above the node, within the "crown group" of tetrapods. Any creature that does not belong to either lies below the node, in the "stem group" of tetrapods.

At this point, the crown group definition would admit only creatures above the node into the formal taxon Tetrapoda, with those falling outside it being relegated to stem-tetrapods, even though they have four legs. The stem-based method, also known as the "total group" method, would admit not only the crown group, but a whole suite of other creatures below the crown on the cladogram, but above the node that admits its nearest living relative. For example, with respect to modern tetrapods, this method would look at the nearest living relative of crown group tetrapods, in this case most likely (but not certainly) the lungfishes, and pinpoint the node at which these groups separated as being the next most significant. All animals belonging to the tetrapod stem lineage and falling above the split between lungfishes and crown group tetrapods—such as osteolepiforms and rhizodonts—would be admitted into the formal taxon Tetrapoda, even though they do not have four "legs" (Fig. 3.20). The conundrum has been partially solved by the decision to call this whole clade the Tetrapodomorpha (Ahlberg 1991a).

Both these methods have consequences that are counterintuitive and confusing for the nonspecialist, and it remains to be seen whether either of them will gain universal acceptance. Thus, for the purposes of this book, and with the recognition that it can be criticized, the term *tetrapod* will continue to be applied to animals with four legs bearing digits.

Historical Background and the Debate about Tetrapod Relationships

Ideas on what tetrapods are, where they came from, and who their closest relatives are have gone through a series of phases over the course of history, echoing the main concerns, interests, and preoccupations of those who studied the problem. It is revealing to study the history of these ideas as much for what it shows about science and its progress as for what it says about tetrapods.

The first phase belongs to the naturalists of the early–mid-19th century, in the days before Darwin introduced his ideas of "descent with modification," or "evolution," as it is now better known. In those days, one of the main jobs of the naturalist, who was often not a professional scientist but might be a cleric, a medical practitioner, an explorer, a military man, or a person (there were a few notable women) of independent income, was to classify the inhabitants of the natural world. The hierarchical ordering of "groups within groups" outlined in Chapter 1 was already well established, and there were few organisms that could not be fitted comfortably into known groups, despite the speed at which new ones were being found. This was a time of expansion and exploration brought about by the Industrial Revolution. Even fossils at that time were recognized as the remains of once living animals, and most of these were readily fitted into the scheme of things. The natural order was considered to be a manifestation of the order in the mind of the Creator when He made the world. The idea that species could become extinct, however, was new, and not accepted by everyone. Thus it was that vertebrates could easily be classified as fish or reptile, mammal or bird. (The distinction between amphibians and reptiles was, however, not yet clearly defined, but for the purposes of discussing the origin of tetrapods as a whole, this is a peripheral issue, and what are now known as amphibians were then included within the term *reptile*. This old amalgamation is reflected in the term *herpetology* for the study of amphibians and reptiles.)

The trouble started with the discovery of the lungfish *Lepidosiren* from South America in 1837. Here was an animal that breathed both with lungs like a tetrapod and gills like a fish, that had little in the way of fins, and that had soft tissue resembling in many ways that of tetrapods. This was followed by the discovery of the African and Australian genera. Where did they fit? The earliest work described them as amphibians, but Richard Owen, at the institution that he largely founded and is now called the Natural History Museum (London), came down on the side of fish. The basis of his decision was the state of the nostrils, because he did not believe that the posterior nostril in lungfishes was equivalent to the choana or internal nostril of tetrapods (Owen 1841). Because Owen was so influential, for a time, this view held sway. The arguments concerned the affiliation of the animals in question, and the problems arose because lungfish did not fit neatly into any of the previously recognized categories. It was a matter of trying to pigeonhole the animals each into its appointed place; the idea of phylogenetic relationships had not yet arisen.

When Darwin published his work on the *Origin of Species* in 1859, it became clear that the hierarchically arranged, nested sets of groups into which the animals fitted was in effect a family tree and resulted from the fact that the animals were related to one another by descent. The word *descent* here is crucial because it produced a search not only for immediate kinship relationships but ancestor–descendant sequences. The search therefore began for tetrapod ancestry. This had two effects. One was to relegate the question of where lungfishes fitted to a side issue: because evolution predicted intermediate forms, it was no longer a surprise to find that some animals could not be slotted into conventional groups. The other effect was to shift the search from among living forms to fossil ones.

One of the first people to classify fishes after the publication of Darwin's book was T. H. Huxley. He coined the term "crossopterygian" for a number of fossil fishes that had fleshy fins, and in 1861 he included fossil lungfishes (not yet recognized as related to living ones), coelacanths (not yet known from living forms), and a number of others. It was the grouping of these forms together that paved the way for the recognition that their fin structure shared certain characters with tetrapod limbs (Fig. 3.21). In 1892, however, E. D. Cope suggested that lungfishes were too specialized to have given rise to tetrapods and that the tetrapods' ancestor should be sought elsewhere among the crossopterygians. However, he spotlighted the crossopterygians as the group from which tetrapods probably arose.

In the early years of this second phase, ideas varied about which animals belonged where among the crossopterygians, and several genera

Huxley's Classification of Fishes (1861)

Ordo Ganoidei
Subordo I - Amiadae (*Amia*)
Subordo II - Lepidosteidae (*Lepisosteus*)
       (These two are now both put among ray-finned fishes)
Subordo III - Crossopterygiidae
    Fam 1. Polypterini
        Dorsal fin very long, multifid; scales rhomboidal
        Polypterus
        (This is also now considered to be a primitive ray-finned fish)
    Fam 2. Saurodipterini
        Dorsal fins two; scales rhomboidal, smooth; fins subacutely lobate
        Diplopterus, Osteolepis, Megalichthys
    Fam 3. Glyptodipterini
        Dorsal fins two; scales, rhomboidal or cycloidal, sculptured; pectoral fins acutely lobate; dentition dendrodont
            Sub-Fam. A with rhomboidal scales
                  Glyptolemus, Glyptopomus, Gyroptychius
            Sub-Fam. B with cycloid scales
                  Holoptychius, Glyptolepis, Platygnathus,
                  [Rhizodus, Dendrodus, Cricodus, Lammodus]
    Fam. 4. Ctenodipterini
        Dorsal fins two; scales cycloidal; pectorals and ventrals acutely lobate, dentition ctenodont
        Dipterus, [Ceratodus, Tristichopterus]
    Fam. 5. Phaneropleurini
        Dorsal fin single, very long, not subdivided, supported by many interspinous bones; scales thin, cycloidal; teeth conical; ventral fins very long, acutely lobate
        Phaneropleuron
    Fam. 6. Coelacanthini
        Dorsal fins two, each supported by a single interspinous bone; scales cycloidal; paired fins obtusely lobate; air bladder ossified
        Coelacanthus, Undina, Macropoma
Subordo IV Chondrosteidae (*Acipenser, Polyodon*)
Subordo V - Acanthodidae ('spiny sharks'; a fossil group now entirely extinct, see
       chapter 4)

My comments are in parentheses.

Note that this scheme places together animals that we would now separate into lungfishes, porolepiforms and osteolepiforms (eg in Fam. 4. the osteolepiform *Tristichopterus* appears in the same family as the lungfish *Dipterus*). However, osteolepiforms, porolepiforms, coelacanths and lungfishes are all included in the crossopterygians.

Figure 3.21. Table showing Huxley's classification of fishes from 1861.

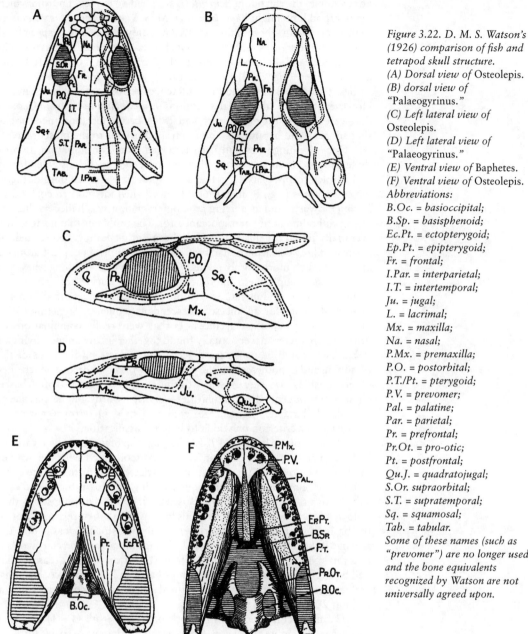

Figure 3.22. D. M. S. Watson's (1926) comparison of fish and tetrapod skull structure.
(A) Dorsal view of Osteolepis.
(B) dorsal view of "Palaeogyrinus."
(C) Left lateral view of Osteolepis.
(D) Left lateral view of "Palaeogyrinus."
(E) Ventral view of Baphetes.
(F) Ventral view of Osteolepis.
Abbreviations:
B.Oc. = basioccipital;
B.Sp. = basisphenoid;
Ec.Pt. = ectopterygoid;
Ep.Pt. = epipterygoid;
Fr. = frontal;
I.Par. = interparietal;
I.T. = intertemporal;
Ju. = jugal;
L. = lacrimal;
Mx. = maxilla;
Na. = nasal;
P.Mx. = premaxilla;
P.O. = postorbital;
P.T./Pt. = pterygoid;
P.V. = prevomer;
Pal. = palatine;
Par. = parietal;
Pr. = prefrontal;
Pr.Ot. = pro-otic;
Pt. = postfrontal;
Qu.J. = quadratojugal;
S.Or. supraorbital;
S.T. = supratemporal;
Sq. = squamosal;
Tab. = tabular.
Some of these names (such as "prevomer") are no longer used, and the bone equivalents recognized by Watson are not universally agreed upon.

Relationships and Relatives • 71

were cast in the role of "model for a tetrapod ancestor" at different times. For example, the rhizodont *Sauripteris* supplied a "model" fin from which the tetrapod limb might have arisen. D. M. S. Watson chose *Osteolepis* as a crossopterygian whose skull appeared not only sufficiently unspecialized to be a tetrapod ancestor, but one that showed a number of similarities in bone pattern and skull construction to the Carboniferous tetrapods that he also described (Fig. 3.22) (Watson 1926). At that time, these were the earliest tetrapods known, and the inferences seemed reasonable to make. However, it was the detailed descriptions of *Eusthenopteron* from Canada that were to make the real impact, and its legacy as a tetrapod ancestor has shaped ideas and hypotheses about tetrapod origins to the present day.

Once the lungfishes had been removed from consideration, two or three groups remained among the crossopterygians: the osteolepiforms and porolepiforms, placed together as "rhipidistians," and alongside them, the coelacanths (e.g., Romer 1966). Hence, when the modern coelacanth *Latimeria* was found as a living crossopterygian, it was hailed as the surviving embodiment of a tetrapod ancestor. The problems that its anatomy eventually posed for this theory have been outlined above. Nevertheless, because of how coelacanths had been classified, for most people, *Latimeria* appeared much more closely related to tetrapods than did lungfishes.

Because people were interested in ancestry during this phase, the emphasis was on finding more and more primitive or generalized forms to provide the base for an ancestor–descendant sequence. Sometimes, this meant erecting hypothetical ancestors that were really combinations of primitive characters never actually found together in any animal, living or fossil. They could be constructed according to taste on the basis of the reasoning and hypotheses that the particular worker favored. The schemes of relationships that were drawn up during this phase were also often based on primitive or generalized similarities—in other words, on the absence of specialized characters. Absence of characteristics, in retrospect, was an unsatisfactory criterion on which to base a classification.

The next phase began in 1981, when a group of workers, who came to be known as the "Gang of Four," became discontented with this way of thinking. As a result of the impact of cladistics, they returned to the question of the relationships of tetrapods to living and fossil forms and subjected it to a radical reassessment (Rosen et al. 1981). The philosophy they followed was to take the living animals first, establish their relationships on the basis of shared derived characters, and then slot the fossils into the scheme. If the relationships between the living animals are correctly worked out, this should be a feasible proposition. (Unfortunately, for reasons that will be explored below, the relationships between lungfishes, coelacanths, and tetrapods are not so easily established as first appeared, and the jury is still out 20 years later.)

The main conclusion of the work of Rosen et al. (1981) was that lungfishes were more closely related to tetrapods than were coelacanths, in an echo of the original ideas of the mid-19th century (Fig. 3.23). More surprising still was that they suggested lungfishes and all their fossil kin to be more closely related to tetrapods than any of the so-called crossopterygians. *Eusthenopteron* was relegated to a lowly and primitive branch on the tree of lobe-fins. The years that have followed have proved this work to be one of the most influential in the subject. Its first effect was to galvanize other workers to spring to the defense of the more traditional views and to reassess their own ideas in cladistic terms (e.g., Panchen and Smithson 1987) (Fig. 3.24). These efforts have culminated in complete reworking of the relationships of lobe-fins, such that terms such as *crossopterygian* and

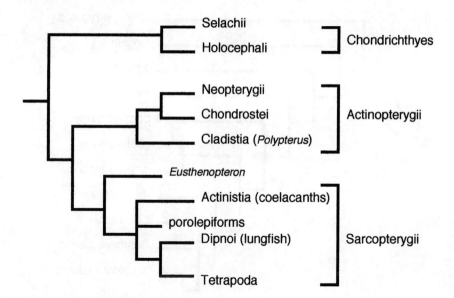

Figure 3.23. Cladogram from Rosen et al. (1981). Note that Eusthenopteron *falls outside the sarcopterygians in this cladogram because it lies below the node that unites tetrapods, lungfishes, and coelacanths, the crown group sarcopterygians.*

rhipidistian are no longer used with their original meanings. As they had come to be defined over the years, they embraced what cladists term *paraphyletic groups* held together only by what they lacked, and part of this problem was caused by the exclusion of the lungfishes. The lungfishes were now returned to the fold, and in most recent cladistic analyses, it is the lungfishes rather than the coelacanths that emerge as the tetrapods' closest living relatives. However, in contrast to the 1981 analysis, a stem lineage of tetrapodomorph lobe-fins, including rhizodonts *Osteolepis* and *Eusthenopteron,* has been established, with many newly described fossil taxa adding support to this idea (Ahlberg and Johanson 1998). Taking account of fossil evidence, many of the lungfishes' soft tissue characters in which they resemble tetrapods appear to be parallel developments, showing not only the danger of the use of only the living forms for analysis, but also the usefulness of fossil ones.

To sound a final note of warning to end this phase, not all cladistic analyses produce the same result; a few have produced the alternative grouping of coelacanths as more closely related to tetrapods than are lungfishes (Zhu and Schultze 1997, 2001) (Fig. 3.25). However, the data

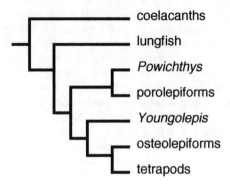

Figure 3.24. Cladogram from Panchen and Smithson (1987).

Relationships and Relatives • 73

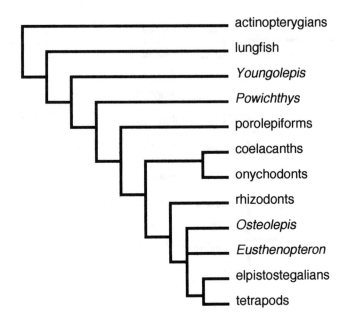

Figure 3.25. Cladogram from Zhu and Schultze (1997). The database supporting this cladogram with coelacanths closer to tetrapods than are lungfishes varies by only a few data points from that of Cloutier and Ahlberg (1997) (Fig. 3.1).

sets that produced this result were different in only a few details from that producing the lungfish–tetrapod grouping and this shows the fragility of both analyses. Part of the problem stems from the fact that all three groups —lungfish, coelacanths, and tetrapods—have a long evolutionary history separate from one another, which means that they are all rather different from their early ancestors. The impact of this fact rebounds into the next phase of inquiry.

With the advent of molecular data analysis, it might have been hoped that the problem of the interrelationships between tetrapods and their two closest living relatives would be quickly and decisively solved. One could have expected that a straightforward comparison between the structure of a molecule such as the DNA or RNA of each would yield definitive results and that there would no longer be any need to resort to fossil data. Events have proved otherwise. There are a number of reasons for this. To begin with, the analyses that have been performed by different research groups have not used comparable molecular data or a comparable set of taxa in their analyses. In order to work out the relationships of tetrapods to lungfish and coelacanths, it is first necessary to compare their molecular structure with those of "fish," to represent the basal or outgroup condition—and here, the choice has varied widely. Then there is the problem of which tetrapod or tetrapods to use as representative, and again, the choice and the number of taxa has varied. No two analyses have used the same set—indeed, most analyses have used either the lungfish or the coelacanth to test against other "fishes" and tetrapods, but not both. Nor have all the analyses used the same molecules; the early ones used amino acids, but some of the more recent studies have used nuclear genes from ribosomal RNA, but again, from different examples. Not surprisingly, not only did the analyses produce different results, but the results were not easily compared with one another. For example, in a study by Stock et al. (1991), the coelacanth and tetrapods emerged as sister groups, but disconcertingly, ray-finned fish and sharks were found nested within tetrapods! This result

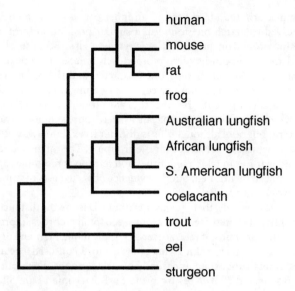

Figure 3.26. Cladogram from Zardoya and Meyer (1996) based on molecular data.

was tested for its robustness, and was found (thankfully) to be only weakly supported by the statistics in any case.

Probably the most comprehensive study in recent years has been that of Zardoya and Meyer (1996), who tested the most representative sample of taxa by means of nuclear genes. That study found lungfishes and coelacanths to be sister groups—in other words, neither one was the sister group or closest relative of tetrapods (Fig. 3.26). This result has some support from a small set of soft tissue characters, but none at all from fossil data. Other studies that used mitochondrial genes suggest lungfishes to be the tetrapods' sister group, but generally, the analyses are poorly supported statistically.

One of the reasons why these molecular analyses have been so inconclusive stems from the fact noted above: that the three groups in question have been separate from each other for a very long time. This can be shown in the fossil record incontrovertibly. Their molecular structure has therefore had several hundred millions of years in which to diverge. Even though coelacanths might be thought to be very conservative anatomically, it has already been shown that at least some profound genetic changes, affecting the fins, have occurred during their evolution. Similarly, the lungfishes have evolved adaptations of the skull extreme enough to debar them for decades even from consideration as part of the tetrapod story. Not only that, but the modern forms are very different in external appearance, and probably internal appearance, from their Devonian forebears. As for tetrapods, whether one takes modern amphibians or any particular amniote, one is dealing with an animal whose ancestors split from the common tetrapod stock more than 300 million years ago. Furthermore, the tetrapods typically used in the analyses (for example, a mammal such as a mouse or a frog such as *Xenopus*) belong to groups that came into existence no earlier than the Triassic.

Thus it seems that on the evidence from molecular studies, the jury is still out. What about analyses of soft tissue anatomy in the three living groups? Does that fare any better or give firmer clues? The problem here is that analyses have tended to concentrate on individual features or systems.

For example, Fritzsch (1987) found an inner ear receptor in coelacanths that appeared to match one found in a similar position in tetrapods and that was not found in lungfishes or anywhere else. Because of this, he suggested that coelacanths were more closely related to tetrapods than were lungfishes. But this is only one character, and the resemblance between the two receptors is not perfect. The two might equally well represent parallel developments. Northcutt (1986) found some features of the brain in which lungfishes and coelacanths were more similar to each other than to tetrapods, and although this is difficult to explain away, it remains one of the few pieces of evidence for the grouping. The range of circulatory and related features in which lungfishes and tetrapods resemble each other has already been mentioned, but the evidence now suggests they must have been developed in parallel.

Recourse to fossils, then, seems inevitable. Despite the distance in time from the split between these three groups, the fossil record is improving all the time with more finds of relevant taxa. For the moment, many of the new discoveries are in fact making things even more confused because they represent animals that span apparently separate groups. For example, the recently discovered *Psarolepis* from the Early Devonian/Late Silurian of China has characters found in placoderms, chondrichthyans, ray-finned fishes, and lobe-fins (Zhu et al. 1999). At present, the balance of evidence is in favor of its being a primitive lobe-fin, but opinions may change as more of the animal, or its contemporaries, are found. Such discoveries will assuredly affect how the relationships between coelacanths, lungfishes, and tetrapods are viewed in the future. For the moment, it is hard to see that opinion will be swayed away from the idea that tetrapods are the descendants of a "tetrapodomorph" lineage including fish such as *Panderichthys, Eusthenopteron,* and *Osteolepis.*

One other aspect of the relationships of tetrapods deserves a mention here. Throughout this book, the assumption is made that tetrapods are what is called a "natural group," that that group had its origins in the mid–late Devonian, and that all modern forms are descended from a unique common ancestor appearing somewhat later, perhaps in the Early Carboniferous. This idea is almost universally held today, but it has not always been so.

In the 1930s and 1940s, thinking about evolution was only just beginning to take the shape that it has today, in a move called the New Synthesis. This brought together the new sciences of genetics and its spin-off, population genetics, with the older discipline of paleontology. Evolution was seen in terms of adaptive radiation, with adaptation to certain environments more important than their phylogenetic descent in determining animal forms. This permitted the idea that, for example, many vertebrate groups could have been descended from more than one ancestor. Mammals, birds, and particularly teleost fishes were considered by some in this way. In this atmosphere, one school of thought became convinced that tetrapods were the product of at least two separate radiations. This school was founded in Stockholm, and its ideas were widely taken up in Eastern Europe. There were two versions of the idea, but both recognized a divergence between salamanders on the one hand, and all other tetrapods on the other. Holmgren (1933) suggested that dipnoans gave rise to salamanders, with all other tetrapods being descendants of osteolepiforms. Indeed, there are an intriguing number of soft tissue and molecular characters in which salamanders and lungfishes resemble each other very closely, although the main thrust of Holmgren's argument was based on the apparently different order in which salamander digits develop compared to those of all other tetrapods. This meant, to Holmgren, that salamanders could not share their limb structure with other tetrapods but must have acquired it separately. Recent reanalysis of salamander embryology has suggested that they have a specialized developmental program for digit production but are not fundamentally different from other tetrapods (see, e.g., Hinchliffe et al. 2001).

By contrast, Jarvik (1942) suggested that salamanders were the descendants of porolepiforms and that frogs and all other tetrapods arose from osteolepiforms. This, he maintained, was supported by detailed resemblances in the structure of the nasal capsules in the two lineages. Not many people outside the Stockholm school could accept this conclusion, not least because it seemed that the soft tissue structures, upon which so much reliance was placed, had been reconstructed in the fossil representatives according to the preferred model. Another curiosity was that Jarvik considered tetrapod diversification, as exemplified by mammals, to have sprung directly from *Eusthenopteron* with no recognizable intervening stages. Thus he viewed the skull and soft tissue of a mammal as having its exact counterpart in *Eusthenopteron,* whose soft anatomy could thus be reconstructed.

Today, the overwhelming majority of workers accept that the long list of characters held in common by all living and, where known, by almost all fossil tetrapods is evidence that they form a monophyletic clade with a single common ancestor. It is now realized that not all these common characters arose at once, and some of the new preoccupations include questions of which arose first, when, and how. Much of this book is devoted to how some of these questions are currently being answered.

Further information on some of the topics covered in this chapter can be obtained from the following books. For coelacanths, Smith (1956) gives his original account of the finding of the first and second coelacanths. Weinberg (1999) updates this with the story of more recent finds. Musick et al. (1991) is a volume covering aspects of the biology of *Latimeria,* and Forey (1998) gives a detailed account of all fossil coelacanths as well as aspects of the biology of *Latimeria* and the problems of determining the relationships of coelacanths to other lobe-fins. For lungfishes, Bemis et al. (1987) give details of the anatomy, physiology, and ecology of modern lungfishes, as well as a bibliography of fossil forms, and for a less technical account of Paleozoic and other fossil fishes, Long (1995) is both entertaining and well illustrated.

Four
Setting the Scene: The Devonian World

Devonian Biogeography and Climate

The Devonian period opened onto a world very different from the present day. In the earliest stages, over 400 million years ago, even the oxygen content of the air was different from today. According to some models, at the beginning of the Devonian, the air contained about half the present levels of oxygen (O_2), but about 10 times the present amount of carbon dioxide (CO_2). The estimates were constructed on the basis of a range of factors, including increasing radiation from the sun over the last 570 million years and how that has affected weathering rates of carbonates and silicates, and how the increase in plant cover changed the rate of weathering as well as the uptake of CO_2 from the air (Berner 1993). The models were backed up by evidence from carbon isotope studies of fossil soils. The effects mean that because carbon dioxide is less dense than oxygen, the total air density was less than it is now. During the Devonian, the proportions of one gas to another changed radically, with oxygen increasing and carbon dioxide decreasing throughout the Devonian and Carboniferous (Graham et al. 1997). It is an example of the influence of the increase in land plants covering the Earth, using CO_2 and giving out O_2. By the end of the Devonian, O_2 had just about reached modern levels, whereas CO_2 had reduced by about half compared to early Devonian levels. Figure

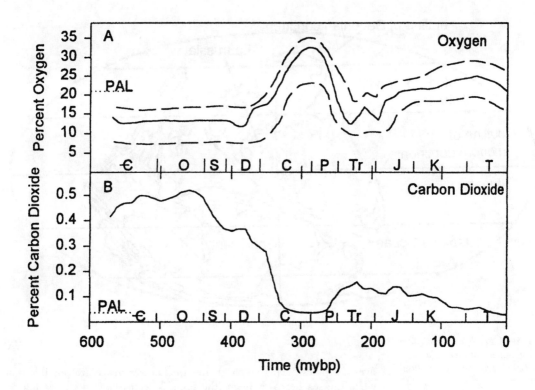

Figure 4.1. Graph of O_2 (A) and CO_2 (B) levels through time (from Graham et al. 1997). C = Carboniferous; $\mathrm{\mathfrak{E}}$ = Cambrian; D = Devonian; J = Jurassic; K = Cretaceous; T = Tertiary; O = Ordovician; P = Permian; PAL = present atmospheric levels; S = Silurian; Tr = Triassic.

4.1 shows graphs of changes in CO_2 and O_2 levels since the beginning of the Paleozoic, according to certain models. The result of these changes profoundly influenced the rest of life on Earth in ways that are only recently beginning to be appreciated (e.g., Dudley 1998).

The continents as they are today did not exist, but separate chunks lay scattered around the oceans. Even the length of the day was different. The Earth has been steadily and gradually slowing down since its origin, and so in the Devonian, it was spinning faster than today, making the days shorter. At that time, however, the sun seems to have been less bright. From essentially bare beginnings in the early Devonian, life on land underwent explosive evolution throughout the Late Paleozoic, so that by the middle of the Carboniferous period, forests with essentially modern ecological structure flourished, and plants and animals showed comparable interactions with each other as they do today, except that until the latest Carboniferous, there were no terrestrial vertebrate herbivores. A comprehensive summary of terrestrial ecosystems through the Paleozoic can be found in the chapter by DiMichele and Hook (1992) in the volume *Terrestrial Ecosystems through Time*. Details have changed since it was published several years ago, but the overall sweep of the conclusions still provides a good picture of the times.

Throughout the period from the Late Devonian to the end of the Paleozoic Era, which closed at the end of the Permian period, the world went through enormous, although gradual, changes. This whole long episode of time is marked by movements of the continents into new configurations, which in turn affected the climate. Different continents experienced glaciations as they slid slowly over the poles. Continents sitting

Setting the Scene • 79

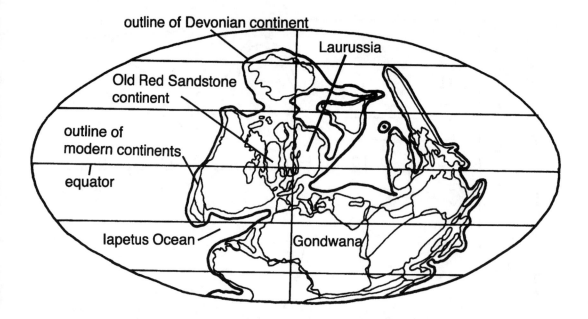

Figure 4.2. Paleogeographic map of Devonian continents during the Late Devonian (363 million years ago).

over the poles means much greater thicknesses of ice are possible than when the sea lies over them (the same is seen today in comparing the present-day Antarctic with the Arctic, the latter covered only by sea ice), and in turn this affects the climate of the whole Earth (Erwin 1995). The changed positions also had profound effects on the circulation of water currents round the globe, which in turn affected the distribution of warm or cold water, thus affecting climate.

By the Early Devonian, the continents had coalesced into two major land masses, Laurussia to the north and Gondwana to the south, separated by a sea that geologists call the Iapetus Ocean (Fig. 4.2). When continents coalesce, the total length of seacoast decreases, limiting the available areas and habitats for marine animals to live in. More communities are at the same time brought into more direct contact. The effect of this is to decrease the diversity of species. While communities are isolated, they tend to evolve away from each other into local specialized and unique species, but when they are brought together, each species has the opportunity to colonize greater areas and communities, become more cosmopolitan. The effect on marine species during the Late Paleozoic was thus a gradual decrease in species diversity, with a corresponding increase in the area of distribution of those remaining (Briggs and Crowther 1990).

Throughout the Devonian, Laurussia and Gondwana were gradually moving together, a process that ultimately ended in the Permian with the formation of the supercontinent Pangaea and the inevitable squeezing out and eventual closure of the Iapetus Ocean. This had a number of profound consequences for life on Earth.

The gradual closure of Iapetus, with the resulting loss of marginal marine or nearshore habitats, may have had the indirect result of forcing many organisms into terrestriality. Certainly more terrestrial habitats became available with the accompanying reduction of aquatic ones. The process of continental collision and loss of coastline habitats seems to have been going on throughout the Devonian, and in the Middle Devonian,

Figure 4.3. Stensiö Bjerg in Greenland. These strata alternate between reddish and greenish. These layers represent repeated cycles of depositional circumstances, probably influenced by globally occurring events such as Milankovitch cycles. Photograph by J.A.C.

present-day North America and northwestern Europe were brought together into what is known as the Old Red Sandstone Continent. Vast areas of terrestrial sediments are associated with this continent, and in them are found fossils of the first known vascular plants and insects. Also, it is here that Devonian tetrapod remains were first found, in Greenland, and more recently, they have turned up in eastern North America. However, tetrapods are now known to have been present in Australia by this time, which was far away, on the opposite end of the southern continental landmass.

With the position of large continental masses (today's Antarctica, South America, and South Africa) in polar regions, the world experienced glaciations on an episodic but gradually increasing scale throughout the late Paleozoic. In their turn, the development of large ice caps tied up water and lowered sea levels generally. This was another cause of reduction of marine habitats throughout the interval. Toward the end of the Permian, the glaciation gradually came to an end, and the supercontinent Pangaea had formed by the coalescence of the northern continents with the southerly Gondwana. This produced a huge area of land that experienced continental climates of extremes of temperature, because land gets hotter than water, and in turn this affected the global climate, producing desert conditions over much of the land surface.

The Earth also experiences periodic fluctuations in its position relative to the sun. Combinations of circumstances in its orbit and declination, added to cyclical changes in the sun itself, combine to create slowly repeating cycles in the Earth's temperature and insolation. These are known as Milankovitch cycles after their discoverer, and they still operate today. They are responsible for cyclical changes to climate with effects such as periodic glaciation, and their periodicity is recorded in the rocks. For example, they can be seen in the cycles of alternating red and green bands in the so-called red-bed deposits of the Devonian of East Greenland (Fig. 4.3) (Olsen 1994). These relatively short-term cycles, over time scales of tens or hundreds of thousands of years, must be added to the effects of changing oceanic currents and continent positions taking place over millions of years during the Paleozoic.

Early Evolution of Plants and Animals

It is hard for people today to imagine the world as it must have been in the middle of the Paleozoic. We can scarcely envisage a world without birds and insects in the air, or grass covering the hills, or even familiar land animals, small or large. These are so much part of everyday life that we take them for granted. But that is what the world was like when the story of tetrapods opens. Nothing moving on land except water and wind through low-growing plants. Nothing in the air except dust and occasional dead plant stems—leaves had yet to evolve. The world would have been not only still, but silent.

Before the beginning of the Silurian, very little except unicellular organisms grew or moved on the land surface. During the Silurian period, organisms such as liverworts, lichens, fungi, and mosslike plants existed that could inhabit humid nearshore habitats (Kenrick and Crane 1997), and there is evidence of terrestrial arthropods from late in this interval (Jeram et al. 1990). Leafless, sticklike plants without true roots, such as *Cooksonia*, could grow around the shores of shallow lakes and rivers. Only close to the water's edge and in shallows would there have been significant vegetation, and here, deposits of decaying plant matter may have formed sufficient organic debris to give footing for plants to send up stems into the air, allowing invertebrates to climb in and around them. The evolution of true vascular plants with water-conducting tissues and pores for gas exchange had happened by the end of the Silurian. Arthropods such as myriapods, scorpions, and trigonotarbids—distant relatives of spiders—appear in numbers in the fossil record during the Devonian and were living among the plants (Fig. 4.4) (DiMichele and Hook 1992; Rolfe 1980). However, none of the plants grew very large, and they formed little cover for animals to hide in. Each type of plant tended to grow in its own small colony, and there was little interaction between plant types, unlike the situation today.

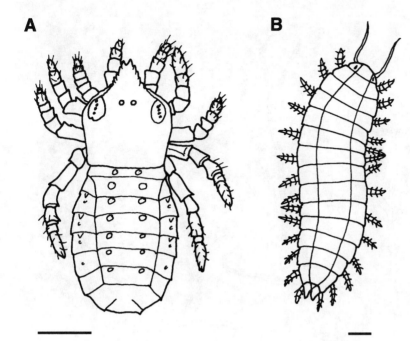

Figure 4.4. Drawings of Early Devonian arthropods. (A) Trigonotarbid from the Rhynie Chert. (B) Arthropleurid myriapod. Scale bars = 10 mm. From Rolfe (1980).

By the beginning of the Devonian and throughout the early part of the period, the vegetation expanded and diversified, and by the middle of the period, the flora can be divided into provinces according to latitude. The story of the Devonian period is one of diversification by land plants and invertebrates so that by the end of the period, a rich and diverse flora and fauna existed. By the end of the period, the increase in plant cover almost certainly had been sufficient to change the balance of the atmosphere so that oxygen reached its present-day levels, while at the same time carbon dioxide decreased as it became tied up in plant tissues.

Early–Middle Devonian Plants and Animals

During the early parts of the Devonian, plants evolved into more complex forms, with at least three new major groups appearing for the first time, such as the extinct zosterophylls and trimerophytes, or becoming more diverse, such as the lycopsids or club mosses that are still around today. Instead of simple stalks, these new forms developed branching stems with small leaflike structures, and some were up to 2 m tall. One of the tallest of these was the trimerophyte *Pertica* (Fig. 4.5). They lived only in wet places, although there is evidence of specialization to different habitats among them. They may even have grown thickly enough to stabilize the courses of streams, limiting flooding (DiMichele and Hook 1992).

At one locality in Scotland, the flora and fauna are spectacularly well preserved because silica-bearing water has mineralized plant and animal remains. The rock formation, known as the Rhynie Chert, has provided an immense amount of information about Early Devonian life. A site at Gilboa in New York State that is a little younger has provided comparable information about the Middle Devonian (Shear et al. 1984).

From these and a few other Middle Devonian localities, it is known that a variety of arthropods were living among these marginal plants.

Figure 4.5. Drawings of Early Devonian plants (not to scale).
(A) Cooksonia caledonica, Late Silurian to Early Devonian, a primitive vascular plant a few centimeters high.
(B) Psilophyton crennulatum, latest Early Devonian, a trimerophyte up to 1 m high.
(C) Sawdonia ornata, Early Devonian, a zosterophyll showing fruiting bodies, stems up to 1 m high.
(D) Asteroxylon mackei, Early Devonian, a prelycopsid, with stems up to 1 m high.
(E) Pertica quadrifaria, a trimerophyte up to 2 m and probably one of the tallest plants of the period.

These included the millipedelike arthropleurids, centipedes, scorpions, pseudoscorpions, mites, and large, predatory trigonotarbids (Rolfe 1980) (Fig. 4.4). The first spiders to show evidence of silk production are known from Gilboa, and the earliest insect is known from the Early Devonian of Canada. Springtails are known from Rhynie, although generally throughout the Devonian, the fossil record of insects is poor. Most of the arthropods were predators that presumably fed on each other, although some plant fossils show damage to leaves and stems, indirect evidence of limited herbivory in arthropods. However, it is also possible that this damage resulted from the plants being chewed after they died and became part of the detritus. True herbivory among arthropods may not have arisen until later, in the Early Carboniferous.

During the early and middle parts of the Devonian, the climate was warm and seasonally dry at least in some parts of the world. It is during this period that large plants first came to dominate the landscape. During the Middle Devonian, several groups of plants achieved bushy, shrublike forms and small treelike statures, including primitive, fernlike types (Fig. 4.6), primitive ancestors of the conifers and horsetails of today. Lycopsids attained treelike morphologies, although not size, for the first time. During the Late Devonian, more recognizable horsetails and fernlike plants evolved. Tree lycopsids of moderate stature appeared. However, the dominant form was the genus *Archaeopteris*, which produced many species throughout the interval and into the Early Carboniferous. These assumed many different habits, the best-known forming extensive forests of trees with woody trunks of up to 1 m in diameter, especially in wetter places such as flood plains and stream valleys.

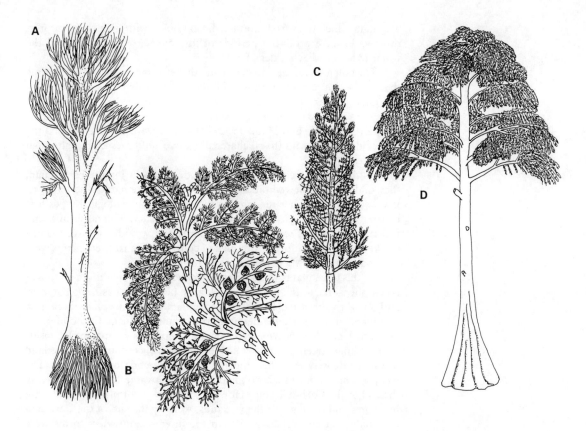

Figure 4.6. Drawings of Late Devonian plants (not to scale).
(A) Calamophyton sp., Late Devonian, a primitive lycopsid several meters high.
(B) Rhacophyton ceratangium, Late Devonian, an early pre-fern, about 1.5 m tall.
(C) Top part of Pseudobornia ursina, Late Devonian, sphenopsid (early horsetail) up to 20 m high.
(D) Archaeopteris sp., Late Devonian, a progymnosperm up to 20 m high.

During the middle of the Late Devonian, at the boundary between the Frasnian and Famennian stages, significant changes are seen in the flora, which preceded a similar change in marine faunas. Many species became extinct, only to be replaced by others, at what is known as the Frasnian-Famennian extinction event (McGhee 1989). This was one of the major extinction events that shaped Earth's history, the most catastrophic of which occurred at the end of the Permian period. The Frasnian-Famennian extinction may have been caused by the movements of the continents so that the large land mass of Gondwana became glaciated, changing the world's climate and lowering the overall temperature.

Some specimens of *Archaeopteris* from the late Frasnian show growth rings, indicating seasonal stress (DiMichele and Hook 1992). Growth rings in trees today form when times of slowed growth during climatic stress (possibly annual, essentially winter) alternate with more rapid growth in better times (spring and summer). It was during the Frasnian that plants first evolved deciduous forms, presumably as another response to the climatic deterioration. In turn, the increase in plant activity along the water margins and on nearby land may have influenced the evolution of air breathing in vertebrates. Several factors may have contributed to this. As plants grew larger, their roots became correspondingly larger, so their root activity increased, forming deeper organic soils. The evolution of resistant seeds allowed the plants to penetrate further inland and increase the ground cover and biomass of plant material. Both these developments led to an increase in leaf litter and decaying plant matter falling into the shallow waters in which the vertebrates lived. Combined with high temperatures, this would have produced periods of low oxygen concentration in the water (Algeo and Scheckler 1998).

The fallen leaves seem to have had other effects as well. It is at this time that fire became a routine aspect of floral ecology, perhaps exacerbated by the fallen leaves and branches, creating a fire hazard in drier regions, especially in view of the high oxygen levels in the Late Devonian atmosphere (DiMichele and Hook 1992). The Late Devonian has yielded the first evidence of fusain (fossil charcoal) (Rowe and Jones 2001), although it is rare.

As the plants recovered from the extinctions during the Famennian, they underwent renewed radiations into many more habits and habitats. It was at this time that the whole range of modern ecological types first became established. Ground cover plants, vines, scramblers, shrubs, and forest trees arose. From the air, the landscape would not have seemed unlike a modern rainforest, although of course, the constituent plants were vastly different (DiMichele and Hook 1992).

By this time, plants were diversifying and adapting to different types of landscape, some to the wetter, waterlogged places, such as *Archaeopteris,* and some to the drier, better-drained uplands. *Archaeopteris* became a dominant form, as it remained until the beginning of the Carboniferous. Ferns such as *Rhacophyton* were common in peaty swamps; the lycopsids evolved huge, treelike forms called lepidodendroids, dominating wetland forests so densely that their remains formed the first coal deposits. The lycopsids maintained their dominance in these swamp forests until nearly the end of the Carboniferous. However, these types of plant, reproducing by spores and confined to the wetlands, were by the end of the Devonian essentially ancient, relict species. In the drier regions, new forms were evolving that reproduced by true seeds with toughened coats. This allowed them to cope with periodic environmental disturbances and reproduce in seasonally dry conditions. These plants were the seed ferns or pteridosperms, which became dominant during the Carboniferous (Fig. 4.6).

Very little is known of terrestrial arthropods from the period between the Frasnian and the middle of the Carboniferous. Their evolution has to be inferred by comparing what is known of the Middle Devonian forms with those from the Late Carboniferous. This huge gap in the invertebrate fossil record is unfortunate, because the radiation of terrestrial invertebrate faunas undoubtedly influenced that of vertebrates.

Most recently, another extinction event has been recognized in the fossil record. It occurred at the boundary between the Devonian and Carboniferous and is called the Hangenberg event. It is manifested as a thin, black shale layer at the stratigraphic boundary (Caplan and Bustin 1999). It seems to have been caused by a rise in water level, disruption of water stratification, increase in organic content, and presumably decrease in oxygen content of the waters. A black shale layer has also been found at the Devonian/Carboniferous boundary in East Greenland (Marshall et al. 1999). This may indicate that the Hangenberg event was much more general and widespread than previously realized. Although most of the localities in which it has been found are marine in origin, the deposits of East Greenland are generally regarded as freshwater sediments.

Late Devonian Fish Faunas

As far as the vertebrates are concerned, the Devonian period is rightly named the Age of Fishes. To give a picture of the kinds of fishes that lived then and the environments they lived in, one of the richest of all localities

Figure 4.7. The locality of Escuminac Bay, where many specimens of Eusthenopteron and other Late Devonian fishes have been found. Photograph by J.A.C.

for Devonian fossil fishes is that of Escuminac Bay in eastern Canada. The locality, near the village of Miguasha, is now a World Heritage Centre in recognition of the importance of its fossil fauna. A museum and interpretive center give insights to its history and paleontology. A section through the sequence of strata has been excavated at one point, shown in Figure 4.7, a photograph taken during an international conference on early vertebrates in 1991. The section dates from early Frasnian times, and recent work has given a very detailed picture of this locality and its fauna (Schultze and Cloutier 1996). Geochemical and faunal studies show that the sediments represent a coastal brackish water to marine environment, although earlier interpretations suggested that it was freshwater (Chidiac 1996).

Representatives of all the major groups of vertebrates except chondrichthyans were present in Escuminac Bay (Fig. 4.8). Jawless fishes, including primitive relatives of today's lampreys, are represented by both the anaspids and the osteostrachans. Anaspids were elongated and rather narrow-bodied forms with downturned tails. They had no paired fins and are thought to have been filter feeders. The osteostrachans such as *Alaspis* had flattened, almost semicircular head shields formed of bony plates up to 300 mm across. Paired eyes were situated close together on the top of the shield, along with a single median nostril. In some osteostrachans, the head shield bore fields of what might have been electric organs. They had a pair of flaplike pectoral appendages attached to the back of the head shield; the rest of the body was covered by scales. It is thought they were bottom dwellers, perhaps detritus feeders.

Very common at Escuminac Bay were the heavily armored placoderms. Placoderms fall roughly into two groups, the antiarchs and the arthodires. Both had heavily armored head shields that featured a joint between the head armor proper and a shoulder region. One of the best known antiarchs is *Bothriolepis*, a moderately sized form about 200–250 mm long in total. The genus is found all over the world at comparable sites, and it is the most common fish at Escuminac Bay. In antiarchs, the head shield was short compared with the shoulder shield, and just by the junction, paired appendages articulated at a complex shoulder joint. The appendages themselves were externally armored and jointed in the middle, giving them the

88 • Gaining Ground

appearance of crabs' claws. These animals had jaws but no true teeth and were probably scavengers. Preservation of some *Bothriolepis* specimens from Escuminac Bay has led to the suggestion that antiarchs had lungs, although this is disputable (Denison 1941).

Arthrodires were, in contrast to antiarchs, major predatory forms. Some grew very large, with heads nearly a meter long, and bodies perhaps a further 2 m long. Their mouths were armed with enormous jaws with "built-in" toothlike pincers or scissors. They had unarmored pectoral appendages and had pelvic girdles, although their pelvic fins are poorly known. Much of the body appears to have been scaleless. At Escuminac Bay, the arthrodire present was called *Pleurdosteus*, up to 250 mm long.

Placoderms are jawed fishes, but their relationships to other fishes with jaws (gnathostomes) are unclear. They are now all extinct, as is another group of early gnathostome, the acanthodians or spiny sharks. There were several genera at Escuminac Bay, some larger (up to 150 mm long), long-bodied, small-finned forms, and some with shorter bodies and longer fins, up to 80 mm long. The fins of acanthodians are their most notable feature. Both midline and paired fins were supported along the leading edge by a bony spine. Some acanthodians had more than two pairs of paired fins ventrally.

Of the true osteichthyan groups, all except tetrapods are represented at Escuminac Bay. Ray-fins are represented by the primitive form *Cheirolepis* up to 500 mm long, whereas of the lobe-fins, lungfishes, coelacanths, porolepiforms, and osteolepiforms are all found there. The lungfishes were short-bodied forms with elongated second dorsal fins. *Fleurantia* at 420 mm maximum length had a long snout, whereas *Scaumenacia*, with a maximum body length of 645 mm, had a short one, indicating different habits and diets in the two. The primitive coelacanth *Miguashaia* (up to 450 mm long) did not show the characteristically modified tail and dorsal and anal fins seen in all other later coelacanths (see Chapter 3). *Holoptychius* is the best-known porolepiform from the locality (up to 470 mm), whereas *Eusthenopteron*, the most common osteolepiform, is one of the most famous and best-known fossil fishes in the world (see Chapters 2 and 3) (Fig. 4.9). This specimen is known from individuals of a wide range of sizes, up to about 500 mm long. Finally, although only three incomplete specimens are known, for this review of tetrapod origins and relationships, one of the most significant fishes in the fauna is *Elpistostege*, the closest Escuminac Bay comes to a tetrapod. Its skull is a little smaller than that of *Panderichthys*, at about 210 mm. For further details of the fishes, see Schultze and Cloutier (1996).

This overview of Escuminac Bay gives an idea of the range of vertebrates around during the middle to late Devonian against which to view the record of the earliest tetrapods.

Figure 4.8. (opposite page) Drawing of Escuminac scene. (1) Anaspid (jawless vertebrate) Endeiolepis. *(2) Osteostrachan (jawless vertebrate)* Escuminaspis. *(3) Acanthodian* Diplacanthus. *(4) Antiarch placoderm* Bothriolepis. *(5) Arthrodire placoderm* Pleurdosteus. *(6) Actinopterygian* Cheirolepis. *(7) Lungfish* Fleurantia. *(8) Lungfish* Scaumenacia. *(9) Osteolepiform* Eusthenopteron. *(10) Fern* Rhacophyton. *(11) "Seed-fern"* Archaeopteris. *(12) Early sphenophyll. (13) Early sphenopsid horsetail* Archaeocalamites. *Illustration by J.A.C.*

The First Tetrapod Tracks, Traces, and Tantalizing Fragments

The Late Devonian provides the first unequivocal evidence of the existence of tetrapods. There are two kinds of fossil material that provide different perspectives on the animals they represent. One is that provided by fossils of the animals themselves (body fossils), and the other is that presented by tracks and trails left by these creatures (trace fossils). This chapter surveys both kinds, starting with the body fossils.

Figure 4.9. Eusthenopteron foordi, *a specimen from Escuminac Bay, showing a dorsoventrally crushed skull (UMZC GN790). Photograph by S.M.F.*

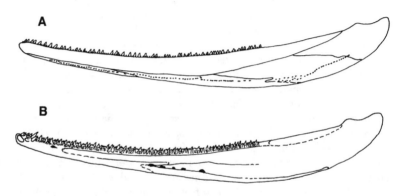

Figure 4.10. Lower jaw of Elginerpeton *in (A) external and (B) internal views. The posterior part of the jaw, including the articular, is mostly reconstructed. From Ahlberg (1998).*

Figure 4.11. The site at Scat Craig near Elgin in Scotland where Elginerpeton *was found. Foreground, Per Ahlberg; at the cliff face, Bob Reekie and Bill Baird from the Royal Museum of Scotland. Photograph by R.N.G.C.*

The Earliest Body Fossils

The earliest occurring tetrapod genera known from body fossils are all represented by fragments, chiefly of lower jaws, and in one of them, some limb and girdle material. They are called *Obruchevichthys* and *Elginerpeton*. They have been recognized only recently, from material that has existed in museum collections since the last century, but which has had to wait for discoveries of other, better-preserved Devonian tetrapod material before its identity could be appreciated (Fig. 4.10).

Obruchevichthys was originally described as a more conventional sarcopterygian fish from the Frasnian of Latvia and western Russia (Ahlberg 1991b). It is known only from lower jaw fragments belonging to an animal with a skull length about 400 mm. The slender, elongate jaws suggest an animal with a rather flattened skull, although the shape of the head can only be guessed at. The remains occur along with many fishes typical of the Late Devonian, such as placoderms, lungfishes, acanthodians, and some other fragmentary lobe-fins. Although no limb or girdle material is known from this animal, its lower jaw shows some features now considered to be characteristic of tetrapods. Even if it proves in the future not to have had true limbs or digits, its position on the cladogram is likely to remain close to animals that do. With *Elginerpeton,* recognized as a tetrapod, it constitutes the very earliest known body fossil evidence of the group.

Elginerpeton is known from a greater number of fossils than *Obruchevichthys,* which allows a more complete picture to be built (Ahlberg 1991b, 1995). Many of these specimens lay in museum drawers in the United Kingdom until 1991, where they had been labeled as "undetermined sarcopterygian" and had not been described or even cataloged in some cases. The material comes from a tiny locality called Scat Craig, where the rocks are exposed in a stream cutting near Elgin in Scotland (Fig. 4.11). An excavation in 1992 collected more material from this locality, but unfortunately, very little of it proved to be of *Elginerpeton*. Like *Obruchevichthys,* it is late Frasnian in age.

The material consists of both upper and lower jaw fragments (Fig. 4.10), including a premaxilla and some limb and girdle elements. Among these is a tibia, a femur, pectoral and pelvic girdle fragments, and, less certainly associated, a humerus (Ahlberg 1998). The skull material shows an animal with a skull length of about 400 mm, with a flat head and a somewhat pointed snout. Both pectoral and pelvic girdle elements are massive and are in accord with the size of the skull. The pelvic girdle in particular is comparable in morphology to that of the better-known *Ichthyostega*. The femur is very short and flattened, and its shape suggests that it was held in such a way as to be unlikely to have supported weight, but was rather used as a paddle. The humerus, assuming it belongs to the same animal (and there is no evidence of any more likely owner), seems to fall somewhere between that of *Panderichthys* and *Acanthostega* in its morphology.

Elginerpeton occurs in coarse sandstone, and the bones are disarticulated and waterworn, so their place of origin cannot be established. It seems certain that they were transported from another locality before the fossilization process. Other remains from the locality include some typical Devonian fishes such as placoderms, a lungfish, a porolepiform, and some jawless fishes. Little information about the habitat or lifestyle of *Elginerpeton* can be gained from the sediments in which it occurs, but the limbs suggest that it was an aquatic animal.

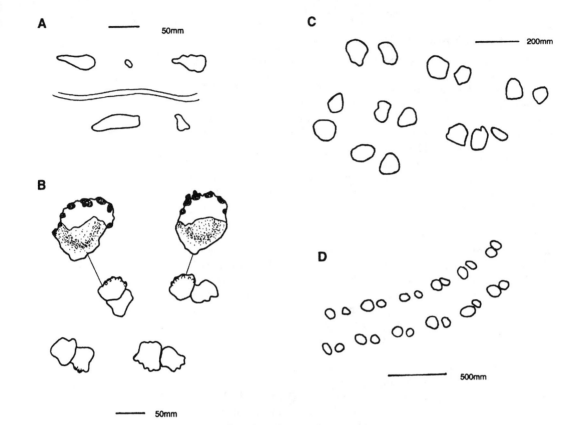

Figure 4.12. Drawings of Devonian tetrapod trackways. (A, B) Genoa River tracks; in B, two of the prints from the second trackway are enlarged. (C) Tarbat Ness tracks. (D) Valentia Slate tracks.

Trackways

There are several trackway fossils described as having been made by Devonian tetrapods. One problem that they all share is that of their dating. Without exception, the dates of these trackways can only be placed within rather broad ranges, so that it is not really possible to treat them in a strict time sequence. Therefore, the easiest way to deal with them is by the date when their descriptions were published.

The first described—and the most firmly dated—are a pair of trackways on the same block from the Genoa River in Australia (Fig. 4.12, 4.13A). These were described in 1972, and the date of the block is given as "probably Frasnian" by the authors (Warren and Wakefield 1972). Thus the trackways are probably contemporary with *Obruchevichthys* and *Elginerpeton*.

The two trackways were apparently made by two rather different animals, or at least differently sized animals moving in contrasting ways. One trackway shows a series of foot impressions in which hind footprints overlap fore footprints in an alternating sequence on left and right sides. This pattern is characteristic of tetrapods and is produced when the animal moves its fore- and hindlimbs alternately, as a dog would do, trotting on the beach. The prints show some digit impressions that are situated to the side of the print rather than in front, as they would be in a modern mammal. There are at least five digits represented by these impressions, although there could be more.

The animal left no tail or belly drag, and this has been interpreted to mean that the animal was walking with its body supported well clear of the ground. A rough calculation based on the separation of the prints gives the animal a body size of about 220 mm, minus the tail.

The second trackway runs parallel to the first, and in this one, a small print from one foot, probably the forefoot, alternates on each side with a larger print, presumably from the hindfoot, which left a drag mark behind it. Down the center of the trackway runs a sinuous body or tail trace, suggesting that in this case, at least part of the body was not being supported. The animal was apparently somewhat smaller than the one that made the first track.

Although these trackways are supposedly from animals walking on the land surface, there is little independent evidence presented to support this assumption, other than the fact that they were made by tetrapods. Other trackways, made by invertebrates, occur on the same bedding plane, but these have not been described. Comparing the footprints with what is known of the earliest tetrapods suggests that none of the known forms is likely to have made these tracks unless they had formed under water (Clack 1997a). The next chapter deals with the morphology of some better-known Devonian tetrapods, and readers can compare them with these tracks for themselves. Nonetheless, to date, these Genoa River trackways constitute some of the best substantiated evidence for possible terrestrial walking by tetrapods in the Devonian.

Recently discovered trackways from the Valentia Slate from the west of Ireland were also certainly made by tetrapods (Stössel 1995) (Figs. 4.12D, 4.13). The trackways are more extensive than those from Genoa River, although the individual prints are less clear. The dating of the rocks is not absolutely secure, but best estimates put the horizon as Frasnian. These trackways consist of several series of footprints made by more than one animal, all preserved on the same bedding plane (Fig. 4.13). Despite having been distorted when the rocks were stretched and pulled by nearby geological events, they still show the typical alternating sequence of prints, with fore and hind footprints distinguishable. Computer techniques have been used to restore the original shape of the prints and trackways.

There is one long, sinuous track and several shorter stretches, clearly made by different individuals. One of the trackways shows broad shallow furrows between the footprints, made by the body that was pulled along the ground and not supported fully by the limbs. If all the trails were made by the same kind of animal, this track is the best evidence that, unsupported by much water, the body dragged on the ground and the limbs lay to the side of it. Another trail shows footprints with a much longer stride length, although the distance between left and right prints remains much the same in all tracks. The footprints show no digit impressions, but they are quite deep and rather egg-shaped, with the pointed end at the outer edge. The suite of tracks suggests that the animals were probably supported to some extent by water, although some were further emersed than others. All the tracks seem to have been made at about the same time, and all the animals were heading in more or less the same direction, but it is not clear whether the differences between them resulted from animals that were moving at different speeds, or animals that were moving in different depths of water, or by differently shaped animals (Fig. 4.13).

No other fossils have been found in the same beds as these tracks that might give clues to dating, nor have any spores been found in them. However, some placoderms have been found at a different point on the western Irish coastline in a nearby formation that probably does not lie too far distant in time from those in which the tracks were found. These placoderms are consistent with a Frasnian age for the rocks, a date that

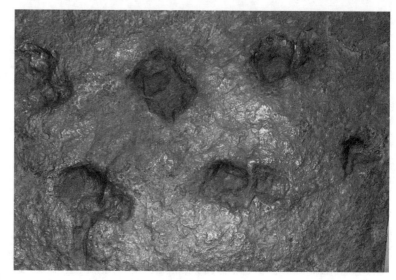

Figure 4.13. (A) Genoa River tracks. Note the digit impressions on the outer edge of the print at top right; this is the print figured at right in Figure 4.12B (from a resin replica; photograph by J.A.C.).
(B, C) Valentia Slate tracks in southern Ireland.
(B) Overview of the main trackway in the background, a shorter trackway in the foreground; this one shows a shallow groove made by the animal's body, whereas the footprints themselves are raised on a ridge where the sediment has been forced upward.
(C) (opposite page) Close-up of some of the shorter tracks showing different print spacing. The track in the foreground is the same as that in the foreground above; note the closer spacing of prints than in the track behind it. Photographs by J.A.C.

would not be unlikely for these trackways. Perhaps they could have been made by an animal such as *Elginerpeton*.

Other tracks that have been described as tetrapod and dated within the Devonian are all subject to problems of interpretation, not only of date but of identity as tetrapod or as terrestrial tracks (Clack 1997a). These include a ladderlike track from the Grampian mountains in Victoria, Australia, which the latest evidence suggests is Late Silurian or Early Devonian. Ladderlike tracks—that is, similar sized individual prints in opposite pairs—are inherently unlikely to have been made by tetrapods, and some recently discovered evidence suggests that some kind of large invertebrate may have made them. Another interpretation, based on comparable ladderlike tracks from the Middle Devonian of East Greenland, is that they were formed underwater by the forelimbs of a placoderm such as *Bothriolepis*. Certainly, a recent study has shown that the Greenland trackways are of exactly the right size to have been made by the fins of the contemporary placoderms. A recently described track from Tarbat Ness in Scotland is more likely to have been made by a tetrapod; it is tentatively dated as Middle to Late Devonian, although it could be as late as Early Carboniferous (Rogers 1990) (Fig. 4.12C). It is notoriously difficult to interpret footprint trails, so trackways cannot be used as reliable evidence of the habitus of the earliest tetrapods.

Origin of Tetrapods: When and Where?

What does all this information suggest about when and where tetrapods are likely to have evolved? The earliest evidence of body fossils of transitional forms between "fishes" and tetrapods are found in the early part of the Late Devonian, the Frasnian. These forms are represented by jaws that either were originally mistaken for fishes or about which there was debate over identity. This was partly because they retain many fishlike characters, but also because the detailed differences between fishes and tetrapods with respect to the lower jaws had not been discovered. Even now, although some of these animals fall within the group usually called tetrapods, it is not clear that they had true limbs with digits in all cases. At this point, the difficulties of definition that were explored in Chapter 2 become significant.

Some things are clear, however. Both *Elginerpeton* and *Obruchevichthys* appear more closely related to tetrapods than was *Panderichthys*. They are also very closely related to each other, sharing some details that cause them to be placed in the same family (Ahlberg 1995). This family was widely distributed in the Frasnian. They were also different from the slightly later Devonian tetrapods, which will be described in the next chapter. They may represent an early and specialized offshoot from the tetrapod branch.

Panderichthys and *Elpistostege* flourished in the early Frasnian and are some of the nearest relatives of tetrapods. But tetrapods appear only about 5 to10 million years later in the late Frasnian, by which time they were widely distributed and had evolved into several groups, including the lineage leading to the tetrapods of the Famennian. This suggests that the transition from fish to tetrapod occurred rapidly within this restricted time span. Neither fishlike tetrapods nor tetrapodlike fish body fossils occur in the record before this (Clack 1997a). Indeed, the osteolepiforms as a whole are not found before the Middle Devonian. This lends weight to the suggestion that the tracks from the supposed Late Silurian or Early Devonian are not those of a tetrapod, and those from the Middle Devonian are unlikely to be so. Given our current understanding of phylogeny, tracks made by a terrestrial tetrapod are unlikely to be found before the late Frasnian, and the body fossil evidence conflicts with the interpretation of any pre-Famennian track as terrestrial.

Inference of a late Middle Devonian or early Late Devonian date for the origin of tetrapods fits very well with what is known of the evolution of land plants and invertebrates. Before that time, plant communities on land were limited in diversity and ecology, as were the terrestrial invertebrates. In the colonization of the land, it seems that the plants led the way, followed by the invertebrates, then the vertebrates. Only when there was sufficient shelter and humidity under the plant canopy and sufficient invertebrates to supply them with food is it likely that vertebrates would have begun to explore the terrestrial environment.

Even though tetrapods had appeared by the Frasnian, it appears that they did not escape the influence of the Frasnian-Famennian extinction event. None of the forms from the Frasnian are yet known to have survived into the Famennian. The tetrapods of the Famennian may represent a renewed outburst of evolutionary radiation from a few surviving Frasnian forms, but only further research and discoveries could confirm this idea. The discovery of the Hangenberg Event further suggests that the Late Devonian tetrapod radiations may have been hit a second time, just as the diversity was increasing again.

The question of where tetrapods evolved is even more difficult to answer than that of when. It falls into two sorts of inquiry. The first is "where" in terms of geography, and the second is "where" in terms of "in what type of conditions." The geographical evidence, so far as it goes, shows that the Frasnian tetrapods that have been found have all come from central Laurussia, but that by the Famennian, they had reached much further west and into what is now Australia. The position of that continent during the Late Devonian is still disputed, but most estimates place it near the equator. However, whether it was attached to other continents by land is not clear, and if so, whether it was attached to the eastward or westward side of Laurussia. Although the record of Frasnian tetrapods is still so poor,

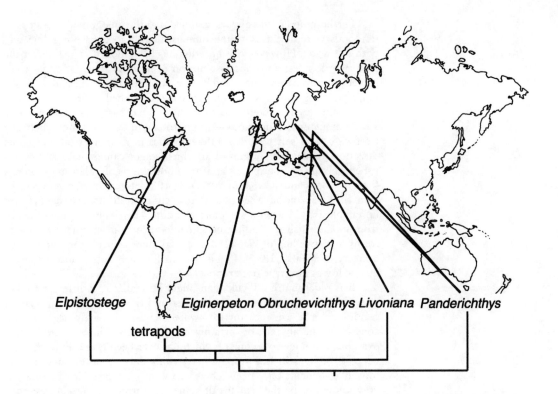

Figure 4.14. Cladogram of tetrapodomorphs showing the largely Laurussian distribution of the closest relatives of tetrapods. From Ahlberg et al. (2000).

it is possible to infer from biogeographical evidence that there was a major radiation of tetrapodomorphs in the Middle to early Late Devonian of Laurussia (Ahlberg 1995; Ahlberg et al. 2000). Most of the significant new finds from this area seem to slot into a reasonably well-resolved stem lineage for tetrapods, with the near-tetrapods such as *Panderichthys* and *Obruchevichthys* cropping up in the Baltic states and Russia (Fig. 4.14). Tetrapodomorphs found in Australia are either rather far removed from the stem, such as rhizodonts such as *Goologongia,* or are rather derived tristichopterids. The recently discovered *Livoniana* from Latvia, by contrast, exemplifies the kind of transitional form that is being found in the Laurussian region (Ahlberg et al. 2000).

Until quite recently, it had always been assumed that the earliest tetrapods—indeed, the earliest lobe-fins—lived in freshwater and that tetrapods made their way onto land from rivers and swamps. In many ways, this is still the most plausible scenario, given the body forms of the best-known genera. *Acanthostega* (Chapter 5) appears to have been adapted for a more or less permanently aquatic existence, and its limbs and digits are thought to have been adapted for use in swampy waters where fins would be a disadvantage. Other strands of evidence that have been used in favor of a freshwater origin for tetrapods include the fact that modern amphibians are, with the exception of a single frog genus, unable to live in saltwater conditions. In the fossil record, only one "amphibian" fossil, *Archegosaurus* from the Permian, has been found in indisputably marine conditions. However, among more recent finds of early tetrapods are some that are now known to be from tidal, marginal marine, or brackish water localities.

Furthermore, it was always assumed that the lobe-finned ancestors of tetrapods were also freshwater animals, unable to cope physiologically with salt water. However, this question has been looked at again over the past two decades, and that assumption at least has been strongly challenged. It now seems more likely that some of the earliest lobe-fins—lungfish, for example—were indeed marine (Campbell and Barwick 1987). Many other early lobe-fins seem to have been euryhaline, or possibly anadromous or catadromous—that is to say, they could move between freshwater and salt water at different times in their lives (although the direction in which they moved at which stages cannot be easily established), because some species appear in both kinds of paleoenvironments (e.g., the coelacanth *Rhabdoderma*, Forey 1981). By contrast, several sharks from the Paleozoic have been shown to occur in freshwater, so that they are unreliable as evidence for salt concentration. There is increasing evidence from modern fish that adjusting salt tolerance from marine dwelling to freshwater dwelling or vice versa is actually not difficult for many vertebrate types (Walsh 1997). If this was so for early lobe-finned "fishes," maybe it was also true of tetrapods.

It is much more difficult than might be realized to differentiate between deposits laid down in freshwater and those laid down in marine conditions. The evidence for freshwater deposits is partly negative and consists of the absence of certainly marine fossils. These would be, for example, echinoderms, none of which has ever been found in freshwater deposits or lives in freshwater today; conodonts, usually found as tiny toothlike elements once regarded as useful only for correlating strata but now known to be the remains of some very primitive vertebrates; acritarchs, microfossils formed from the shells of tiny marine organisms; the shells of sedentary polychaete worms; or even sharks and other fishes that are associated with known marine faunas. Vertebrates, although one of the most widely used indicators of salinity conditions in the past, are now seen as the most subject to the dangers of circular reasoning. Other more independent evidence springs from geochemical analysis, in which the isotopic ratios of elements such as oxygen and carbon and the amounts of boron are found to be characteristically different in marine and freshwater deposits. The isotopic ratios of strontium within the bones of some fossil vertebrates can also be used in this way (Schmitz et al. 1991). Understanding the sedimentary sequence in which the fossils occur as well as their wider geological context also provides clues about their environmental origins.

As an example of how well-known Devonian deposits can be reinterpreted, Escuminac Bay is one such locality that was previously assumed to have been freshwater in origin (see above). One reason for this assumption was the high proportion of lobe-fins in its fauna, combined with an absence of sharks or any specifically marine invertebrates.

Looking at the localities in which early tetrapods were found provides some food for thought. The deposits of East Greenland where Devonian tetrapods were first found have almost universally been regarded as freshwater in origin. They are interpreted as having been laid down in a great river basin, surrounded by mountains and bounded by faults. No evidence of marine influence has ever been found there. The East Greenland faunas were highly significant in influencing the growth of ideas about the origin of tetrapods, and the next section outlines some that resulted from the discovery of *Ichthyostega*. Other Devonian localities are also firmly interpreted as freshwater. The Australian localities yielding their fragmentary tetrapods are continental in origin; that is to say, they were laid down in small lakes or rivers in areas far distant from any marine influence. The Famennian tetrapods of the Catskill Basin in the United States are associated with sequences of sediments representing a large river valley and its meandering channels, in many ways similar to those in which the East Greenland faunas occur (Daeschler et al. 1994). However, by contrast, the locality of Tula, with its complement of tetrapods and fishes, has also yielded charophytes and stromatolites, strongly suggesting at least some marine influence, and the geological conditions suggest that the site was rather far from the nearest land (Lebedev and Clack 1993). The Baltic Frasnian sites from which animals such as *Obruchevichthys* and *Panderichthys* come are interpreted as marginal localities at the edge of a large marine basin (Kurss 1992).

Taking the tetrapod sites worldwide, one thing is obvious: they lie scattered over the globe in places that were remote from each other on separate continents, even in the Devonian. It is hard to see how the tetrapods could have reached these widely dispersed continents unless they could negotiate marine conditions (Thomson 1980). The alternative, that tetrapods radiated independently from lobe-fins that had originally been euryhaline and subsequently lost their salt tolerance, seems even more unlikely and is countered by the detailed similarities that are found in the tetrapods that are now known from over the world. The shallow, swampy waters of marine lagoons, newly populated by emergent plants, might have been the breeding ground for the earliest tetrapods, an idea further explored below.

Theories and Speculation: Why Did Tetrapods Evolve?

One of the questions that interests many people is, "Why did the tetrapods come out of the water onto the land?" Over the years during which people have been developing ideas about evolution of tetrapods from lobe-finned fishes, many theories have been proposed as possible answers to this question. These ideas, most of which were put forward at a time when information was even scarcer than it is now, have often found their way into popular imagination, not to say mythology. It is time to examine some of these critically, not really to see which is correct, because there probably is no single answer to the question, but to see how many there have been, to examine some of the suggestions, and to see how they stand up to criticism and to recent discoveries about actual fossil animals.

Scenarios to explain the evolution of limbs with digits were intimately tied to those speculating on the environment of the earliest tetrapods, and it is difficult to tease the strands apart. Some of the earliest theories go back to the early years of the 20th century, when the observation was made that sediments from the Late Devonian period, the time when tetrapods were assumed to have evolved, consisted largely of red-beds (Barrell 1916; Lull 1918). These are layers of sandstone that often have a reddish color due to the presence of iron, and such sediments are found all over the world from this period. The eastern United States, northern and eastern Europe, Australia, and China all have them. They have usually been interpreted as the result of arid or semiarid conditions. Indeed, comparable red-beds are also found from other time periods such as the Permian and Triassic, which are likewise associated with deserts or semideserts, and the plant fossils they preserve support this interpretation.

Early theories of the origin of tetrapods suggested that these arid conditions caused a general drying up of the pools and lakes in which the lobe-finned ancestors of tetrapods lived, leaving the creatures stranded.

The suggestion was that those that had lungs and were able to breathe air were the most likely to have been able to survive. One problem with this idea is that many lobe-finned fish types survived past the end of the Devonian into the Carboniferous period, and even into the Permian. Another is that not only are all of the early lobe-finned fishes thought to have had lungs, but it is probable that early ray-fins did too.

Building on the red-beds scenario, the next suggestion, by A. S. Romer (1933, 1945), was that those fishes whose fins were strongest and most resembled the structure of limbs, such as the lobed fins of *Sauripteris* or *Eusthenopteron,* were favored by a strong selective pressure on the animals to get back into the water. Those with more limblike legs were better able to struggle over the dry surface and so were more likely to reach another pool. According to this idea, limbs actually arose to enable the animals to get back into water, not to be better able to leave it, and Romer argued that it is not until the Carboniferous that truly terrestrial tetrapods are found (Romer 1958).

This idea was widely accepted for many years. It gave rise to a variety of scenarios about how the changes to the limbs might have occurred, assuming that those changes took place between a fin like that of *Eusthenopteron,* and a pentadactyl limb like that of the Permian temnospondyl *Eryops* (see Chapter 9). Some of these ideas are dealt with in more detail in the section on the evolution of limbs in Chapters 5, 6, and 10. One of the threads that unites ideas from this time is that because *Eusthenopteron* and its allies had fins with a bone structure like those of tetrapods, they must therefore have used them in a similar manner. Changes to the locomotory pattern were thought most likely to have evolved while the animals were still in water, but only excursions onto land would provide the selective pressure to lose the fin rays (Eaton 1951).

Over the middle decades of the 20th century, particularly during the 1950s, a whole series of articles addressed the question of the origin of limbs and terrestriality, putting forward objections to old ideas and coming up with new ones. Looking at modern amphibians gave one worker the idea that limbs might originally not have been for walking at all. Rather than walking away from drying pools, many modern amphibians, and indeed crocodiles as well, bury themselves in the mud at the bottom of the pool, where they stay until the rains come again. It was therefore suggested that the limbs of early "prototetrapods" might have been evolved for burrowing into the mud (Orton 1954). However, even the proponent of this idea recognized that many animals that move or even burrow on land do not use limbs to do so, and some other explanation must be sought for why the animals eventually became terrestrial. It may also be pointed out that the lungfishes *Protopterus* and *Lepidosiren,* two of the tetrapods' closest living relatives, are noted for their ability to estivate in mud burrows. They are the lungfish genera with fins least like those of tetrapod limbs—they do not use them for burrowing.

Arguments about the origin of limbs and terrestriality continued in the literature with the observation that most modern amphibians, rather than seeking new pools when their own dries up, congregate in the drying pools and often die there. If they are disturbed for any reason, they may leave the pool, but the directions they take are random (Ewer 1955). Some may find new pools, but the selective pressure would be on those that could best withstand desiccation, not those with the strongest legs. The same would have been true in the Devonian. Population pressure in existing pools today may sometimes stimulate migrations in humid conditions, and in this case, the selective pressure may have been to favor those with limbs. If limbs had originally evolved for digging estivation burrows, Ewer (1955) pointed out, it is hard to see why they would ever have been used for walking.

One contribution envisaged the pressure being on the larvae of these prototetrapods: "To be a successful frog, one first has to be a successful tadpole" (Warburton and Denman 1961). It was suggested that the animals laid their eggs in shallow water that then dried up, and the conditions favored those that could survive in temporary pools. The problem with this suggestion is that it is simply not known how these early prototetrapods reproduced. Study of modern amphibians suggests that their modes of reproduction are strongly tied to their ecology, and they show a greater variety of strategies than any other vertebrate group. Although the majority of frogs lay large numbers of small eggs in "spawn," which they then abandon, this is not necessarily a guide to what Devonian prototetrapods did. The most primitive frog genus has internal fertilization, as do all salamanders and caecilians, and most members of these two latter groups produce their eggs and young internally. This idea will be developed further in Chapter 6, but it shows that theories about pressure on larvae may all be quite wrong.

This argument also fails to take into account that predatory forms too must have laid eggs and had young that sought small food, so that they too could well have been part of the marginal pool fauna, exploiting the prototetrapods that sought refuge there.

One of the most telling objections to the "drying pool" idea is that on closer examination, red-beds are found not to be invariably correlated with arid climates. In modern times, red-beds are found in low-latitude, tropical climates, and they are always associated with oxidizing conditions. This was realized as long ago as 1957 (Inger 1957) and has been corroborated many times since. The red-beds may be associated with rainforests or places where the rainfall is high but confined to monsoonal periods. The conclusion from this is that red-beds are not really a good indicator of arid environments, even though they are frequently found along with evaporite deposits.

A good model for red-bed-producing conditions today is the Amazon basin. Many of the fishes that live in the rivers there are air-breathing forms. As the season becomes drier, they simply move downriver to deeper water. This model prompted one worker to look at the behavior of air-breathing fishes today (Inger 1957). Inger pointed out that in many cases, air bladders allow the fishes to remain in the water when the water becomes anoxic, rather than to allow them to make excursions on the land. Some of these fishes do feed on land, and some do leave the water to find other pools, but only when the climate is humid, and only to seek pools where the population is less dense (Goin and Goin 1956). Therefore, Inger suggested that one possible stimulus for early tetrapods to leave the water might have been population pressure. In addition, he pointed out that if the conditions in which the early tetrapods lived were indeed humid, then limbs with digits are unlikely to have evolved to dig estivation burrows but could have helped them to migrate to new locations.

There is another point that the red-bed theory fails to take into account. Even if it were correct that the Devonian red-beds were indicative of desert or semiarid conditions, it does not necessarily mean that these conditions are the ones in which the earliest tetrapods arose. By no means do all Middle or Late Devonian sediments consist of red-beds. Some represent river and lake basins or nearshore deposits from lagoons that are rich in organic material, suggesting nearby forests. The earliest known tetrapod remains are Late Devonian in age, but presumably they actually evolved somewhat before that time, perhaps before the red-beds were deposited. Furthermore, it is quite possible that they evolved in places other than those that the red-beds represent. The kinds of ecological conditions in which early tetrapods first arose are not known, but if some recent hypotheses are correct, they may have lived not in purely freshwater habitats but in brackish or lagoonal ones, within the influence of the sea. In modern ecosystems, the intertidal zone is the richest in terms of ecological niches and diversity of forms, from the species to the phylum level, of any that exist. This is because the conditions are constantly changing, influenced by the tides and exposure to air, and this produces environmental stresses that put great evolutionary pressure on the animals that live there. It would not be surprising to find the same diversity among intertidal animals in the Devonian, and it could be that a tidal influence, perhaps in an estuary or lagoon, might have been important in the evolution of air-breathing, land-capable vertebrates in those days, as it is today in the form of teleosts such as mudskippers. If so, the deposition of red-beds elsewhere is of no consequence.

Two major objections to the fin-to-limb scenarios have only recently become apparent. First, the modern coelacanth, with the bony fin structure most like that of tetrapods among living fishes, does not use its fins for walking (Fricke et al. 1987, 1991). Before it was studied in its natural environment, many people predicted that its fin structure would mean that it used the fins for walking on the bottom of the sea or among the coral reefs where it lived. However, film of the fish in action shows that this is never the case, and it uses its paired fins for slow paddling, albeit in an ipsilateral manner, with one fore fin moving in concert with the hind fin from the opposite side. This is the same sequence used by tetrapods, but not for the same purpose.

The second major objection to the fin-to-limb scenarios is that recent discoveries show that the earliest tetrapods were not pentadactyl after all, and the evolution of joints and digits did not proceed in the order or fashion that early theoretical studies assumed (see Chapters 5 and 6). The ideas that were put forward about how the changes took place do not fit the facts.

A third objection to this idea is that there are many modern fish that have not only lost most of their fin rays and all of their fin webbing, but have modified the remaining fin rays into bendable structures that can be used for clinging to vegetation in the water, or for feeling around in the sand, or for walking on the sea bottom. Conversely, one of the fishes that makes the longest excursions over land is the eel, which has very reduced paired fins and an elongate body (see below). Logically then, there need be no connection between loss of fin rays and webbing, evolution of digits, and excursions onto land.

Many of these ideas about the origin of tetrapods have assumed that the origin of limbs was intimately connected with the origin of terrestriality. Some have tried to separate the two, such as Romer, who suggested that the limbs helped the animals get back into the water. However, the problem actually consists of three parts: origin of limbs with digits, origin of walking, and origin of terrestriality. No one yet knows which came first, or which order they occurred in. Each aspect may require a quite separate explanation that is not necessarily dependent on the others. (Furthermore, the origin of the clade known as the Crown Group Tetrapoda is a different issue again, not necessarily connected with limbs, walking, or terrestriality at all. Some of the problems associated with this are dealt with in Chapter 9.)

Some ideas about the acquisition of terrestriality and limbs may be

Figure 4.15. Drawing of a frogfish, a ray-finned fish in which the pectoral and pelvic fins are adapted into jointed, digitlike "fingers" and "toes." Illustration by J.A.C.

gained from looking at modern ray-finned fishes that do show some adaptations for terrestrial excursions, as well as those that have evolved digitlike equivalents from fin-rays. The first thing to notice is that these two groups are almost mutually exclusive—that is, those with digitlike fin-rays by and large are not those that venture onto land, and conversely, those that do venture onto land do not necessarily do so by means of digitlike or limblike fins.

Fishes that have evolved limblike adaptations are often those that live in deep waters or that are habitual bottom dwellers. The Sargassum frogfish has developed perhaps the closest analog to tetrapod digited limbs (Fig. 4.15). It uses its jointed, grasping fin rays to cling to the sargassum weed among which it lives, remaining stationary by this means until potential prey passes. Notably, there are eight of these jointed rays on each "foot." Batfish, with comparable adaptations, use their limblike fins for bottom walking, sometimes at abyssal depths. Gobies and blennies also develop digitlike fin-rays, although these animals are often tide pool species that do sometimes emerge from the water. However, their appendages are used for walking on the pool bottom, not on land (Fig. 4.15). Eels, which make long journeys overland to reach their spawning grounds, not only have nothing in the way of limblike appendages, but their paired fins are reduced in size. *Periophthalmus,* the mudskipper, of which there are many species in different parts of the world, does use modified fins for walking, but walking catfish may use modifications of the opercular series. Once again, there is little relationship between the origin of limblike appendages and terrestriality.

Fishes leave water and spend time on land under a wide range of conditions and circumstances (Sayer and Davenport 1991), but there is no link between those that do leave the water and those that have evolved digitlike structures. The emergence of tetrapods onto land may have involved any one or a combination of the sorts of conditions that induce terrestrial excursion in modern forms. Some of these are seen in marine fishes, some in freshwater forms, and some in intertidal or brackish water forms. Lowered oxygen concentration is one such factor. Low oxygen levels may themselves occur for a variety of reasons, such as high daytime temperatures in tropical shallow freshwaters, low activity of photosynthetic organisms at night in tide pools, or decaying vegetable matter being oxidized by bacteria. Under these circumstances, many air-breathing fish increase the incidence of air gulping or the amount of body exposed to the air. Temperature may also be an influence in another way—some mudskippers, for example, only leave the water when the temperature on land is high enough, whereas some killifish will leave the water when the water temperature gets too low.

Biotic factors probably play a generally larger role in why fish leave the water. Increased competition for food or space sometimes plays a part, as suggested for amphibians above, as does avoidance of predation or aggression, but one of the most common reasons is for feeding on terrestrial or semiterrestrial food sources. Associated with this, it has been shown that fish that feed on land and then stay there for a while actually experience an increased rate of digestion; indeed, some aquatically feeding fish leave the water after feeding to take advantage of this effect. Raised ambient temperature appears to be the key factor here. A very few fish leave the water in order to spawn. Some will leap out of the water to attach egg masses to leaves overhanging the water surface, whereas others deliberately seek places to lay their eggs that are exposed to the air for at least half the time. There may be several reasons for this behavior; raised temperatures may speed development, or such places may be inaccessible to the most common predators, or the eggs may get an improved oxygen supply. The plethora of reasons for why modern fish leave the water demonstrate how difficult it would be to cite any one as the main stimulus for the tetrapods to have done so.

As far as the origin of terrestriality in early tetrapods is concerned, two explanations are currently the main contenders. One is that tetrapods, either the adults, their eggs, or their young, were being pressured by predatory fish or arthropods to seek more and more marginal habitats where the predators could not follow (McNamara and Selden 1993). Certainly there were very large predatory fish around during the Devonian, but they would not have been able to penetrate the shallow, vegetation-choked water where the tetrapods are assumed to have lived. This explanation uses the timing of a rapid expansion in the diversity of fishes during the Devonian to suggest that competition for resources in the water grew rapidly, with the land offering new niches free from predation pressures.

During the Devonian, increase in diversity of fishes in the water occurred at the same time as the radiation of plants onto land, followed by an increase in the variety and numbers of arthropods that fed on them. By the Late Devonian, vegetation at the margins of lakes and rivers was dense and lush, with the appearance of forests whose structure resembled that of modern forests. The many arthropods living in and on this vegetation would have provided a rich and unexploited resource for any vertebrate that could reach them. No competing vertebrate forms were on the land before them, so that any vertebrate that could in any degree survive out of, or partially out of, water would find itself at an advantage, independent of what predation pressures might have existed in the water.

This brief survey is meant to show how difficult it is to tease apart all the influences that were at work in the Devonian to produce the appearance of tetrapods. It shows how ideas are numerous, but evidence is equivocal. Most ideas can be countered by objections or alternative suggestions, but without a much more complete fossil record, the truth can only be guessed at.

Other reviews of the questions and problems associated with the origin of tetrapods can be found in Thomson (1991, 1993) and in Daeschler and Shubin (1995).

Five
The First Feet: Tetrapods of the Famennian

By the latter part of the Late Devonian, the Famennian, vertebrates with indisputable limbs bearing digits—tetrapods—had appeared. Some remarkably well preserved material from several localities provides details of their anatomy and lifestyles. This chapter examines each of these in turn, and then goes on to see what pointers they may give to the origin of the group.

Famennian Tetrapods from East Greenland

East Greenland has provided the most detailed knowledge of Devonian tetrapods. East Greenland has been studied by geologists for many decades. One of the reasons is that the terrain lies within the polar semi-desert, so that it is relatively sparsely vegetated and the geology can be seen easily during the short summer season when the ice and snow around the coast melt. At the same time, in contrast to many desert areas, availability of water is not a problem for the visiting scientists, although accessibility to the area is often hampered by bad weather. Most of the sites are usually only reached by helicopter, although in the past, ice breakers carried the scientific teams to the fjords, from where they reached the sites by inflatable dinghies. Recent work has been carried out in collaboration with the Denmark and Greenland Geological Survey, who use as their base camp the Danish Air Force airstrip of Mestersvig.

Figure 5.1. Flying over Kejser Franz Joseph Fjord in a Twin Otter in July 1987. Ymer Ø and Celsius Bjerg in the foreground with some of the classic Ichthyostega *sites on its northeastern side and Gauss Halvø in the background. Photograph by R.N.G.C.*

Part of the central area of East Greenland is composed of rocks representing the Middle and Late Devonian period, and the area around Kejser Franz Joseph Fjord, which lies 400 km north of the Arctic Circle, has been noted for yielding fossils of Devonian vertebrates since the early years of the 20th century. The main localities are centered around Celsius Bjerg on Ymer Ø and the mountains of Sederholm, Smithwoodward, Stensiö, and Wiman Bjerg that make up the peninsula of Gauss Halvø (Fig. 5.1).

From the late 1920s until the mid-1950s, a series of expeditions undertaken jointly by Danish and Swedish paleontologists brought back remarkable finds of fishes and what were then the earliest known tetrapods. Two genera of tetrapod from this area have become particularly well known from extensive collections. *Ichthyostega* was the first genus of Devonian tetrapod to become established in textbooks, scientific articles, and in popular and children's books (Fig. 5.2). It is sometimes referred to as the "four-legged fish." Many specimens of this creature were found during the early years of exploration of East Greenland, and these are vividly described by Jarvik (1996) in his monograph on the animal. *Acanthostega gunnari* was named and described from two specimens in 1952 (Jarvik 1952), but since then, it has become much better known as a result of collections made during the summer of 1987, when a joint expedition from Cambridge, England, and Copenhagen, Denmark returned to the area (Bendix-Almgreen et al. 1990; Clack 1988). Most recently, an expedition in 1998 led by the author has collected new ichthyostegid material and has discovered a bit more about the circumstances in which the tetrapods were deposited (Clack and Neininger 2000). This expedition was supported by the National Geographic Society. Figure 5.3 shows three of the expedition members on Gauss Halvø on a reconnaissance trip, and Figure 5.4 shows the author seated at the 1998 campsite under the brow of Stensiö Bjerg.

When *Ichthyostega* and *Acanthostega* were alive, Greenland would have been in tropical latitudes. The climate of the region seems to have been monsoonal, and the animals lived in an extensive river basin fed by meandering streams flowing from mountains to the south. These fed into a large freshwater lake or inland sea whose margins were determined by a series of

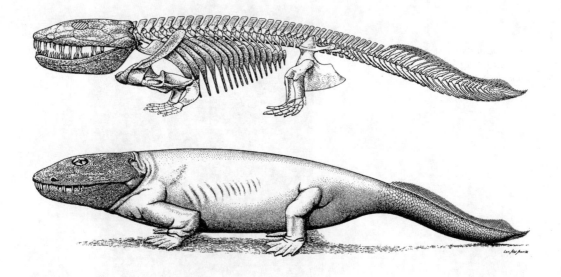

Figure 5.2. (above) Jarvik's (1996) reconstruction of Ichthyostega.

Figure 5.3. (left) Three of the 1998 expedition members: from left to right, Becky Hitchin, Sarah Finney, and the author. The photograph was taken by the fourth expedition member, Sally Neininger (S.L.N.).

faults, cutting a broad parallel-sided valley into the mountains of the Old Red Sandstone continent (Olsen 1993; Olsen and Larsen 1993).

Along the peninsula now called Gauss Halvø, tetrapods have been found in two formations, a lower level called the Aina Dal Formation and an upper one called the Britta Dal Formation (Fig. 5.5). The formations represent two periods of seasonal deposition in the river basin. They are separated by another formation, the Wimans Bjerg, which has yielded very few vertebrate fossils, although trace fossils and burrows have been found, indicating a rich fauna of bottom-dwelling invertebrates (Clack and Neininger 2000). These three formations constitute the Celsius Bjerg Group. The Wimans Bjerg Formation is thought to represent a drier period when

Figure 5.4. The author seated in front of Stensiö Bjerg in 1998. Photograph by S.L.N.

Figure 5.5. Photograph of Smithwoodward Bjerg and Stensiö Bjerg on Gauss Halvø showing the three formations of the Celsius Bjerg Group. At their base, the top of Aina Dal Formation is marked by an outcrop of harder rock forming a small cliff, running across from about 300 m at the left of the photograph and dropping to about 100 m on the right. The Wimans Bjerg Formation is the middle, slightly paler gray layer, whereas the uppermost Britta Dal Formation is marked by another slightly darker red layer. Photograph by J.A.C.

a shallow lake or playa was all that remained of the once deeper lake of Aina Dal times. In succeeding Britta Dal times, the return to a wetter climate filled the lake and rivers once more. In other parts of the Kejser Franz Joseph area, these separate formations cannot be distinguished so readily, suggesting slightly different geological conditions (Fig. 5.5) (Olsen 1993; Olsen and Larsen 1993).

Poorly preserved remains show that there were large plants growing in

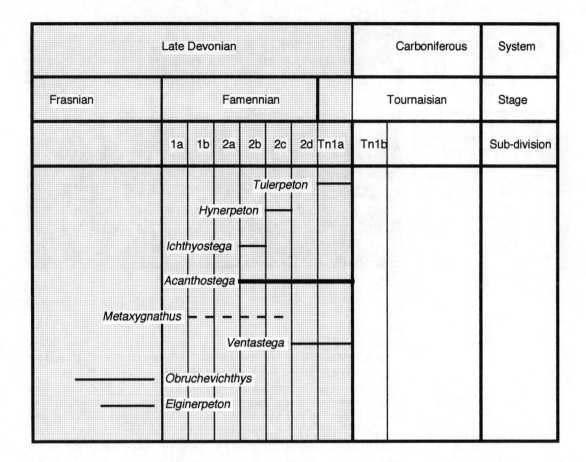

Figure 5.6. Chart showing relative dates for Devonian tetrapods based on spore analysis. The lines indicate error bars, not ranges, except in the case of Acanthostega. The dating of Metaxygnathus is problematical. From Marshall et al. (1999).

the region, probably along the riverbanks, but they do not give enough detail to be able to identify them precisely. The sediments have been oxidized in most regions, and most of the organic matter of the plants, including their spores, has been destroyed. Plant spores are useful for dating the sediments in which they occur, and until recently, the lack of spores in these Devonian rocks hampered precise dating. Recent work has fortunately unearthed spores in sediments above the Britta Dal Formation and below the Aina Dal formation, demonstrating them to be unquestionably Famennian (Marshall et al. 1999). *Acanthostega* and *Ichthyostega* are now the most securely dated of all Devonian tetrapods, with their occurrences bracketed between two marker bands defining quite a narrow range. Now that these spore dates have been worked out, they can be compared with similar spore data associated with other Devonian tetrapods. Other genera have much broader margins for error. Even though each occurrence of other tetrapods is only a single locality, the spore assemblages found there can still only give a time range, rather than a precise date. Taking these ranges into account, *Acanthostega* and *Ichthyostega* have been found to be the earliest of the Famennian forms (Fig. 5.6).

From study of the sediments in which it was found, it has been suggested that *Acanthostega* inhabited actively flowing river channels (Bendix-Almgreen et al. 1990). One small lens of rock contained numerous articulated skeletons of this tetrapod, and layers above and below con-

Figure 5.7. The 1987 team on Stensiö Bjerg when they first encountered the Acanthostega-*bearing horizon. The large block to the right of the author is that which provided the most complete articulated individual. At the base of the mountain, the campsite can just be seen. From left to right in the photograph: the author and Per Ahlberg (kneeling), Birger Jørgenson, and Svend Bendix-Almgreen (standing). Photograph by R.N.G.C.*

tained disarticulated bones. Figure 5.7 shows the 1987 expedition at the small outcrop that yielded most of the *Acanthostega* material, about 800 m up Stensiö Bjerg, which is reached only after several hours' climb. This deposit was interpreted as an accumulation of sediment on the bank of a river at the inside of a bend, called a point bar, where maybe during a flash flood, many individuals were heaped up together after they had died. Animal bones and carcasses were dumped on the bar formed of fine, silty sand, along with coarser sediments carried by the fast-flowing floodwater. From the good states of preservation of some of the skeletons, it does not look as though they had been transported far before they were fossilized, so it may be that they lived in the river where they died. Other interpretations have suggested that the channel in which the animals were deposited was formed during a sheet flood event that carried the carcasses, or indeed the living animals, suggesting that in fact they had not been living in the area but had been swept there by the water. Work in progress on these deposits should help to clarify the picture.

The site that yielded most of the *Acanthostega* specimens also produced numbers of scales of fishes such as the porolepiform *Holoptychius* and a lungfish, but the fish remains were all isolated, disarticulated elements. They almost all represented much larger animals than *Acanthostega*, and they could either have been living upstream in a lake of deeper water, or they could have fetched up where they did from a river or lake that flooded downstream.

The sediments in which *Ichthyostega* has been found vary from coarse red to fine-grained, blackish-red sandstones, and they frequently contain fish scales and plates of placoderms. *Ichthyostega* is the most frequently found tetrapod in the Aina Dal formation, and is usually encountered in massive blocks of ankle-breaking talus slope (Fig. 5.8). Although fish scales such as those of *Holoptychius* and lungfish are found alongside remains of ichthyostegids, placoderms are as often as not found in different lenses of sediment from those in which tetrapods are preserved. It is often the case that both placoderms and *Holoptychius* occur in lenses or bedding planes containing only remains of those animals. Most ichthyostegid specimens are disarticulated and appear to have been preserved in flood deposits. A

Figure 5.8. Aina Dal scree at the base of Stensiö Bjerg, where many of the Ichthyostega *specimens were found in the 1930s and in 1987. There is a figure in the picture to give some idea of scale. Photograph by J.A.C.*

few specimens held together better and are partly articulated, and these may have been carcasses that dried out and became mummified before being caught up in the flood. It appears that the ichthyostegid specimens have all been carried quite a long way from their original habitat (Clack and Neininger 2000).

Ichthyostega and *Acanthostega* represent two very different kinds of animals, not particularly closely related to each other, and adapted in very different ways. (The tendency in some literature to refer to them both as "ichthyostegalians" is mistaken and misleading.) Although they were contemporaries, they have not been found in exactly the same deposits, and they most likely exploited contrasting environments. They show that by the end of the Famennian, tetrapods were already diverse in their body forms and modes of life.

Ichthyostega

Ichthyostega was first described in 1932 by Säve-Söderbergh and by Jarvik in a series of papers over many years, the most recent being his monograph of 1996. Originally, four species were created, although their status as valid taxa is uncertain. Jarvik (1996) suggested that the name *I. stensioei* had priority over the others. Until further work is done on the material, this seems reasonable because the type specimens of these species all came from the same locality and the same altitude. However, there may still be more than one species of *Ichthyostega* present in the Celsius Bjerg Formation, to judge from the variability seen in the skull material.

The skull of *Ichthyostega* ("fish-armor") had a solid roof, like most

Figure 5.9. Photograph of one of the skulls of Ichthyostega *worked on by Jarvik (MGUH VP 6158). Skull about 200 mm long. Photograph by J.A.C.*

vertebrates of its time. The head of the largest specimen is about 250 mm long (Figs. 5.9, 5.10). It had a bluntly rounded snout, with a rather flattened profile, although at the back, the cheeks were quite steeply angled away down from the top of the head (Figs. 5.2, 5.10). The skull table, like those of many early tetrapods, had a hole for the pineal or parietal organ. The nostrils were near the edge of the snout, hardly separated from the mouth. Running through tubes in the bone of the head shield were the nerves and canals of the lateral line system, which opened to the outside by small pores. In this respect, *Ichthyostega* more closely resembles its fishy relatives than later tetrapods.

Although the braincase retains some primitive features, in others, it resembles that of no other known creature. Some recent work by the author and colleagues shows that its braincase was extremely narrow compared with any other contemporary fish or tetrapod. The regions to each side of the braincase appear to have been hollow chambers whose sides were specially strengthened by bony walls (Fig. 5.11). The contents and function of this unusual construction are still a mystery. The stapes has been identified as a thin, oval-shaped, curved lamina of bone, convex ventrally, and attached to the braincase by a fairly narrow neck. Unlike all other tetrapod stapes, it does not appear to have penetrated the braincase wall but attached to it anterior to a vestibular fontanelle similar to that found in *Eusthenopteron* (Figs. 2.1, 2.11) . Thus it could represent an early stage in the evolution of the characteristic tetrapod stapes. More work remains to be done to understand this curious and unique structure (Clack and Ahlberg 1998).

Where the cheek met the skull table, there was a deep embayment, bounded along the top edge by a downwardly turned flange from the skull table. The purpose of this notch has been the subject of speculation, which

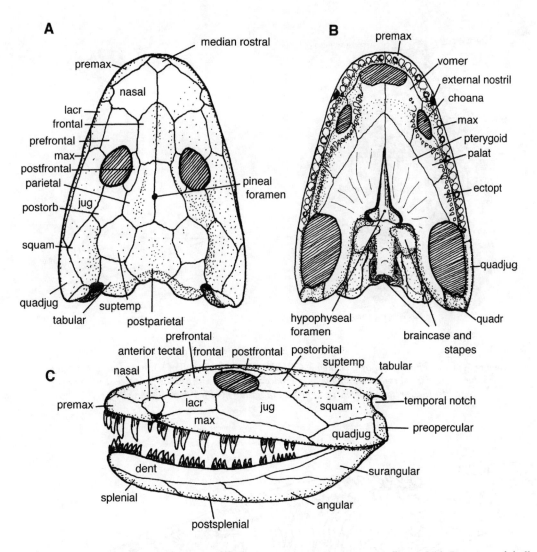

Figure 5.10. Drawings of skull of Ichthyostega.
(A) Dorsal view of skull roof.
(B) Ventral view with braincase.
(C) Lateral view of skull roof and lower jaw. Abbreviations as for Figures 2.1 and 2.10. Based on Jarvik (1996) and the author's personal observations.

will be dealt with in a later chapter. No element of a gill-bar system has yet been identified for *Ichthyostega*.

The mouth was armed with many sharp conical teeth. In the upper jaw, there were two rows, an outer row of about 27–30 large teeth and an inner row of about the same number of smaller ones, borne on the solid bony palate. The upper teeth contrast with those in the lower jaw, where surprisingly, in the outer row, rather than matching the upper teeth in size and number, there were more, although smaller, teeth. The reason for this discrepancy is not known, but *Ichthyostega* must have been capable of eating fairly large food items. It is likely to have fed on fish and invertebrates, although it is possible that it scavenged dead food as well as capturing live prey.

One of the more peculiar features of *Ichthyostega* was its ribs (Figs. 5.2, 5.12). It was a specimen containing the ribs of *Ichthyostega* that first gave paleontologists a clue that there were tetrapods to be found in the rocks of East Greenland, but it caused a great deal of puzzlement initially.

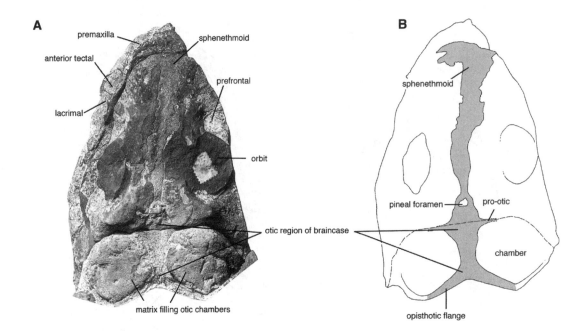

Figure 5.11. (A) Photograph of specimen MGUH VP 6158 of the skull of Ichthyostega, *showing otic chambers. The dermal skull bones have been eroded away over most of the dorsal part of this skull. (B) Interpretation of the specimen. Photograph modified from Jarvik (1996).*

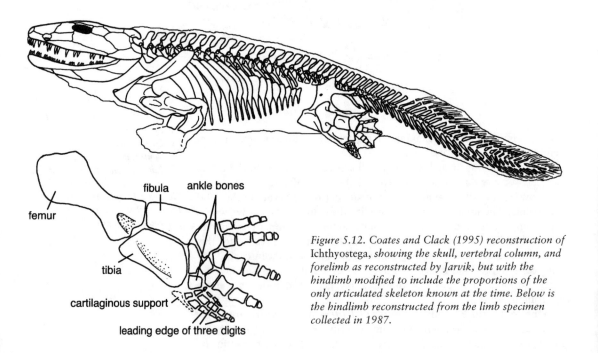

Figure 5.12. Coates and Clack (1995) reconstruction of Ichthyostega, *showing the skull, vertebral column, and forelimb as reconstructed by Jarvik, but with the hindlimb modified to include the proportions of the only articulated skeleton known at the time. Below is the hindlimb reconstructed from the limb specimen collected in 1987.*

In the trunk region, its ribs were greatly expanded, so that one rib would have overlapped three or four of the more posterior ones, forming a heavy, almost inflexible corset around its middle. The cervical or neck region ribs were relatively slender, and the ribs grew narrower and shorter toward the pelvis. The arrangement would not have allowed the animal to move in a sinusoidal motion, as might have been expected in an early tetrapod and as can still be seen in lizards and salamanders. Toward the rear of the body, the ribs seem to have become narrower and angled more steeply backward.

It is not known what purpose the expanded ribs served. One suggestion is that they served to support the body in place of having well-ossified vertebrae (Thomson and Bossy 1970). Although *Ichthyostega* had vertebrae with quite well formed neural arches, the pleurocentra on which they rested were rather small. Another possibility is that the ribs were there to support the internal organs and to stop them from collapsing when the animal came onto the land. They are unlikely to have been of use in helping to ventilate the lungs, as those of mammals are, for two reasons. First, the overlap would effectively have prevented this. Second, although ribs are usually thought of as closely linked with ventilatory movements, in reality, it is mainly in modern amniotes (mammals, birds, and reptiles; see Chapter 10) that ribs are used to bring about inhalation. They were not necessarily used like this in *Ichthyostega*.

The shoulders of *Ichthyostega* were massive for the size of its body, and recent restudy of the specimens suggests that Jarvik's (1996) restoration underestimates their relative size. In early tetrapods, the shoulder girdle still retained parts of the dermal skeleton, which was inherited from the fish, including a cleithrum (Fig. 2.5). This bone in *Ichthyostega* was large and fused with the cartilage–bone scapulocoracoid where the articulation for the forelimb was placed. Judging from the size and position of this articulation point, the forelimb would have been fairly restricted in its range of movement. For example, it might not have been able to move forward further than about 90° to the body, although it could have been pulled back into the side of the body when the animal was swimming.

The upper arm bone or humerus was L-shaped and fairly flattened, with large processes forming attachment points for muscles to the shoulder girdle. At the outer end of the humerus were the attachment points for the forearm bones, the radius and ulna. In some individuals, the radius pivoted at a bulbous condyle on the underside of the humerus, whereas the ulna swung around a strap-shaped condyle at the end, although in others, the radius was placed more toward the leading edge of the humerus. The radius and ulna were rather short, stubby bones, although the ulna bore a process at the elbow, called the olecranon, to which muscles attached for raising the forearm. Because of the way the radius and ulna attached, the elbow could probably not have been extended much further than a right angle. Unfortunately, nothing is described of the wrists or hands of *Ichthyostega*, although some specimens do preserve incomplete material that remains to be investigated.

When the postcranial skeleton of *Ichthyostega* was first described in detail, it was the tail that aroused the greatest interest because it was more fishlike than tetrapodlike. In many other respects, *Ichthyostega* is a very strange animal, and parts of it are like no other known tetrapod or fish. But the tail was unequivocally fishlike. Most obviously, it had bony fin rays supporting a web, as is found in fishes, but not at the time in any other known tetrapod (Figs. 5.2, 5.12). Extra supports called radials attached to the tips of the neural arches and helped to stiffen the web. The fin web

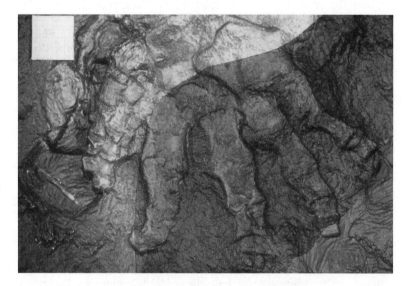

Figure 5.13. (Top) Hindlimb, pelvic girdle, and tail found in 1987, with the counterpart of the four largest toes placed at right (MGUH f.n. 1349). (Bottom) Part and counterpart toes put together in a photographic transparency by Mike Coates. Scale bar = 10 mm.

extended not just dorsally but continued around underneath the end of the tail. Clearly, this tail was used in water to aid swimming.

Several specimens with preserved tails are known, and some of them also have preserved elements of the hindlimb and pelvic girdle. Probably the best of these was that found in 1987 (Coates and Clack 1990), whose significance is discussed further in Chapter 6. The pelvic girdle was a large structure, robust, and formed from a single bone on each side. It bore both a horizontally directed process oriented backward and a shorter, stouter one directed upward. It probably had a rib that was especially adapted to join to the pelvis on the internal face of this process, although this rib has not been found. The backwardly directed process almost certainly bore muscles that helped to power the tail.

The hindlimb is known from three or four specimens, the best preserved of which was found by the 1987 expedition (Figs. 5.12, 5.13). The femur was flattish in cross section but had a large flange on the side nearest the body; the flange carried muscles that moved the limb inward and downward. At the outer end, the bone widened and provided attachment for broad and flat lower leg bones, the tibia and fibula. These in turn were attached to what seems to have been a rather inflexible ankle joint composed of fewer bones than those of modern tetrapods, although these, like those of the lower leg, were flat and quite large.

Perhaps the most surprising feature of all in *Ichthyostega* is the complement of toes. The 1987 specimen clearly shows that at the leading edge of the limb were three tiny toes, the smallest of them being the third one from the edge. Then followed four quite stout toes, for a total of seven. The leading edge also seems to have been strengthened by a strip of a cartilagelike substance that was preserved along with the toes (Fig. 5.13).

The picture that emerged from putting together the material collected in 1987 with some of the specimens collected by earlier expeditions was surprisingly different from the view that is found in textbooks and the popular literature for children and nonspecialists. What is more, over the years, the perception of *Ichthyostega* has changed. Although in 1952 it was restored as a semiaquatic animal with paddlelike hindlimbs, by 1980 (Jarvik 1980), it was portrayed as a solidly terrestrial animal, with four stout, weight-bearing legs. The material collected in 1987 suggested that the earlier reconstructions of 1952, as well as a more recent one by Bjerring (1985) of a more aquatically adapted animal, were more correct. New material collected in 1998 will be crucial in revising the image of *Ichthyostega* and improving its accuracy. It may not be the stumpy-legged creature lumbering over the sand dunes, as it is sometimes portrayed (Fig. 5.2).

Although *Ichthyostega* is represented by many specimens, especially skulls, until recently there has been only one in which a head, shoulders, forelimb, and trunk are thought to be associated with a femur and part of the presacral column, and this is the one on which the reconstruction is

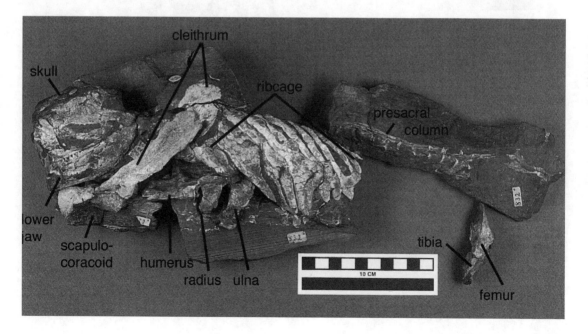

Figure 5.14. Articulated specimen of Ichthyostega *(MGUH VP 6115). The skull is broken off just behind the orbit. Half the left lower jaw can be seen. The specimen preserves both humeri, cleithra, and scapulocoracoid, and this photograph shows the left shoulder girdle with the humerus tucked under the ribcage. The radius, ulna, and one wrist bone (ulnare) were preserved in articulation with it. The broad overlapping ribs of the anterior part of the body can be seen. At the hinder end, part of what may be the presacral column is shown, although it cannot be fitted onto the main specimen, nor can the femur with its associated tibia. This femur, even allowing for breakage and crushing, is much shorter than the humeri and may not belong to the same individual. Photograph by S.M.F.*

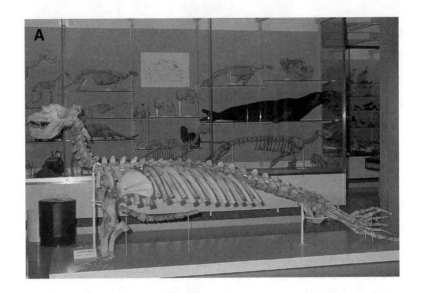

Figure 5.15. (A) Skeleton of an elephant seal in UMZC. (B) Photograph of dolphin paddlelike forelimb (UMZC specimens; photographs from UMZC archives).

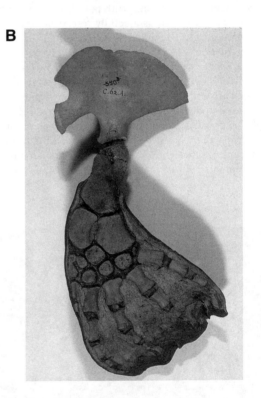

based (Fig. 5.14). However, the association of the femur is dubious, and the supposed presacral column section may not in fact belong to one individual. Thus, an accurate picture of its proportions is not yet possible, although the separate parts are known in some detail. However, taking the proportions of this single most complete specimen and combining them with knowledge of the other hindlimbs, pelvic girdles, and tails gives the best estimate, with the caveat that there are numerous problems still to solve with respect to the body form of *Ichthyostega*. The picture that emerges from a strict adherence to these specimens is a rather unexpected one for such an early tetrapod (Fig. 5.12), and this in itself may provide grounds for questioning its accuracy.

The femur associated with the most complete body is only about half the length of the humerus, making the forelimb much larger and stouter than the hindlimb. This is unusual in tetrapods, whose hindlimbs are, as a rule, larger than the forelimbs because the hindlimbs provide the power in walking. In *Ichthyostega*, the hindlimbs were not only diminutive compared with the forelimbs, but with their flattened bones and inflexible ankles, they were much more paddlelike than leglike. By contrast, the massive shoulders and forelimbs look as if they were adapted for supporting the head and lifting the front end of the body off the ground. It appears that in *Ichthyostega*, the forelimbs provided the motive power on land, although in the water, it would have been the tail. The paddlelike hindlimbs might have aided steering and braking, and perhaps stabilization or providing purchase against the ground on land.

The skeleton of a modern seal, such as an elephant seal, compared with that of *Ichthyostega* shows a number of parallels that may provide clues about *Ichthyostega*'s lifestyle. In the elephant seal, the elbows are held at right angles when on land (Fig. 5.15). The forelimbs are about twice the size of the hind, which in turn are paddlelike. All the limbs can be tucked into the sides for speed swimming powered by the tail, or they can be used to steer the animal in direction changes. When the seal comes onto land, the forelimbs drag the body, hind legs, and tail, and they also tend to move together. It is possible to imagine *Ichthyostega* progressing in much the same way.

The skeleton of the paddlelike hindlimb also has another parallel among modern forms in its similarity to that of the forelimb of some dolphins, especially a river dolphin. In these animals, the bones are flattened, and there is a reduced number of them in the wrist. At the leading edge of the paddle is a small digit; the other four are stout, and all are bound together by a web (Fig. 5.15). It is not known whether the digits of *Ichthyostega* were embedded in a web, but all of the known specimens of hindlimbs have their digits present, tending to suggest that they might have been.

Figure 5.16 gives a life reconstruction of *Ichthyostega*, one based on the most recent information. In summary, *Ichthyostega* is a curious mixture of features, some of them primitive but some of them specialized and unique. Because of this, after its discovery, it did not lead to many new insights into the origin of tetrapods. New material currently being studied should go a long way to rectifying this, as well as to putting the animal into a more detailed biostratigraphic and paleoecological context.

Acanthostega

Acanthostega was the second Devonian tetrapod to be described from East Greenland (Jarvik 1952). For many decades, all that was known of this animal were two partial skull roofs, and from these it was clear that it

Figure 5.16. (above) Life reconstruction of Ichthyostega *incorporating some of the most recent observations by the author. The number of fingers on the manus is unknown. Illustration by J.A.C.*

Figure 5.17. Holotype specimen (MGUH 6033) of Acanthostega *described in 1952 by Jarvik. Skull about 140 mm long. Photograph from UMZC archives.*

was a very different animal from its contemporary. Figure 5.17 shows the holotype specimen on which the original description of *Acanthostega* was based. The specimen shows the right side and midline bones of the skull more or less complete, with the tooth-bearing maxilla somewhat detached from the rest of the cheek. The pattern of bones in the skull roof was in some respects different, in particular at the back edge of the skull table, where *Acanthostega* had a tabular bone that formed both a deep embayment and long, pronglike, tabular "horn." It is this distinctive structure that gives it its name, which means "spine-armor."

When following up some accidental finds of *Acanthostega* material made in 1970 by a sedimentologist, the joint Copenhagen-Cambridge expedition succeeded in collecting what has turned out to be spectacular new material of this creature (Figs. 1.3, 1.5, 5.18). Remarkably, among the

Figure 5.18. (left) The most complete of the Acanthostega *specimens found in 1987, after preparation was finally completed in 1998 (MGUH f.n. 1227). A second skull is seen to the top left. The forelimb with its eight digits lies at bottom left. Photograph by S.M.F.*

Figure 5.19. (bottom) Drawing of Acanthostega *skeleton (modified from Coates 1996).*

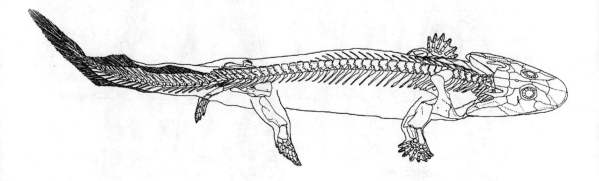

specimens they collected in 1987 were the bodies of the animals whose skulls had been found in 1970 (see Fig. 5.21). An account of this story is entertainingly told by Zimmer (1998). From these accumulated finds, almost the whole skeleton of *Acanthostega* is now known and has been described in a series of articles, listed as they are discussed below. The animal has emerged as much more what would be expected of a primitive tetrapod (Fig. 5.19). Its study has led to some changed ideas about what early tetrapods were like and has also led to some new ideas about the development of limbs, the evolution of air breathing, and the subsequent radiation of tetrapods from the Devonian to the present day.

Apart from primitive similarities, *Ichthyostega* and *Acanthostega* have almost nothing in common. As in *Ichthyostega,* the skull of *Acanthostega* was quite shallow, with a bluntly rounded snout (Clack 1994a, 2002) (Fig. 5.20). Its palate was likewise formed of a bony sheet, although not so solidly attached to the rest of the skull as in *Ichthyostega*. As in *Ichthyostega,* the external nostril lay very close to the edge of the mouth, and the lateral-line organs were in most cases housed in tubes through the bone, although these merge into grooves in some places, and the pores through which they open to the outside are more conspicuous than in *Ichthyostega*.

The First Feet • 121

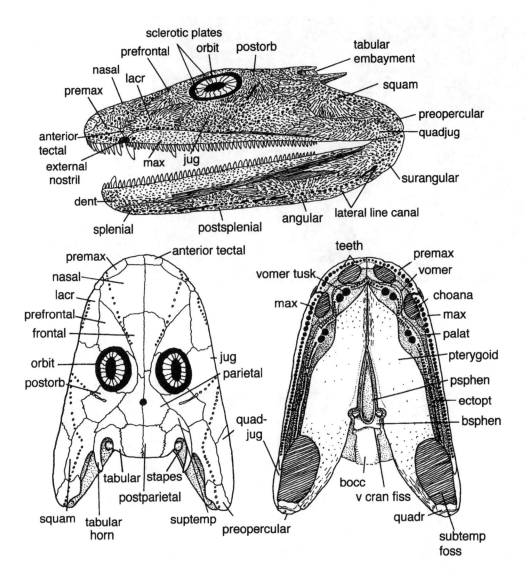

Figure 5.20. Drawings of skull of Acanthostega: *top, left lateral view; bottom left, dorsal view; bottom right, ventral view. Abbreviations as for Figures 2.1 and 2.10.*

The skull and the dermal shoulder bones of both animals had a sculptured surface of pits, ridges, and furrows, in common with other early tetrapods and many contemporary fishes (Figs. 1.3, 6.5). The function of this ornament is unknown, but something similar is seen in certain modern fishes, especially catfishes. In the skull roof, the suturing of the skull table bones to each other is particularly secure, with interdigitations interfingering in three dimensions. In the snout, there are broad areas of overlap where the sutures are designed to resist the twisting of the snout in feeding (Clack, 2002).

Acanthostega was a rather smaller animal than *Ichthyostega*, with a skull of a maximum size of about 200 mm and a dentition suggesting a different diet. The outer row of marginal teeth were much smaller than in *Ichthyostega*; the inner row on the palate consisted of both much smaller toothlets and denticles, plus some large fangs. The contrast in upper and

Figure 5.21. (left) Acanthostega specimen showing the skull in ventral view, with the gill skeleton underneath and shoulder girdle elements lying more or less in place (UMZC T1300 and MGUH f.n. 1258). The skull was found in 1970 and the body in 1987.

Figure 5.22. (below) Drawing of head and shoulders of Acanthostega. (A) Relationship of head to shoulder girdle.
(B) Skull roof showing arrangement of gill skeleton and stapes inside.
(C) Anterior view of left half of shoulder girdle showing the postbranchial lamina.

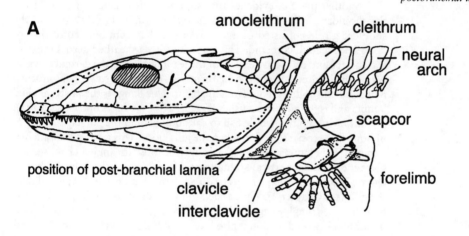

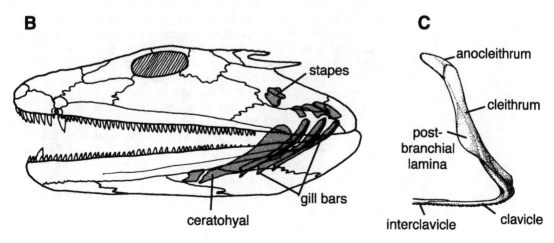

The First Feet • 123

lower tooth size seen in *Ichthyostega* is less noticeable in *Acanthostega*. Almost all the features of *Acanthostega* that have been discovered suggest that it was likely to have been totally aquatic, so its dentition, resembling in some respects that of many of the contemporary lobe-finned fishes, was probably adapted for feeding in water.

Because of the remarkable preservation of several of the *Acanthostega* specimens, parts of the skeleton, particularly the more delicate ones, are better known than in *Ichthyostega*. A good example of this is the preservation of the gill skeleton, not often discovered in any fossil tetrapod (Figs. 5.21, 5.22). In *Acanthostega,* the gill skeleton proved to be remarkably fishlike (Coates and Clack 1991). Each of the branchial elements was strongly ossified and had a deep groove running down the back. Elements from at least three pairs of gill arches have been found. When compared with modern fishes in which the gill structure is known, the gill bars of *Acanthostega* most closely resemble those of the Australian lungfish *Neoceratodus*. This fish uses internal gills for breathing in water like most fish, but it also uses lungs for breathing air. There are three functional pairs of gill arches, as found in *Acanthostega*. The groove along the individual elements accommodates the afferent branchial artery, which takes blood to the gills for oxygenation. This is one piece of evidence that suggests that *Acanthostega* used internal gills like *Neoceratodus* as part of its breathing mechanism.

Another piece of evidence that supports this conclusion comes from the shoulder girdle (Fig. 5.22). In *Acanthostega,* as in *Ichthyostega,* the shoulder girdle consisted of a stout dermal cleithrum bonded to the endoskeletal scapulocoracoid. The whole unit was sheathed across the ventral surface and up the lower part of the leading edge by the dermal clavicle. Along the leading edge of the cleithrum in *Acanthostega* was an inturned flange, a little hollowed out from the front. This flange is the postbranchial lamina and is found in most fishes (Fig. 5.22B, C) (Coates and Clack 1991). (Next time you buy a whole fish to eat, look behind the gill flap at the rim of bone that receives the flap when the gills close. This is the postbranchial lamina.) This lamina performs a couple of jobs. Its angled cross section gives the shoulder vertical strength, and it also forms the back of the gill chamber, where its shape helps to direct water out once it has passed over the gills (see Chapter 2). This, added to the evidence from the gill elements themselves, strongly suggests that *Acanthostega* still used internal gills for breathing. It would no doubt also have used lungs as well, as did *Neoceratodus,* because lungs or their equivalent appear to have been a common heritage of all bony fishes. *Acanthostega* also retained another fishlike character of the shoulder girdle: the presence of the anocleithrum, a remnant of the more extensive dermal shoulder girdle of fishes (Coates 1996). There was no postbranchial lamina in the shoulder girdle of *Ichthyostega,* and no gill skeleton has been found, although this may have more to do with the degree of ossification of the elements, or the state of preservation of the specimens, than with their actual absence in the living animal.

In contrast with *Ichthyostega,* in *Acanthostega* the surface of the glenoid facet was not strongly contoured but was a relatively flat oval, directed to the side of the animal. This suggests that its forearm also was directed almost entirely sideways, although it may have had more freedom of movement to the front than that of *Ichthyostega*. The forearm of *Acanthostega* has revealed a number of the most remarkable new facts about Devonian tetrapods (Coates and Clack 1990).

The humerus of *Acanthostega* is shaped like a broad L, as in most other early tetrapods, although as in *Ichthyostega,* a number of foramina

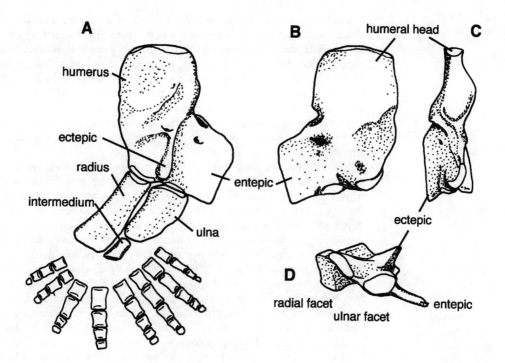

Figure 5.23. Drawing of forelimb of Acanthostega. (A) Complete left forelimb. (B) Humerus from the ventral side. (C) Anterior view of the humerus. (D) Distal view of the humerus.

and processes were present that disappear during the evolution of tetrapods as a whole. However, the bones of the lower arm, the radius and ulna, are different in several ways from those of any other tetrapod (Figs. 5.22, 5.23). First, the radius was an elongate element, cylindrical in section near the articulation with the humerus, but flattened and spatulate at the lower end. It was also nearly twice as long as the ulna. The ulna was a short, flattened bone, with a strap-shaped articulation where it joined the humerus but a very narrow articulation where it joined the wrist. The leading edge facing the radius was straight, but the back edge was curved. There was no olecranon process like that in *Ichthyostega*.

These proportions and shapes are unique among tetrapods, but they also reflect those of a lobe-finned fish such as *Eusthenopteron* remarkably closely. Therefore, the most plausible interpretation is that the shapes and proportions are those of a very primitive forelimb that retains several fishlike features, rather than one that has been adapted from that seen in a more "conventional" tetrapod. This conclusion is important for interpreting many aspects of the morphology of *Acanthostega*, including what is probably the most surprising aspect of the animal: the hand and digits.

Only one true wrist bone is known, that attached to the ulna, called the intermedium. Although a number of other possible elements were discovered that may be wrist bones, they were jumbled up and it is not possible to say where they were positioned. The wrist itself must have been broad and inflexible because of the discrepancy in length between the radius and ulna. It is difficult to see how that arrangement could have functioned as a weight-bearing joint.

More surprising are the digits. One specimen shows the digits in full articulation at the end of the arm. There are eight of them. Each is complete, and none is the duplicate of any of the others. They are arranged in a regular array, with three small ones at the leading edge and two very slender ones at the rear. It is not clear whether these digits were enclosed in any kind of webbing, although the fact that they are still in articulation suggests that they might have been, at least to some extent. Taking into account the structure of the joints and the arrangement of the digits, the limb seems unquestionably to be adapted as a paddle rather than a walking leg. This discovery has proved significant in several ways, not only with respect to the original function of limbs with digits but also for ideas on the embryonic development of limbs. These implications will be discussed in later chapters.

Almost the complete vertebral column of *Acanthostega* is preserved, including the most anterior elements forming the neck and those in the region where the pelvis attaches (Coates 1996). Also preserved are many of the ribs of the neck and those just behind the pelvis. One of the striking things about the column is how little variation there is along its length. In this respect, it was similar to a fish such as *Eusthenopteron* or an aquatic tetrapod from the Carboniferous such as *Greererpeton*. The processes on the neural arches, the zygapophyses, were poorly developed in *Acanthostega,* although in a few cases the arches had additional articulations between them instead. The centra were essentially husks surrounding the notochord and were formed from two halves that remained unfused in the midline, except for the atlas and sacral intercentra. This lack of differentiation and lack of zygapophyses is consistent with an animal whose head and body were routinely supported by water, rather than by the limbs and girdles that transmitted forces to the vertebral column by specialized muscles.

Although the cervical ribs of *Acanthostega* are well preserved and found in articulation with the vertebral column, most of the trunk ribs are missing or disarticulated. The cervical ribs were scooped out at the ends and broadened to provide seating for neck muscles and those that supported the shoulder. The trunk ribs were about the same length as the cervicals; they were short and straight, and their heads were only weakly differentiated into upper and lower process for articulation with the vertebrae. The were quite unlike the heavy, flanged, overlapping ribs of *Ichthyostega,* and this probably constitutes the clearest evidence of the differences between them in body form and lifestyle.

The hindlimb and pelvic girdle of *Acanthostega* are also known from a number of specimens (Fig. 5.24) (Coates 1996). The pelvic girdle was relatively small for a tetrapod, compared both with *Ichthyostega* and other somewhat later species from the Carboniferous. However, it was still large by comparison with that in any fish and was more tetrapodlike than fishlike in other respects, too. It was a single element as in *Ichthyostega* and similarly bore two dorsal processes, a short, stout, upwardly directed one and a long, slender, backwardly directed one. The backwardly directed one lay at an angle of about 45° to the vertebral column. The more dorsally directed process would have been attached on its inner surface to the vertebral column, but there was no facet for the sacral rib to have articulated with it. Therefore, it seems as though the attachment must have been formed by ligaments. The sacral rib that formed the link between girdle and vertebrae was scarcely modified for the purpose, although it had a slightly expanded end to give a greater surface area for the attachment.

The femur in *Acanthostega* was somewhat longer than the humerus, contrasting with the short femur of *Ichthyostega*. On the surface nearest the body was a large rectangular adductor blade (Fig. 5.23). This flange

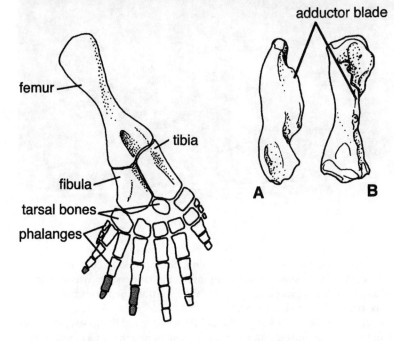

Figure 5.24. (Top) The hindlimb of Acanthostega *(MGUH f.n. 1375). (Bottom) Reconstruction of the hindlimb of* Acanthostega *by Coates (1996). Shaded elements are reconstructed. The position, but not the number, of tarsal bones has been inferred from the disarticulated specimen. To the right of the reconstruction are the femur in (A) side and (B) internal views, to show the adductor blade. From Coates (1996).*

would have carried the adductor muscles to pull the leg inward and backward. The size of the flange suggests large, powerful muscles, but the position of the adductor blade suggests that they were used in swimming. The lower leg bones, the tibia and fibula, like those of *Ichthyostega*, were flattened with a slight overlap between them along their length. As in *Ichthyostega*, the ankle was formed of flattened elements. Although only known from one disrupted specimen, the number of digits is likely to have been eight, as in the forelimb, on the basis of a rather conservative judgment. In this case, very small digits are placed one at each edge of the foot, with a second small one contributing to the leading edge (Fig. 5.24).

This structure strongly suggests that as in *Ichthyostega*, the hindlimb was used largely as a paddle, and the powerful action produced by the

Figure 5.25. Life reconstruction of Acanthostega *by J.A.C.*

adductor muscles was used in swimming, not in walking. Support for this idea comes from some modern aquatic salamanders, in which a large adductor blade is placed further down the shaft of the femur than it is in terrestrial ones (Coates 1996).

Although in *Ichthyostega* the tail fin was relatively modest, in *Acanthostega*, it was substantial, most closely resembling that of a lungfish (Fig. 5.19). The fin rays were longer, more numerous, and reached further around beneath the tail than in *Ichthyostega*. This is the tail of a lurking predator, which was able to produce and control fine rippling movements of the fin by means of the fin rays while the animal lay stationary in the water, but could have produced a fast, powerful thrust to give rapid acceleration when it lunged to catch prey. This is not the tail of an animal that ventured regularly onto land. Moreover, the existence of the dermal fin rays suggests that it was not descended from an animal that had been more terrestrial. During the evolution of tetrapods, there is a steady trend toward reduction of the dermal skeleton, including the fin rays, which, once lost, is not redeveloped. More details of the postcranial skeleton can be found in Coates's monograph of 1996.

The postcranial skeleton of *Acanthostega* shows numerous features that clearly mark it as an aquatic animal that rarely, if ever, made forays onto dry land and whose legs would have almost certainly been incapable of supporting its body had it done so (Fig. 5.25). However, the more important question is whether this aquatic adaptation was a primitive relict of its fish ancestry or was secondarily derived from more terrestrial tetrapod ancestors (Clack and Coates 1995; Coates and Clack 1995). This question will be discussed further in the last section of this chapter.

In addition to the hyobranchial and postcranial skeleton of *Acanthostega*, there is another key region of the skeleton that has provided important insight into the events of the fish–tetrapod transition, and this is the braincase and associated structures. In contrast to that of *Ichthyostega*, the braincase of *Acanthostega* is directly comparable to that of related fishes such as *Eusthenopteron* and *Panderichthys*. The similarities and differences between these fishes and *Acanthostega* are particularly instructive and will be considered in Chapter 6.

Famennian Tetrapods Worldwide

Tulerpeton

Tulerpeton curtum is the third Devonian tetrapod genus for which articulated postcranial remains are known (Fig. 5.26). It comes from the Late Devonian, from a site called Andreyevka, near Tula in Russia. The site has yielded many taxa of fishes and invertebrates and is unique among localities that have yielded Devonian tetrapods in showing clear evidence of marine influence in the sediments. Serpulid worms and stromatolites, the latter of which are accretions of calcite and limestone laid down in layers by algae, are characteristic of marine or brackish sediments. Charophytes, which are a form of primitive plant probably intermediate between vascular plants and algae, are also known from this site. The chemical makeup of the sediments shows signs that the salinity of the water fluctuated widely, and the geological evidence suggests that the Andreyevka site was situated about 200 km from the nearest landmass (Alekseyev et al. 1994).

Tulerpeton is represented by a nodular block containing the belly scales, right hind- and forelimbs, and left shoulder girdle of a single individual, along with a premaxillary bone and its attached vomer. This specimen is the only one from the site that was preserved as an articulated series of bones. Many other vertebrate bones have been recovered from the site, but all of them have been found by using acid to digest the rock away, leaving the bones behind. Other tetrapod elements are among these bones, including further isolated skull bones, vertebrae, ribs, interclavicles, and part of a pelvic girdle. It is not certain that all these belong to the same taxon because there are two patterns of tabular bone in the collection that has been made from Andreyevka.

When *Tulerpeton* was first found and described, it was considered aberrant in having six digits on the forelimb, and almost certainly six on the hindlimb (Lebedev 1984, 1985). Now that other Devonian tetrapods have been found similarly showing more than five digits, *Tulerpeton* is seen as fitting into a pattern of polydactylous (having more than five digits) early tetrapods. The implications of this are discussed in Chapter 10. In addition to the development of digits, several other important aspects of the early evolution of tetrapods are shown by *Tulerpeton* (Lebedev and Clack 1993; Lebedev and Coates 1995).

The first impression of its limbs is that, apart from the digits, they resemble those of more conventional and later tetrapods. The humerus resembles most closely that of a group called anthracosaurs known from the Carboniferous (see Chapters 7 and 8). The radius and ulna, and to a lesser extent the tibia and fibula, are relatively more elongate and slender than those of the other known Devonian tetrapod limbs, those of *Ichthyostega* or *Acanthostega*. These give it the appearance of a more terrestrially adapted animal. However, the ankle construction bears some resemblance to those of *Ichthyostega* and *Acanthostega* in having fewer bones than those of later tetrapods. The tibia and fibula still retain some of the flattened structure seen in the two other Devonian genera, making the foot more of a paddlelike appendage rather than one used for walking. Furthermore, as in *Ichthyostega* and *Acanthostega*, the femur bore a large quadrangular adductor blade placed about halfway down the bone, and this kind of construction is found today associated with swimming muscles in aquatic salamanders.

The shoulder girdle shows several primitive features including retention of the anocleithrum, but unlike those of *Ichthyostega* and *Acanthostega*, the dermal cleithrum had become separated from the scapulocoracoid. *Tulerpeton* lacks a postbranchial lamina that was still present in *Acanthostega*, suggesting that *Tulerpeton* had lost the use of internal gills for breathing.

Although very little of the skeleton of *Tulerpeton* has been found, it has proved possible to generate some hypotheses about where it fits phylogenetically among other early tetrapods. By use of postcranial characters, Lebedev and Coates (1995) found that it appeared more similar to the lineage of animals normally associated with the origin of amniotes than to the lineage usually thought to have given rise to modern amphibians. The implications of this are discussed in Chapter 9.

Ventastega

Ventastega curonica, one of the most recently discovered of the Devonian genera, comes from Latvia near the Venta River, from which its name derives (Ahlberg et al. 1994). This is another example of an animal that was first recognized as a tetrapod from bones in a museum collection labeled as "unidentified sarcopterygian." Study of *Acanthostega* allowed its lower jaw to be picked out of its drawer and relabeled as a tetrapod.

The bones occur as beautifully preserved three-dimensional specimens, embedded in a light, sandy matrix that is easily washed off, revealing remarkable detail. There is enough of the skull and the limb girdles present to be sure that they are those of a tetrapod. Exploration of this site is only just beginning, and only a few parts of the skeleton of *Ventastega* have yet been found. In the past few seasons, more bones have been found, but they tend to be duplicates of what is already known rather than supplying new information. However, further expeditions to this and other potentially productive sites are planned for the future, with the hope of finding more clues.

The skull and palate of the animal resemble in many ways those of *Acanthostega*, although details show that it is not the same animal (Fig. 5.27A). For example, in the lower jaw, the inner row of tooth-bearing bones, the coronoids, each carry a fang pair as well as a row of smaller teeth (Ahlberg et al. 1994). These are similar to, although smaller than, those seen in fish such as *Panderichthys*, but they are absent in *Acanthostega* and *Ichthyostega*. Fang pairs are also carried on the palate in *Ventastega*, a feature also seen in *Acanthostega* and many later tetrapods. Their presence seems to be a fishlike and primitive character, perhaps associated with feeding in water, and gradually lost among tetrapods. For this reason, although only the lower jaw has been considered, it appears in a phylogeny

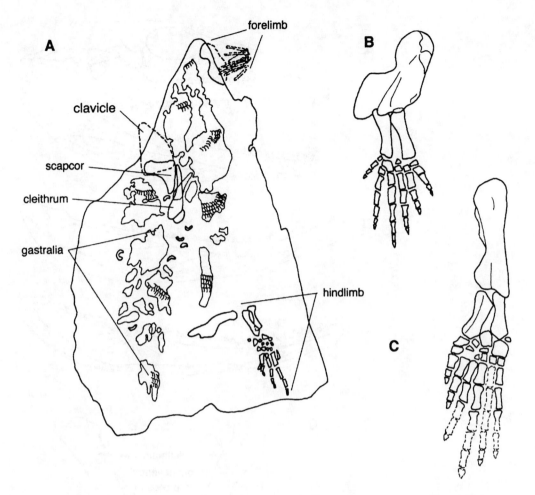

Figure 5.26. Tulerpeton curtum. (A) Diagram of specimen showing position of elements. (B) Forelimb and (C) hindlimb reconstructions. Based on information in Lebedev and Coates (1995).

as being a more primitive animal than *Acanthostega* (Ahlberg and Clack 1998) (see Chapter 10). The really informative parts of the skull, such as the skull table with the tabular and supratemporal, have not yet been found.

Parts of the postcranial skeleton that have been found include the cleithrum with a small portion of the scapulocoracoid integrated at its base, showing the condition found in most other Devonian tetrapod shoulder girdles. Not much of the scapulocoracoid portion is preserved, unfortunately. The clavicle is well represented by several specimens, and it shows a broad, triangular base with the typically narrow stem angled at about 90° to the basal plate. The interclavicle was large and lozenge-shaped, as in *Acanthostega*, with similar dermal ornament. Thus, it will be possible to reconstruct not only the complete shoulder girdle, but also to get some idea of the body profile. A radius has also been found, but nothing else of the postcranial skeleton.

Hynerpeton and *Densignathus*

The discovery of *Hynerpeton bassetti* demonstrates that Devonian tetrapods are waiting to be found in areas that have apparently been thoroughly surveyed geologically many times. It comes from the middle of a popular tourist region of Pennsylvania in the United States. However, the sediments are only visible in road cuts, and are not readily accessible. The

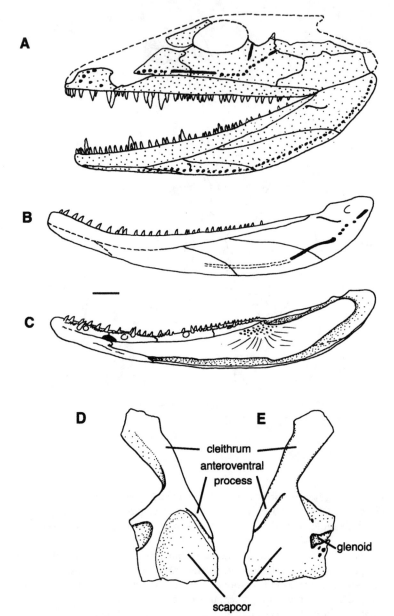

Figure 5.27. (A) Reconstruction drawing of skull of Ventastega in lateral view (from Ahlberg et al. 1994). Lower jaw of Metaxygnathus in (B) outer and (C) inner views (from Campbell and Bell 1997). Drawing of the shoulder girdle of Hynerpeton in (D) inner and (E) outer views (based on Daeschler et al. 1994).

rocks are of Late Devonian age and also contain several taxa of fish, including lungfishes and osteolepiforms. They are thought to represent sediments laid down in a large river valley system with meandering streams cutting channels and carrying remains of carcasses into the area.

So far, two examples of shoulder girdles (Daeschler et al. 1994), two lower jaws (Daeschler 2000), a jugal bone, and some gastralia are all that have been found of any tetrapod. However, the two lower jaws are different from one another in size and dentition and are thought to represent two different genera. One of the jaws has been named *Densignathus,* and the smaller is attributed to *Hynerpeton.* Excavations are still in progress

Figure 5.28. The locality of Red Hill near Hyner, Pennsylvania, where Hynerpeton *was found, with its discoverer Ted Daeschler standing at the level from which the specimens came. Photograph by J.A.C.*

at several sites in the area with a view to finding more material. Figure 5.28 shows Ted Daeschler, the discoverer of *Hynerpeton,* standing on the site where the specimens of both animals were found.

The shoulder girdles are both apparently from the same species and were found in the same part of the same bed as the *Hynerpeton* jaw, as well as being of about the right size to go with it. They are robust and clearly carried large, strong muscles for powering the forearm (Fig. 5.26C). However, no limb bones are yet known, so it is not possible to say whether *Hynerpeton* walked or swam using these muscles. Taking the shoulder girdle alone, *Hynerpeton* shows some differences from both *Acanthostega* and *Ichthyostega,* including some in which it is more advanced. For example, it lacks a postbranchial lamina, but the cleithrum and scapulocoracoid are still bonded together as in these two forms, and this seems to be a primitive feature of the earliest tetrapods.

Metaxygnathus

Metaxygnathus was described in 1977 (Campbell and Bell 1977), but until recently, there was debate about whether or not it really was a tetrapod. Because the fossil consists only of a single lower jaw, it was natur-

Figure 5.29. The locality near Forbes, New South Wales, Australia, where Metaxygnathus *was found. Photograph by J.A.C.*

ally difficult to confirm its identity for a number of years. However, with the description of the better preserved lower jaws of *Acanthostega* and *Ventastega*, paleontologists have realized that there are some characters of the lower jaw unique to tetrapods that can be used to identify isolated jaws as tetrapod or not (Fig. 5.27B).

Metaxygnathus comes from the Late Devonian of New South Wales, Australia, at a locality that has yielded placoderms but no other tetrapod remains (Fig. 5.29). The dating of the rocks from which it comes is not very precise, giving a range from between the Frasnian-Famennian boundary and later in the Famennian than the East Greenland taxa. The specimen occurs along with plates of placoderms and the lungfish *Soederbergia* and offers no clue about the ecology or lifestyle of the animal, and almost none about the anatomy. Its importance lies in showing that tetrapods were widely distributed by this time.

Significance of *Acanthostega*

The above survey outlines the whole range of fossils available to represent the fish–tetrapod transition. The next task is to interpret this evidence. To do this, the first task is to establish which of the taxa most closely represents a truly primitive tetrapod, so that it can be compared usefully with its closest nontetrapod relative. This can then be used to infer the sequence and timing of changes that took place during the origin of tetrapods. There are two lines of evidence that can be used in this study.

The first and most important step is that of working out the phylogeny of the taxa involved. By use of the methods of computer-aided cladistic analysis, the state of each of a wide range of characters is itemized and stated in as unambiguous a way as possible. These are then compared to those found in an outgroup. In the case of early tetrapods, this usually means *Eusthenopteron*. This exercise has been done several times for early tetrapods in different ways by different people and for different regions of the body. Some used a range of features from the whole skeleton, some used only the postcranial skeleton, some the skull roof with the palate, and one used only the lower jaw. Almost all analyses produced the result that *Acanthostega* was the most primitive of the Devonian forms, with *Ichthyostega* being more derived (Clack 1998a, 1998b, 2001; Coates 1996; Lebedev and Coates 1995; Ruta et al. 2002). Only in the lower jaw did *Ichthyostega* appear as more primitive than *Acanthostega* (Ahlberg and Clack 1998). Only one analysis (Laurin and Reisz 1997) placed them together as sister groups.

There is another way to argue the case, however, on the basis of the anatomy of the forms involved and taking into account some more subjective arguments about what people might expect to have occurred at the fish–tetrapod transition. In this argument, there are two sides to the problem. Either *Acanthostega* is truly a primitive form whose aquatic adaptations result directly from its fishy ancestors, or it is a secondarily aquatic form derived from earlier but more terrestrial tetrapods. In that case, its fishlike characters might either be reversals to an earlier condition or derived in parallel with other lobe-finned fishes. Resolving the issue is important because if *Acanthostega* is truly primitive, it can help to infer more directly the sequence and timing of changes during the fish–tetrapod transition. If it is secondarily aquatic, there may have been a whole range of earlier, more terrestrial forms that are currently unknown (Clack and Coates 1995). If so, neither the sequence nor the timing of the changes is any nearer to being understood.

Whichever conclusion is drawn at this stage could of course be overturned sooner or later by new discoveries of fossil tetrapods from the Devonian, but it is only possible to work with what is available and deal with the evidence as it really exists.

The evidence that may suggest that *Acanthostega* is descended from more terrestrial ancestors rests on a number of different lines of evidence. Perhaps the strongest and the most difficult to refute is the fact that it has a tetrapodlike pelvis and a large femur articulating with it. It seems hard to imagine what the pelvis and femur evolved for if not to produce a walking leg. A similar argument is used with respect to the appearance of digits. As explained in the last chapter, some have suggested that the evolutionary pressure to lose the fin webbing and evolve jointed radials could only have occurred in response to a terrestrial mode of life.

Another region of *Acanthostega*'s anatomy that might point to the existence of terrestrial forebears is the hind part of the skull, where *Acanthostega* has lost the operculogular series of bones. The argument runs that if the animal were still as aquatic as a fish, these bones would still be present and functional in the movements associated with gill ventilation. In this view, the gill skeleton and associated shoulder girdle characters that are seen in *Acanthostega* would be primitive, but no longer functional, relicts.

As far as loss of the operculogular series in *Acanthostega* is concerned, in fact there is a very similar situation in lungfishes. They have lost this series yet retain the use of gills, although they complement gill-breathing with lungs and air breathing, just as envisaged for *Acanthostega*. Like lungfishes, *Acanthostega* and other early tetrapods probably gulped air too, by raising the snout out of the water and swallowing the air. Loss or reduction of the operculogulars is found in a variety of modern fishes, some that breathe air and others that do not, such as the Moray eel.

To counter the arguments concerning the limbs, the anatomy of the forearm and the tail is particularly persuasive. The proportions of the bones of the forearm, as shown above, are, among all known vertebrates, most similar to those of *Eusthenopteron*. No secondarily aquatic tetrapod has a radius that is substantially longer than the ulna, but it is the rule in osteolepiforms. Thus its condition in *Acanthostega* is far more likely to be primitive than secondary.

The existence of the large, deep tail webbed with long lepidotrichia is another feature that suggests that the animal is primitively fishlike. As outlined above, and as is discussed further in the next chapter, the dermal fin skeleton seems unlikely to have reevolved once it had been lost, but in addition, it seems equally as likely that any more terrestrially adapted form would have lost such a fin. It is at least as difficult to imagine a terrestrial animal with a tail fin like *Acanthostega* as it is to imagine a primitively aquatic one having evolved a large pelvis and femur. It can be argued that a finned tail would be a liability on land because it would be eroded or torn by contact with the ground and subject to drying out or infection as a result. Again, counterarguments can be put forward. Any early tetrapod of the Devonian presumably produced young that lived at least the early part of their lives in water, whatever their adult status might have been. Possibly the juveniles of even a fully terrestrial adult might have retained fin rays until metamorphosis. Therefore, it is not impossible to imagine a pedomorphic juvenile of such a more fully terrestrial adult evolving into an adult animal like *Acanthostega* with a fully finned tail. This is discussed in more detail in Chapter 6.

The evolution of a pelvis and femur can be accommodated in a number of imaginative scenarios. A predator lurking in underwater weeds may require thrust by a hind paddle to force it through the plant growth, as in the Sargassum frogfish or one of the gurnards (which have their pelvic girdles attached to their shoulder girdles) (Fig. 4.15); it may need increased power and more accurate changes of direction that an enlarged hindlimb might produce. The pelvis was the site of origin for muscles controlling tail movements used in swimming, so the elaboration of the girdle may be related to increasing use of the tail to provide rapid thrusts that a lurking predator employed.

If *Acanthostega* is truly primitive, it can provide clues to the problem of how limbs with digits might originally have been used before they became adapted for walking on land. It would certainly suggest that they evolved while tetrapods were still aquatic for most of the time, and so their major function was an aquatic one. Several models for what such a use might have been can be put forward based on modern fish with digitlike fins. Some of these have been suggested in Chapter 4, such as clinging onto weed to help maintain position in vegetation-rich waters, or slow bottom walking. Other possible uses include pushing through weed-choked swamps or feeling around the plant stems and debris to rake up food items. Finally, as an example of the lengths to which such scenarios can be taken, if the spawning behavior of fishes and amphibians are compared, one of the most conspicuous differences lies in the amphibians' use of amplexus, in which the male's limbs are wrapped around the female's body to ensure that fertilization is completed. Perhaps this was the reason for the invention of limbs. Other ideas to explain invention of limbs have been put forward and discussed in the previous chapter, but of course, none of them is currently testable as a scientific hypothesis. However, if the phylogeny is sound, then one conclusion seems to fit best: that limbs evolved first in water and only later became modified for locomotion on land.

If *Acanthostega* really is a good model for a very primitive tetrapod, then its anatomy can provide a great deal of information about the order in which some events took place throughout the transition. For example, limbs with digits appear to have evolved before walking. *Acanthostega*'s wrist was quite unlike the wrist of subsequent tetrapods. Because the radius and ulna were such different lengths, the ends could not have formed an effective bearing surface on which the animal's weight could be balanced. This can be said with some confidence, even though the wrist bones themselves are not ossified. However, *Acanthostega* obviously had digits, if somewhat unusual ones, so digits must have evolved before wrists. Similarly, the ankle joint was also unsuitable as a weight-bearing joint, being rather inflexible, so hindlimb digits also evolved before the ankle joint.

It is sometimes said that in tetrapods, the wrist and knee joints are simple hinges, whereas the elbow and ankle joints are rotatory. This observation is based mainly on modern amniotes, although it also applies to modern amphibians. However, it does not apply to the earliest tetrapods. The wrist of *Acanthostega*, far from being a hinge, occupied a long arc from front to back, whose range of movement is hard to interpret. The knee joint was also unlike that of later tetrapods except *Ichthyostega*, in that tibia and fibula seem to have overlapped each other a little where they joined the femur. Both bones were rather flattened, and as with the wrist, it is hard to envisage their possible range of movements. This is discussed in more detail in Chapter 10.

Apart from the possession of limbs with digits, a number of the supposed tetrapodlike characters found in *Acanthostega* that are often linked with terrestriality can actually be found among modern fishes that have no reputation as land dwellers. The air-breathing *Arapaima* of the Amazon Basin is one such fish. It has a substantial endoskeletal coracoid portion to its shoulder girdle, there is an enlarged pelvis (although it is not attached to the vertebral column), there are zygapophyses in the well-ossified vertebral column and large ventral ribs, and it has a tail fin of a very similar shape to that of *Acanthostega*.

For the pattern of other skeletal changes during the fish–tetrapod transition, the skeleton of *Acanthostega* can be taken to represent primitive members of the group and can then be compared with those of *Panderichthys* and *Eusthenopteron* to represent tetrapodlike fish, with animals such as *Balanerpeton* or *Dendrerpeton* to represent later tetrapods.

Devonian Tetrapods: An Overview

The tetrapodomorphs of the Frasnian, *Obruchevichthys, Elginerpeton,* and *Livoniana*, although not known from extensive material, hint at a level of diversity that exceeds that found in the tetrapods known from the Famennian. This is shown by the differences among their dentitions, with patterns not seen anywhere else among the lineage. The tetrapods of the Famennian, however, show a conservatism in jaw, palate, skull roof, and girdle structure that is quite striking. They are all about the same size and have similar broad, flattened head shapes. This contrasts with the elongate shape seen in *Elginerpeton*. Only subtle differences mark the lower jaws of, for example, *Acanthostega, Ventastega, Metaxygnathus,* and *Hynerpeton*, although they are still clearly different animals. The minor differences in tooth complement suggest a degree of divergence in their habits and food preferences, although these cannot have been great. The probability is that they were all fairly slow-moving animals most of the time, but capable of bursts of energy when catching moving prey. All probably fed in water (see Chapter 10). This conservative radiation of tetrapods nevertheless succeeded in colonizing widely separated parts of the world.

Some of the ideas that these recent finds have generated will soon be tested by new finds reported but not yet analyzed. At least two new sites in Russia are beginning to yield Devonian tetrapods, and a possible third taxon from East Greenland is also being worked on.

Six
From Fins to Feet: Transformation and Transition

Reading the Evidence:
Eusthenopteron–Panderichthys–Acanthostega

Chapter 2 looked at opposite ends of a spectrum. At one end was the structure of a fish such as *Eusthenopteron,* and at the other that of a tetrapod such as *Dendrerpeton,* an early tetrapod belonging to the group known as temnospondyls (see Chapters 8 and 9). The problem embodied in the phrase "the fish–tetrapod transition" is this: how did evolution get from one to the other? One of the ways to study this is to look at intermediate forms, but what makes a suitable "intermediate form"? In the past, a temnospondyl such as *Eryops* would have been featured in the role of "primitive tetrapod," and *Ichthyostega* would have been seen as an intermediate between *Eusthenopteron* and *Eryops.* Recent analyses, however, have suggested not only that *Ichthyostega* has some highly specialized features that may make it unsuitable as a representative Devonian tetrapod, but it is now also clear that *Eryops* is a highly specialized and unrepresentative temnospondyl, and *Eusthenopteron* is not as close a relative of tetrapods as are some other lobe-fins. In the last few years, with the discovery of animals such as *Panderichthys* and *Acanthostega,* it may be possible to get a much more reliable view of the transition from fins to feet. Although *Panderichthys* may be closer to tetrapods than is *Eusthenopteron,* it remains much less well known, and for the meanwhile, *Eusthenopteron* still provides good information about tetrapodomorph structure.

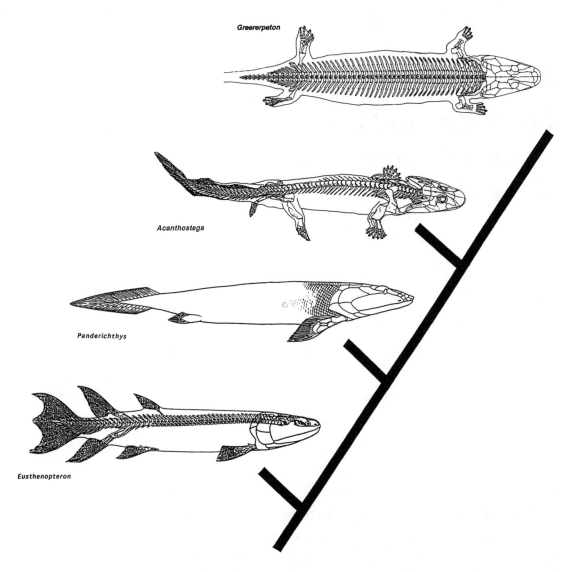

Figure 6.1. Illustrated cladogram with Eusthenopteron, Panderichthys, Acanthostega, *and* Greererpeton.

In the first part of this chapter, *Eusthenopteron, Panderichthys,* and *Acanthostega* are examined more closely for the light they shed on skeletal changes across the transition between fish and tetrapods. These three animals and another later tetrapod, *Greererpeton,* are featured in the cladogram in Figure 6.1. This sequence may justifiably be used to infer the order in which some of the changes occurred because it reflects their phylogenetic relationships as inferred from most modern analyses. In turn, understanding the order of events may provide clues to the underlying processes. The second part looks at what can be inferred about the changes to soft tissue anatomy across the fish–tetrapod transition from studying modern fish and also from the differences between living amphibians and amniotes. Aspects of "soft" anatomy and other nonfossilizable information such as sensory systems, respiration and breathing, the origin of limbs and digits, locomotion, and reproduction fall into this category.

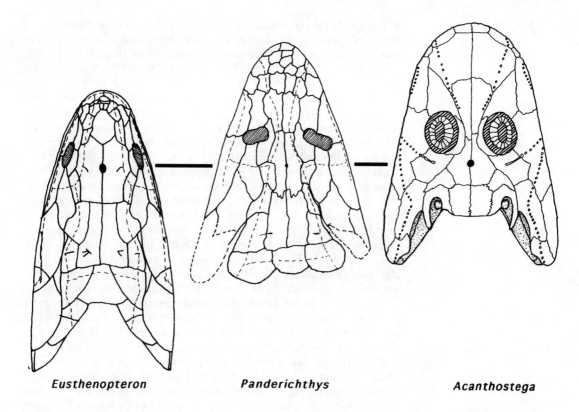

Eusthenopteron *Panderichthys* *Acanthostega*

Figure 6.2. Skulls of Eusthenopteron, Panderichthys, *and* Acanthostega *drawn to the same length from snout to quadrate and lined up on the parietal foramen.*

Skull

Skull Roof and Lower Jaw

Figure 6.2 presents the skulls of *Eusthenopteron, Panderichthys,* and *Acanthostega* drawn to the same snout–quadrate length and lined up on the position of the parietal foramen. Some of the changes to the skull can be seen immediately; others are less obvious. To begin with, changes to the external part of the skull, the dermal skull roof, will be considered.

Early studies of the fish–tetrapod transition had few data on which to base ideas. Skull structure in *Eusthenopteron,* the tetrapods of the Late Carboniferous and the Early Permian, and the little that was known of *Ichthyostega* were all that were available. One of the problems that exercised several workers on the subject in the middle years of the 20th century was the problem of bone homologies in the skull roofs of these animals. In fishes and tetrapods, although the pattern of dermal bones in the skull roof of each is broadly similar, it differs in detail. Another aspect of the problem stemmed from the fact that workers whose interest was mainly fish tended to use a different system of nomenclature from those whose interest was mainly tetrapods. Many articles tried to deal with the questions of which bones of fishes became which bones of tetrapods and argued about the rationale behind the different answers that were produced. The problem arose partly because of the different proportions of parts of the skull between "fishes" and "tetrapods," and because some of the bones appear to have been lost in the transition, making it difficult to draw up a one-to-one correspondence between them. Other factors in the argument were differences in the position of the lateral line grooves and the placement of the pineal foramen, often with either or both being used as landmarks from which to assess the relationships of the surrounding bones (Jarvik 1944, 1980; Klembara 1992, 1994; Parrington 1956, 1967a; Watson 1926; Westoll 1938, 1943a, 1961).

The arguments were never really reconciled, and although several attempts were made to bring the two systems together, none was generally taken up (e.g., Borgen 1983). However, more recently, the whole issue has subsided in view of advances in developmental genetics. It is now clear that strict one-to-one homologies between two such disparate systems are very difficult to draw and may not be meaningful. As will be shown in the case of limbs, what probably does underlie the development of the skull roof in both fishes and tetrapods are fundamental ground rules of formation, applied in different circumstances to produce different end results. In fact, only a few skull regions are affected by such problems—the snout, midline, and temporal series—and for most other bones, it is fairly clear where the equivalents lie in fishes and tetrapods, even though their shapes have changed, often quite radically. This in itself might be food for thought about the developmental processes controlling different parts of the skull.

As was suggested above, one of the most conspicuous differences between the skull of *Eusthenopteron* and that of *Acanthostega* is that of proportion. In *Eusthenopteron,* the snout region is relatively short, with the eyes placed nearer the front than the rear, and the region behind the eyes and overlying the braincase and otic region is relatively long. The eyes are placed laterally and are relatively small. In *Acanthostega,* the opposite is true: the snout is long and the otic region short, and the eyes are large and placed dorsally. *Panderichthys* shows an almost exactly intermediate condition. Snout and postorbital region are about the same length as each other, and because the skull was flattened, the eyes looked dorsally. They have a curious elongated shape, rather than being circular, which may suggest they looked laterally as well as dorsally. *Eusthenopteron* shows the mosaic of snout bones common to most osteolepiforms, but *Panderichthys* has a single pair of enlarged bones making up much of the snout and interorbital region, the equivalent of the tetrapod frontal bones. It is the only osteolepiform to show these bones, and this is thought to be a character shared with tetrapods (Vorobyeva and Schultze 1991). There are fewer bones in the rest of the snout mosaic in *Panderichthys* than in other osteolepiforms, although there are still more than in tetrapods.

In these features, then, it appears that the transition between fishes and tetrapods occurred in a gradual manner, with *Panderichthys* representing an intermediate stage. The inference is that the snout mosaic bones gradually became consolidated from the central midline region of the frontals outward toward the nasal bones. This may be related to the overall lengthening of the snout, in which the presence of fewer larger bones might have been stronger than a collection of many smaller ones. If the snout were habitually raised out of the water, either to feed or to gulp air, this may have been an advantage as the impact of gravity was brought to bear.

The external nostril of *Panderichthys* is larger than that of *Eusthenopteron* and lies closer to the jaw margin. It is still bounded by the anterior tectal and lateral rostral bones, as in *Eusthenopteron. Acanthostega* has lost the lateral rostral, and all other tetrapods seem to have lost both bones. But the position and size of the nostril in *Panderichthys* is very similar to that in many early tetrapods. It can be inferred that the nostril assumed the tetrapodlike condition—that is, larger in size and low on the snout—before the bones surrounding it changed their relationships, with some of them being lost. It is difficult to explain why this change in topology took place in the tetrapod external nostril because it seems likely that it was doing the same job in *Panderichthys* as in *Acanthostega*—that is to say, as an intake for water or air for chemosensing and olfaction, rather than for intake of air for breathing.

The internal nostril—the choana—has an almost identical arrangement of bones defining the opening in all three, although in *Acanthostega* the hole is somewhat larger. Although many people have assumed that *Acanthostega*, being a tetrapod, used its nose to take in air for breathing, the external nostril is small enough to suggest that this was not the case; rather, it remained fishlike in its nasal function. The next section discusses further aspects of the nose and its functioning during the transition.

Early tetrapods are consistent in the pattern of bones of the cheek region and are consistently different from osteolepiforms. In osteolepiforms, the squamosal at the back of the cheek meets the maxilla, and this pattern prevents the quadratojugal from contacting the jugal. In tetrapods, the quadratojugal contacts the jugal so that the squamosal cannot meet the maxilla (see Fig. 2.10). In one specimen of *Panderichthys*, one cheek is described as showing the osteolepiform pattern, whereas the other shows the tetrapod pattern (Vorobyeva and Schultze 1991). This seemingly insignificant change in cheek pattern cannot easily be explained in functional terms, although it correlates with a great increase in the size of the jugal in tetrapods, which also impacts the size of the orbit. An isolated jugal bone from an early tetrapod can be distinguished from that of a fish by its size and the proportion of the orbit margin that it provides.

Some regions of the skull roofs of modern fishes frequently show a range of variation within species that does not seem important to the fishes themselves (Grande and Bemis 1998). Perhaps the variation in cheek pattern was one of those that became variable among the tetrapods' closest relatives and later stabilized in a different pattern among tetrapods themselves. Until more is known about tetrapodlike fish and fishlike tetrapods such as *Obruchevichthys, Elginerpeton,* and the others that are known only from scraps, it will not be possible to test this idea.

In the skull roof, one feature of most osteolepiforms conspicuously absent in *Panderichthys* is the hinge between front and rear halves (see Figs. 2.10, 6.2). In *Eusthenopteron,* this lies at the suture between the bones known as the parietals and postparietals. Although in *Eusthenopteron* it may not have been functional as a hinge, the position of this feature is clear. In *Panderichthys,* there is no sign of it, and the parietals and postparietals are joined by an interdigitating suture, just as they are in tetrapods. In all primitive tetrapods except *Ichthyostega,* in which there is the anomalous situation of having a single ossification there, the postparietals are shorter than the parietals, but in *Panderichthys,* these two pairs of bones are about the same length, as they are in *Eusthenopteron.* This can be read to mean that the hinge line was lost before the rear of the skull became shortened.

Some workers have suggested that the loss of the cranial hinge line was connected with emergence onto land. Once the head was out of the water, it becomes effectively much heavier, and to shift the bulk of the components closer to the neck joint would make it easier to lift. While maintaining the length of the tooth row, it would bring it to a position in front of the eyes, allowing the animal to see more fully what it was catching and allowing the tooth row to be longer relative to the length of the head (Thomson 1966). The proportions of *Panderichthys,* an almost exact intermediate between

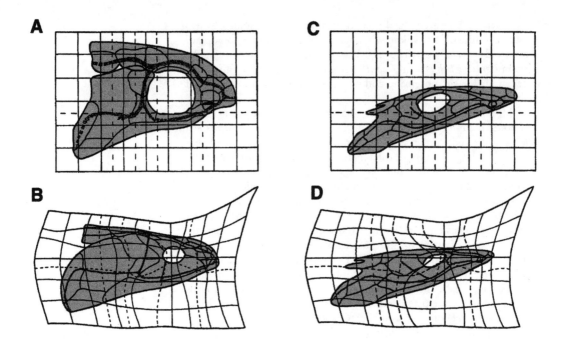

Figure 6.3. Diagrams to show grid transformations of juvenile to adult Eusthenopteron *applied to the skull of* Acanthostega.
(A) Juvenile Eusthenopteron *placed on a plain grid.*
(B) How this grid is distorted by growth into the adult form.
(C) Skull of Acanthostega *placed on a plain grid.*
(D) Skull of Acanthostega *after it has been subjected to the same distortion as from A to B. From Coates and Clack (1995).*

Eusthenopteron and Acanthostega in this respect, do not really help answer this question because nothing can be known about how much time it spent feeding with its head out of water.

Thomson (1967) made a study of the effects on intracranial kinetics of lengthening the anterior part of the skull at the expense of the posterior. He showed that as the front part of the skull became elongated, so the range of movement possible at the intracranial joint must necessarily have been reduced, until at a certain point it became completely ineffective. This kind of change in proportion is that seen across the fish–tetrapod spectrum, with *Eusthenopteron, Panderichthys,* and early tetrapods showing a progressive increase in snout length and a corresponding loss in the features associated with intracranial movement.

Changes to the size of the orbit may be explained in a quite different way. In a series of *Eusthenopteron* skulls representing increase in size during growth, a sequence of changes can be seen. The eyes in the smaller, more juvenile individuals are relatively much larger than in the adults, although the snouts are relatively shorter. The relative size of the orbits in juveniles is much more like that of tetrapods than that of the adult fishes. This has led to the suggestion that increase in eye size at the fish–tetrapod transition may be caused by retention of juvenile proportions.

This idea was explored by Coates and Clack (1995). In an earlier study, Schultze (1984) had examined skulls of *Eusthenopteron*. The Escuminac Bay locality has provided a great size range of specimens, from small juveniles to large adults. He examined the proportional changes they underwent during growth by placing the outline drawing of a juvenile skull onto a grid. Points where the drawing and grid intersected were regarded as fixed. Then the positions of these same points were observed when the outline of an adult skull was placed on the grid. In order to make the grid lines pass through these same intersections, it has to undergo distortions. The amount and shape of the distortions gives a clue to the degree of proportional changes at various points in the skull. Coates and Clack employed this technique in reverse. They placed an outline of the skull of *Acanthostega* on the undistorted grid, then subjected it to the same transformation that *Eusthenopteron* underwent in becoming adult. The skull shape that results looks very much like that of *Panderichthys*. In other words, the skull shape of *Acanthostega* is very much what would be expected if it had been the juvenile of a hypothetical *Panderichthys*-like ancestor (Fig. 6.3).

At the back of the skull of any osteolepiform, *Panderichthys* being no exception, there is a notch situated between the cheek plate and the skull table. The notch, or cleft, is continued backward by the gap between the extrascapulars and the opercular series (see Figs. 2.1, 2.10). The apex of the notch is assumed to be the site of the opening of the spiracle, the external exit of the spiracular pouch. The spiracular pouch is the gill pouch associated with the hyoid arch and hyomandibula, and although no fully formed and functional gill has ever been found associated with the spiracular pouch, a rudimentary gill is sometimes found there in some modern forms. Other functions of the spiracular pouch (confusingly also sometimes known as the spiracular cleft) are as a chemosensory organ detecting the amount of oxygen or carbon dioxide in the water, or as an outlet for exhaled air in air-breathing forms such as *Polypterus*. It has also been suggested that in *Polypterus*, the spiracle can be used for taking in air without the rest of the fish breaking the water's surface (Magid 1966, 1967). This observation has been disputed and is hard to verify. A third use to which the spiracle is put is found in skates and rays, flattened chondrichthyans whose main gill openings lie entirely ventral to the animal. In these animals, the spiracle is the opening through which water enters the pharynx, to exit via the main gill openings in the usual way.

It is usually assumed that all early bony vertebrates had air bladders or functional lungs (Romer and Parsons 1986), and the possession of a functional spiracle is also attributed to them, especially because there is one in *Polypterus*. Early ray-finned fishes also show a groove in the side of the braincase wall, interpreted as the position where the duct leading to the spiracle would have run. The same is true of many early lobe-fins.

In *Acanthostega,* and indeed many other early tetrapods, a notch appears in the back of the skull in what is positionally the same place, between the skull table and the cheek. The function of such a notch has been disputed over the years. In earlier studies it was called an "otic notch" and was assumed to have been part of an aerially adapted ear, holding the tympanic membrane. However, the braincase of *Acanthostega* has a groove in the same position as the spiracular groove in early ray-fins (Clack 1998c). The evolution of this region of the skull and its underlying braincase region have been subjected to much change during the fish–tetrapod transition, and the structural and functional modifications that the region has undergone are discussed in much more detail later in this chapter and in Chapter 10.

Moving back from the main part of the skull, one or two more things can be seen in the contrast between *Panderichthys* and *Acanthostega*. *Panderichthys,* although retaining a full set of operculogular bones, nevertheless had these bones much reduced in anteroposterior length compared with *Eusthenopteron* (Fig. 6.4). This may be tied up with its much more consolidated skull and possibly indicates less complete reliance on gills for breathing. Modern fishes that habitually breathe air generally have reduced operculogular bones, although not all fishes that have reduced this series breathe air regularly (Coates 1996). However, lungfishes, which are

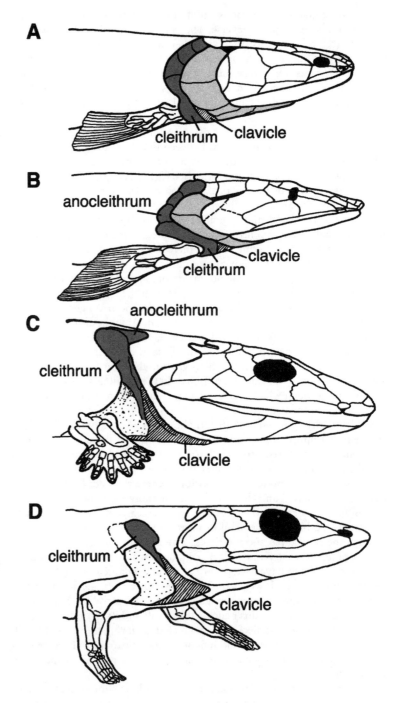

Figure 6.4. Skulls and shoulder girdles of (A) Eusthenopteron, *(B)* Panderichthys, *(C)* Acanthostega, *and (D)* Dendrerpeton, *to show changes to the opercular bones and to the dermal and endoskeletal parts of the girdle. Dermal parts of the girdle are dark-shaded and hatched, the opercular series is light-shaded, and the scapulocoracoid is stippled.*

specialist air gulpers, have lost this series of bones altogether. It may be that *Panderichthys* was a more regular air breather than *Eusthenopteron*. However, this is speculative, because except for the hyomandibula, which is as long and stout as that of *Eusthenopteron*, its gill skeleton is not known. But the sequence may be taken to indicate at least that the opercular series became gradually less important and reduced in size, rather than there being a sudden loss of all the bones together.

One of the bones present in the fish cheek, but absent in most tetrapods, is the preopercular (cf. Fig. 2.10 with, e.g., 7.8, 9.7, 9.10). In osteolepiforms, this is an elongate bone clasping the rear of the squamosal and forming part of the rim against which the opercular series abutted. Its disappearance can be followed among early tetrapods. It remains present in *Ichthyostega* and *Acanthostega,* where it is short and polygonal in shape, and it is also found in an Early Carboniferous form called *Whatcheeria* (see Chapter 7). However, in the latter two animals, it is a rather superficial element of the cheek, and in *Whatcheeria,* it is very thin. Sometimes it is difficult to see at all, and in some specimens, it appears to have been absent (Lombard and Bolt 1995). Loss of this bone was apparently gradual; its place was taken by the meeting of the squamosal and quadratojugal bones underneath it, not its sudden disappearance or its fusion with other bones.

The extrascapular and supracleithral series join the shoulder girdle to the head in fishes and thus are not strictly postcranial (Fig. 6.4). These series are both lost in tetrapods. A remnant of the supracleithrals, the anocleithrum, is still present in *Acanthostega,* but it had no bony connection with the head. That means the head was effectively disconnected from the body except via the occiput and surrounding fleshy tissue. However, a true neck joint or occipital condyle had not evolved in *Acanthostega* or *Ichthyostega,* which were still fishlike in this respect. The notochord still ran through the vertebrae and into the braincase in these animals, meaning that the neck must have had very restricted movement. The notochord is bendable but springs naturally back to a straight line unless muscular effort is exerted. The earliest tetrapods probably had little control over the attitude of the head and probably had to expend considerable energy to move it sideways or up and down.

Muscles that move the head run along above and below the spine and insert onto the occiput and braincase in modern tetrapods. The story of the tetrapod neck is one of consolidation of the bones at the back of the skull, increasing the surface area onto which to fasten these muscles and specialization of the joint surfaces. These did not happen all at once, and recent work suggests that they occurred differently and separately among amphibians and amniotes (see Chapter 10). None of these processes had yet begun in the earliest tetrapods so far known.

For many years, the lower jaws of early tetrapods and their close relatives were little studied, even ignored as uninformative. More recently, a series of characters has been found in which even the earliest tetrapods differ consistently from osteolepiforms. These characters include the presence of teeth versus denticles on the parasymphysial plate, the presence or absence of a couple of foramina around the symphyseal area, and the amount of ossification of the Meckelian bone in the body of the jaw. The characters mainly show differences in the dentition and structural make-up of the jaws and must be related to changes in diet and the stresses under which the jaws operated. These changes show what appear to be an orderly sequence of appearance, and the significance of this will be dealt with in the last chapter (Ahlberg and Clack 1998).

That brings the discussion to the overall skull structure and the way it reflects differences in function at the fish–tetrapod transition. In fishes such as *Eusthenopteron,* the dermal bones are generally joined to one another rather loosely, by narrow under- and overlapping flanges. The margins of the individual bones appear straight or as simple curves in surface view. In

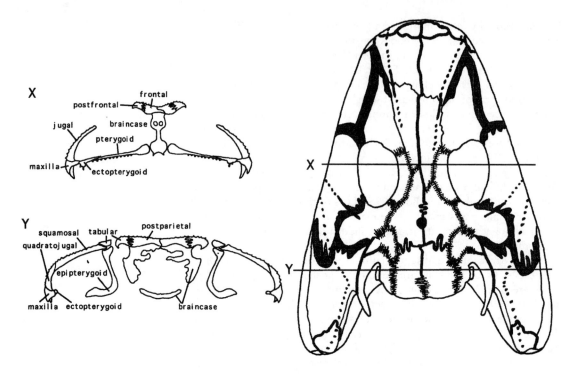

Figure 6.5. Skull roof of Acanthostega *to show sutural types. Heavy lines indicate butt joints with no overlap; gray-shaded areas represent areas of sutural overlap (scarf joints); interdigitated joints are shown by hatch marks. (Left) Sections through the skull at levels X and Y, illustrating some of the sutural types in section. From Clack (2002).*

some places, however, the bones fuse altogether and the junctions between them are difficult to see at all. In early tetrapods, many of the bones are interdigitated—joined together with fingerlike interlinking processes—so that in surface view, the junctions are often highly irregular zigzags. The effect of this is to make many of the sutures between bones into very strong junctions. Even in early forms such as *Acanthostega,* the range of sutural types seems to have increased, so that butt joints, scarf joints, and tongue-and-groove joints, as well as three-dimensionally interdigitating joints, are all found at different points in the skull (Fig. 6.5). This must reflect different stresses experienced by the skull at different points and may indicate that already the head was exposed to the air and thus to the effects of gravity. One of the regions to be most strongly integrated is the skull table, where tabular, postparietal, parietal, and supratemporal have the most complex mutual sutures in the skull. This part of the skull would have been to a great extent involved for the attachment of muscles from the vertebral column to hold up the head. One possible function for the tabular horn of *Acanthostega* is as an attachment point for a ligament or muscle joining the head to the shoulder girdle. Other parts of the skull of this animal show extensive overlapping sutures, good for resisting torsion perhaps created by catching or lifting prey out of the water (Clack, 2002). The contrast in skull integration between fishes and tetrapods may also indicate different feeding strategies, although this is not plainly reflected in jaw structure and dentition. *Panderichthys* shows an intermediate form of construction and is one of the few osteolepiforms to show interdigitating sutures between the skull bones, visible in external view (Figs. 3.16, 3.17, 6.2) (see also Chapter 10).

With a bit of experience, early tetrapod paleontologists can usually

Figure 6.6. Close-up of the dermal ornament of Acanthostega. *Photograph by S.M.F.*

distinguish isolated dermal bones of fish from those of early tetrapods, even when they are incomplete. The clues lie in the external surface of the bone. Most dermal bones carry sculpturing or "ornament"—that is, the surface is marked by wrinkles, pits, pustules, ridges, or grooves. By and large, in fishes, there is a recognizably different type of ornament. It tends to be a quite fine, evenly distributed, "pimple and dimple" texture, so that there is no obvious center of radiation. In tetrapods, the ornament tends to be coarser, to be higher profile, and to show a definite growth center (Figs. 1.3, 6.6). Usually toward the middle of the bone, the ornament is arranged in regular pits and ridges, often described as "honeycomb," but toward the edges, the ridges are aligned more or less radially, with grooves fanning out between them. There are exceptions to the rule, but generally, in early tetrapods, the ornament can be described as "starburst." It was the discovery of this type of ornament pattern that first alerted paleontologists to the existence of Devonian tetrapods in Scotland and Latvia. The differences in ornament between fishes and tetrapods probably reflect differences in growth pattern, but there may be other functional differences as well. In later tetrapods, there are characteristic differences between temnospondyls, in which the ornament tends to be quite regular, and anthracosaurs, in which it becomes rather irregular or almost absent altogether (see Chapter 9 for more on these animals). Again, reasons for this are unknown but could reflect physiological differences. Where the ornament is clear and regular, it has been interpreted as indicating that the outer skin was quite closely applied to the surface, but where it becomes shallower in profile, the bone was more deeply buried below the surface.

From Fins to Feet • 149

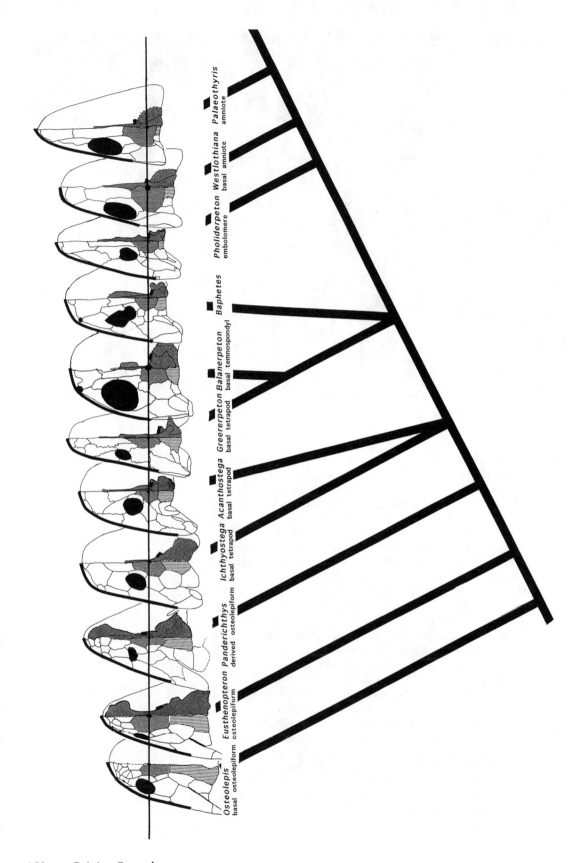

Braincase and Occiput

A number of changes to the outer skull bones are evident across the series *Eusthenopteron–Panderichthys–Acanthostega*, but even more profound ones are seen internally, in the braincase. Although changes to the skull roof appear gradual in this series, changes to the braincase seem to have occurred rapidly between *Panderichthys* and *Acanthostega* (Ahlberg et al. 1996). Some of the changes are subtle, but they have important consequences for the subsequent evolution of the tetrapod skull.

One of them concerns one of the points at which the braincase and palate are fastened together, the basal articulation. In fishes such as *Eusthenopteron* and *Panderichthys*, this is an S-shaped joint in which the braincase and palate contribute more or less equally to the joint surfaces, each with a concave and convex component. There might have been some form of movement at this point, to compensate for movements elsewhere in the skull between the skull roof and palate. In tetrapods, the form of the joint changes, so that even in *Acanthostega*, the braincase contributes a conspicuously bulbous, faceted process that fits into a recessed socket on the palate (Figs. 2.10, 2.12). Early tetrapods maintained this peg-and-socket arrangement, and although it was sealed up in at least one lineage, in another, it remained throughout evolution and can be seen even in modern forms such as some lizards. No one is really sure why the joint became as elaborate as it did, but it does seem to have allowed, or possibly constrained, some kind of skull kinesis.

Apparently occurring at the same time as this change was another, affecting much the same braincase regions. In tetrapods, the two halves of the braincase, the sphenethmoid in front and the otoccipital at the rear, instead of being separate as in *Eusthenopteron* and *Panderichthys*, became intimately joined to each other. The basisphenoid and basioccipital regions became integrated just at the point where the notochord would have ended inside the braincase. In *Acanthostega* and *Ichthyostega*, a suture can be seen at this point, although in all other known early tetrapods, the dermal parasphenoid has expanded over the ventral surface of the braincase and hides this suture from view.

In the restructuring of the braincase during the transition, two other regions are of special interest, and they both concern sensory systems.

In all osteolepiforms, including both *Eusthenopteron* and *Panderichthys*, the nasal capsule is ossified in the adult, meaning that its fossilized remains can be found in good specimens. In all tetrapods, including both *Acanthostega* and *Ichthyostega*, the nasal capsule never ossifies, and so there is no information about its construction. However, this does not necessarily mean a complete lack of data. The inference is that for some reason, the former bony capsule was left in what may have been a more juvenile state, remaining as cartilage in the adult. A number of possible reasons can be suggested for this. Perhaps it was connected with weight reduction in the elongated, stronger snout of tetrapods, like the reduction in the number of bones seen earlier, in a snout that may have been habitually raised out of the water. If so, the condition in *Panderichthys* suggests that elongation and consolidation of the snout roofing bones began before the loss of bone from the nasal capsule.

Reduction of ossification in skeletal structures is often associated with retention of the juvenile condition in animals, a process known as pedomorphosis. The nasal capsule is a second instance, as suggested for changes to the proportions of the skull and orbit, of possibly pedomorphic pro-

Figure 6.7. Cladogram of skulls of fish and tetrapods to show the changes in proportion to the snout, otic region, and eyes. In each skull, on the right, the ossified portions of the braincase are shown shaded. On the left, the corresponding bones in each are hatched.

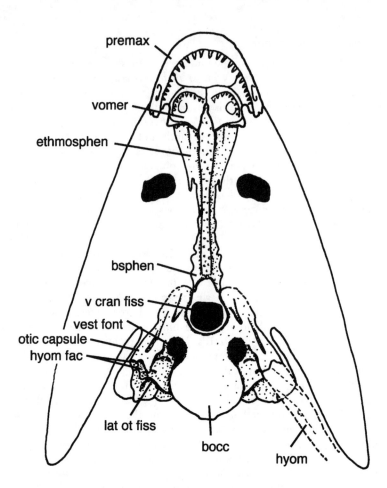

Figure 6.8. Drawing of braincase of Panderichthys *in ventral view, with the skull outline and orbits indicated. From Ahlberg et al. (1996).*

cesses occurring at the origin of tetrapods. A third may be connected with changes to the otic region of the braincase.

Panderichthys has a braincase construction essentially the same as that of *Eusthenopteron,* despite the tetrapodlike features of its skull roof (Fig. 6.8). One of the most surprising things is that its braincase still shows many of the features associated with the division between the ethmosphenoid and oticoccipital parts that are seen in *Eusthenopteron.* It has the fishlike type of basal articulation and size and position of parasphenoid. It even shows some of the processes and facets that in *Eusthenopteron* are associated with the hinge that exists between the braincase regions, despite lacking the corresponding hinge in the skull roof. It appears that although changes to the skull roof appear to have been gradual, changes to the braincase were much more rapid, and occurred, like changes to the nasal capsule, between *Panderichthys* and *Acanthostega* (Ahlberg et al. 1996).

In all tetrapods (with the possible exception of *Ichthyostega,* where the region is poorly known and difficult to interpret), the hyomandibula, which in fishes functioned as a palate, jaw, and gill arch support, changed form and function and became the stapes. Although the bone is the same, the name is changed in response to a very important and radical restructuring of the proximal end (Fig. 6.9). Although in fishes the head of the hyomandibula articulates against the side wall of the braincase at a bridge

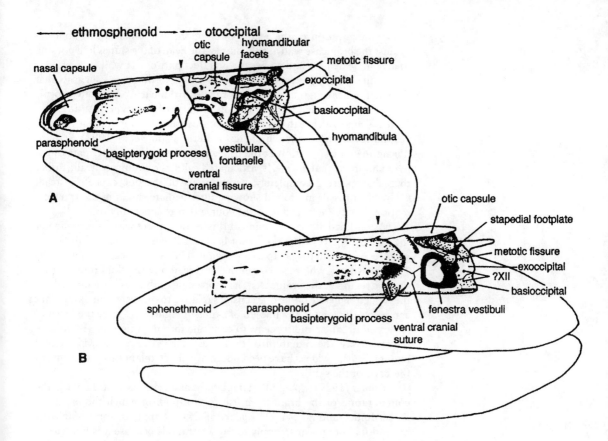

Figure 6.9. Braincases and hyomandibula or stapes of (A) Eusthenopteron and (B) Acanthostega to show contrast in otic region construction. The outline of the skull roof is superimposed.

of bone called the lateral commissure, in tetrapods, the head or footplate of the stapes fits into a hole in the braincase wall formed where this bridge used to be, in combination with an existing hole in the braincase wall known in fishes as the vestibular fontanelle (see also Fig. 2.12). In later tetrapods, this hole reduces in size as the stapes itself becomes more slender and the footplate diminishes. The hole, covered by a membrane, is termed the *fenestra vestibuli* or *fenestra ovalis*, and this and the stapes change during evolution from a supporting bone to part of a hearing mechanism. The story of the evolution of the ear region is told in greater detail in Chapter 10.

Panderichthys shows no sign even of the beginnings of this profound change, whereas *Acanthostega* is in most respects a "good" tetrapod. Crucial changes to the location of the stapedial head and the side walls of the braincase had already taken place in *Acanthostega*, with the footplate of the stapes inserted into the braincase wall. The stapes in *Acanthostega* and other early tetrapods was nevertheless still a stout bone with connections to the palate. It formed the only link between the braincase and the palate except for that at the basal articulation (see Chapter 2), and so its main function could still have been as a support for the braincase.

The inside of the otic region and the side walls of the braincase are invariably poorly ossified in early tetrapods, and it is tempting to think that this might be a comparable pedomorphic state to the lack of ossification in the nasal capsule. The poor ossification may in some way be connected to

From Fins to Feet • 153

the reconfiguration of the whole otic region, where instead of abutting against the braincase wall at solid facets, the head of the stapes had nothing to articulate against and so became in effect part of the wall.

In many living tetrapods, the embryology of the stapes suggests that the portion bearing the footplate is formed from parts of the otic capsule that become attached to the more distal parts of the stapes. At any rate, there is some evidence that part of the otic capsule surrounding the fenestra vestibuli is derived from neural crest material (Couly et al. 1993). Perhaps during the reorganization at the fish–tetrapod transition, that part of the hyomandibula that formerly articulated with the braincase wall, rather than disappearing completely, became part of the otic capsule instead. Only when fossils are found showing intermediate stages in the transformation may there be a clue to the course that events really took.

As outlined at the beginning of this section, there was a shortening of the posterior part of the skull roof between *Panderichthys* and *Acanthostega*, especially in the postparietal region. This was noticed as long ago as 1937 by Romer, who pointed out that this region of the skull roof overlies the otic region of the braincase. The shortening of the postparietals reflects the reduction in size of the otic capsules relative to the rest of the skull. That can be clearly seen in the line-up of skulls in Figure 6.7. Here the "trend" is very noticeable, and it seems to continue into the origin of amniotes— that is to say, the radiation of tetrapods that led to reptiles, birds, and mammals. Many ideas have been put forward to explain this shortening in the back of the skull.

Romer (1937) thought that it might be associated with a change in the configuration of the brain from fish to tetrapod, in which the brain becomes folded and takes up less space. This may apply to some amniotes, but it does not appear to apply to early tetrapods because it is not true of all amphibians and amniotes. The shortening does seem to have occurred at the time of tetrapod origins because *Acanthostega* shows the typical tetrapod proportions, with an otic region markedly shorter than that of *Panderichthys*.

Another suggestion was that it might have been related to the development of a neck in tetrapods and the development of strong neck musculature on the occiput (Westoll 1943a). The shortening of the occipital region might have provided a mechanical advantage in a skull that was habitually raised on the occipital joint. The discovery that *Acanthostega* shows the shortened tetrapod condition while retaining a fishlike, notochordal occiput and braincase argues strongly against this idea. The shortening had already occurred before the occipital condyle had evolved.

Yet a third suggestion was that the semicircular canals, which monitor posture and balance, became smaller and less sensitive once the head was separated from the body by the neck (Bernacsek and Carroll 1981). Jones and Spells (1963) had shown that sensitivity of the canal system was directly related to the cross-sectional radius of the semicircular canals and also to their radii of curvature. They showed that these dimensions increase only slowly compared with the body mass in living animals. They also showed that fishes tended to have larger, and thus more sensitive systems than those of the tetrapods that were measured at the same time, namely a selection of mammals and reptiles. Bernascek and Carroll (1981) added to this original data measurements of some fossil osteolepiforms and two fossil temnospondyl amphibians (see Chapters 8 and 9). In the fossil amphibians, the canal sizes were slightly larger than those of amniotes but significantly smaller than the osteolepiforms, which clustered with the living fishes. They suggested that the size decrease corresponded to a decrease in the sensitivity of the canals and their sensory apparatus at the transition. One possible explanation might be that once on land, the semicircular canals of tetrapods only had to monitor the head and not the body as well. At the initiation of terrestriality, the body was effectively moving only in two dimensions instead of the three exploited by animals living in water, so the semicircular canals would be doing a slightly different job. Once on land, the animals would have experienced the effects of gravity and would have needed to elicit a "trip response" to imbalance or falling. Fish don't fall. Perhaps the semicircular canals became less sensitive in order to edit out some of the noise that terrestrial experiences produced. The observations of Bernascek and Carroll (1981) need to be augmented by measurements from the otic capsules of some of the more recently discovered early tetrapods to test the hypothesis of reduction across the transition.

If modern fishes are compared with modern tetrapods, broadly speaking, the difference in proportions is discernible. However, the rule is not very general, and many exceptions can be found. Among modern fishes, it is true that in some cases, those that engage in "walking" on land have smaller or narrower semicircular canals than those that do not (Gauldie and Radtke 1990). However, it is very difficult to judge from the fossils what the diameter of the semicircular canal tubes might have been, nor is it strictly possible to predict their sensitivity by knowing their diameter.

In this context, the condition in *Ichthyostega* is particularly intriguing. Its phylogenetic position is equivocal. On postcranial anatomy, it consistently appears above *Acanthostega* on a cladogram, but on braincase and jaw anatomy, it appears to be more primitive. Figure 6.7 shows this uncertainty by placing it at the same node as *Acanthostega* with respect to other tetrapods. According to its skull proportions, it indeed shows a much more fishlike condition and has front and rear portions in much the same relation as *Panderichthys*. Like *Acanthostega,* it has a fishlike notochordal occiput, corroborating the suspicion that tetrapodlike limbs at least evolved before the occipital condyle and that otic shortening only occurred later in tetrapod evolution, after the development of limbs. However, its braincase is very peculiar, in particular the otic region. It is not obvious how large the otic capsules were because they were very poorly ossified, although it might be tempting to interpret them as similar in size to those of *Panderichthys*. Recent work (Clack and Ahlberg 1998) suggests that it did not have the tetrapod type of stapes inserted into the braincase side wall seen so clearly in *Acanthostega,* and that if anything, the semicircular canals might have been smaller, rather than larger than expected. The difficulties associated with interpreting this region of *Ichthyostega* might be related to its transitional nature.

A possible confounding factor in the whole debate is that it is difficult to tell whether the reduction in relative size of the ear region is a real effect or whether it results from the increase in the length of the snout. Only a full morphometric study could really tease apart the separate influences on proportional changes.

In its display of tetrapodlike characters, *Panderichthys* shows that many of the changes in the skull roof seen between fish and tetrapods actually arose among animals that would still have been called fish—that is to say, in animals that still had fins and that were still aquatic. Other tetrapod characters, however, seem to be unique to that group, so far as is known. These are, for example, changes to the braincase and limbs.

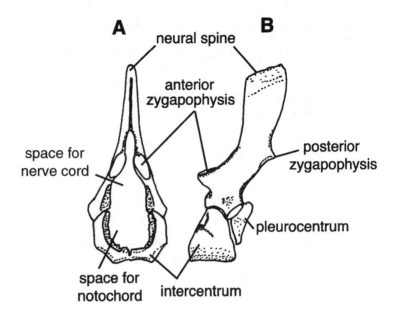

Figure 6.10. Diagram of Acanthostega *trunk vertebra.*
(A) Anterior view.
(B) Left lateral view.

The reasons for the changes to the braincase wall are unknown, but they are profound and characteristic of tetrapods; apparently they occurred at the same time as those to the limbs that produced digits (Ahlberg et al. 1996). Only further researches in paleontology and developmental biology will show whether these two events could be in any way connected, but it is tempting to speculate whether they, perhaps occurring simultaneously, account for a sudden evolutionary event that resulted in the appearance of tetrapods.

Postcranial Skeleton

Unfortunately, the postcranial skeleton of *Panderichthys* is still not very well known, but there is sufficient knowledge about some parts to allow inferences about the order of appearance of some tetrapod characters. This section describes the changes that can be seen to the postcranial skeleton in the sequence *Eusthenopteron–Panderichthys–Acanthostega*.

Vertebral Column and Ribs

The vertebral column is one of the regions to show the least radical change between *Eusthenopteron* and *Acanthostega*. The columns of *Panderichthys* and the similar *Elpistostege* are poorly known and difficult to interpret, although some things can be said about them—for instance, that they seem to have been more robustly built than those of *Eusthenopteron*. The arrangement of centra in *Eusthenopteron* is that called rhachitomous, with horseshoe-shaped intercentra, often formed from paired halves, and small, dorsally placed pleurocentra (Fig. 2.4). The centra were essentially husks around the notochord, which was still patent even in the adults. There was little to distinguish the atlas or axis centrum from any other. The centra of *Acanthostega* were similar to those of *Eusthenopteron* in almost every detail (cf. Fig. 2.4A and Fig. 6.10), except that the pleurocentra were slightly larger and the axis intercentrum was about twice the length of the other intercentra. No features distinguished the sacral centra from the others in *Acanthostega,* even though there was a recognizable, if only slightly distinctive, sacral rib (Coates 1996).

As far as the neural arches are concerned, a few more differences can be found. The atlantal and axial arches in *Acanthostega* are somewhat different from the trunk arches, the atlas arch being rather smaller and the axis arch rather larger than those of most trunk vertebrae. The fourth cervical arch is rather reduced too. None of these distinctions can be made in *Eusthenopteron* (cf. Fig. 2.3 with Fig. 5.19; see also Fig. 10.22). From about the sacral region to the end of the tail in *Eusthenopteron*, the arches become longer and more slender and take on a more acute posterior slope, and in *Acanthostega*, this shape is achieved at about the 12th arch past the sacrum. The arches of *Eusthenopteron* do not interarticulate except for a few at the anterior end of the column that touch each other. They do not have defined points of articulation. In *Acanthostega*, the arches of the trunk region do interarticulate, via poorly developed zygapophyses and at an additional articulation more dorsally. However, even these are poorly defined and would hardly contribute to making the column a weight-bearing structure.

Additional bony spines articulate at the end of the arches in the tail of both *Eusthenopteron* and *Acanthostega*, called caudal supraneural spines or radials, a fishlike feature retained in this early tetrapod (cf. Fig. 2.3 with Fig. 5.19; see also Fig. 10.22). They are present also in *Ichthyostega*. Until recently, supraneural radials were not known to occur in any other early tetrapod, but a specimen has now come to light (Carnegie Museum specimen CM 34638) that shows clearly that at least one embolomere (see Chapter 9) had them. They may have been a primitive feature retained more widely in early tetrapods than had previously been realized.

The presence of substantial dorsal ribs articulating with the intercentrum and neural arch is one of the features that most clearly distinguish tetrapods from fish in general. *Panderichthys* and *Elpistostege* appear to have had some kind of accessory structure attached to the centrum, although it is not absolutely clear they were the equivalent of tetrapod ribs (Schultze and Arsenault 1985; Vorobyeva and Schultze 1991). *Eusthenopteron* has short, dorsally pointing spinelike processes spanning the gap between the intercentrum and neural arch, but there was no equivalent of a rib shaft. *Acanthostega*, by contrast, certainly had ribs in the tetrapod sense of a shaft articulating to the vertebral column at two points proximally, but the shafts were very short compared with almost any other tetrapod (Fig. 5.19). The cervical ribs had spoon-shaped expanded ends, but the trunk ribs were simply straight (Coates 1996). They form a very good intermediate morphology between *Eusthenopteron* and the ribs of later tetrapods such as the aquatic *Greererpeton* (Chapter 7). They were almost as different as could be from the massive overlapping ribs of *Ichthyostega* (Figs. 5.2, 5.12). The functional correlates of rib morphology will be discussed in Chapter 10.

Shoulder and Hip Girdles

One of the most striking changes to the dermal shoulder girdle is the development in tetrapods of the interclavicle (see Fig. 2.5). It is an important bone in even the earliest tetrapods. In *Acanthostega*, it is shaped like an old-fashioned kite, or a lozenge in heraldic terms, whereas in *Ichthyostega*, it has a rounded anterior component with a long, narrow, posteriorly pointing process. No modern tetrapod has an interclavicle of this type, but the function may in some ways have been similar to that of the sternum in modern animals, providing attachment for shoulder–girdle muscles; or it may have spread the weight of the animal across the chest in an animal that crawled on its belly, taking pressure off the heart.

Again in the shoulder girdle, there is a change in emphasis during the transition from fish to tetrapods in the contribution made by the different parts (Fig. 2.5). In fishes such as *Eusthenopteron,* the emphasis is on the dermal components, with a large, dermal, platelike clavicle strongly attached by a suture to the cleithrum. The ventral part of the clavicle consists of a broad plate, tapering into a dorsal process by which it is attached to the cleithrum. The cleithrum bears a small tripod-shaped scapulocoracoid on its inside surface, carrying the socket where the fin articulated. A small midline dermal element, the interclavicle, was present in some forms.

In tetrapods, the clavicle remained roughly the same general shape as in fishes, although the dorsally tapering process became rather more slender in most species, and the cleithrum was correspondingly narrower. Often in Carboniferous forms, the exact way in which the cleithrum and clavicle were joined is not obvious because they are usually found disarticulated in the specimens. Another significant difference between fishes and tetrapods, probably related to the changes to the cleithrum, concerns the scapulocoracoid. This became greatly increased in size to accommodate the enlarged glenoid at which the forelimb articulated. Figure 6.4 shows some of these changes to the shoulder, including that of the Carboniferous *Dendrerpeton*.

In *Panderichthys,* the enlargement of the scapulocoracoid had begun, and this element was attached over much of its surface to the cleithrum, in contrast to *Eusthenopteron,* in which it attached at only three points. The cleithrum itself, however, was still much as it was in *Eusthenopteron.* In tetrapods, the glenoid was not only enlarged but changed its orientation to face laterally. Strengthening buttresses ran up its posterior margin, and excavations on the internal surface provided housing for chest muscles. From Carboniferous forms onward, the scapulocoracoid acquired a dorsal component that occupied a similar area to that which the cleithrum did in fishes.

The Devonian forms (with the exception of *Tulerpeton*) are rather different in construction from either the fishes or the later Carboniferous forms and may give clues to how the changes came about in terms of function, although it will require a lot more study to understand them fully. In *Ichthyostega* (Figs. 5.2, 5.12) and *Acanthostega* (Fig. 5.19), and also in *Hynerpeton* (Fig. 5.27C) and *Ventastega,* the cleithrum was relatively reduced, although it was still a substantial dorsally extensive blade. Rather than a clavicle firmly sutured to the cleithrum, as in fishes, these animals had the cleithrum and scapulocoracoid fused into a unit so that the joint between them can sometimes only be seen in a cross section through the bone. The scapulocoracoid itself had a substantial platelike form equivalent to the coracoid portion of later tetrapods, but there was no endochondral component further dorsally equivalent to the scapular blade. This region was still the province of the cleithrum. However, the cleithrum was a much less superficial element than in fishes. In fishes, it carried surface ornament characteristic of dermal bone that is close to the animal's surface, whereas in the Devonian tetrapods, the ornament is almost totally absent, suggesting a layer of thicker skin or even muscle external to the bone. At the same time, the clavicle had lost its firm connection to the cleithrum and the dorsal process had taken up the more gracile form found in later Carboniferous forms. *Tulerpeton* shows a more typical tetrapod-type relationship between the scapulocoracoid and the cleithrum. The bones were separate as they are in later forms, but there was still no scapula blade. The inference is that the expansion of an ossified scapula blade followed the acquisition of the tetrapod-style anatomical relationship between the two.

Some of these changes must surely have been connected with a shift in the function of the shoulder girdle during the transition and in the changes in stress patterns imposed on it by the enlargement and changes to the use of the limbs. However, some of them may also have been associated with changes in breathing patterns and the reduction of reliance on gill breathing. Muscles used to operate the gills that would once have been attached to the shoulder girdle at some stage must have been reduced in importance and in the positions of their attachments.

Even though the connection between the shoulder girdle and the head was lost in tetrapods, remnants of the supracleithral series of bones remained in some species. The anocleithrum is a small oval bone now known in a few early tetrapods. It attached to the dorsal end of the cleithrum and internal to it, so that it lay below the skin and bore no ornamentation. *Acanthostega* and *Tulerpeton* both retained the anocleithrum, and the bone has also been found in a few Carboniferous forms. The embolomere *Pholiderpeton* and the seymouriamorph *Discosauriscus* (Chapter 9) have both recently been discovered to retain this bone (Clack 1987a; Coates 1996; Klembara 2000) so that, like supraneural radials, it may be a primitive feature more widely retained in early tetrapods than was previously realized.

Chapter 2 outlined the profound differences in the construction of the pelvic girdle between fishes and tetrapods. Even in the earliest tetrapods known, such as *Elginerpeton,* this change had taken place, and that animal is known to have had a pelvic girdle very like that of *Ichthyostega*. However, the reason for having this large pelvis is less clear. Although in later tetrapods the hindlimbs came to be substantially larger than the forelimbs, the evidence from the earliest tetrapods suggests that this discrepancy between front and rear had not yet evolved. If the femur of *Elginerpeton* is correctly interpreted, it was short and flattened, and the limb that it operated was paddlelike, whereas the humerus appears if anything to have been a little longer than the femur (Ahlberg 1998). Similarly in *Ichthyostega,* although there is some doubt about the proportions of front to rear limbs, what evidence there is at present suggests again a larger (in this case much larger) forelimb than hindlimb. In *Acanthostega,* where both limbs are known in detail, the hindlimb and forelimb are of about equal size. Both limbs in this animal were paddlelike. As outlined in Chapter 5, the early function of limbs with digits may have been more associated with aquatic habits, in which case the increase in size of the pelvis is puzzling.

Fins and Limbs

Panderichthys had paired fins broadly comparable in construction to those of *Eusthenopteron,* but they were different in detail. For example, the ulna in *Panderichthys* was rather larger than in *Eusthenopteron* and had a D-shaped outline (Fig. 6.11A) (Vorobyeva 2000; Worobyeva 1975). Fewer ossified elements are found distal to the radius and ulna in *Panderichthys,* so in some ways it appears less tetrapodlike than the fore fin of *Eusthenopteron*. The humerus, however, does show some very tetrapodlike features and shows an almost perfect intermediate morphology between *Eustenopteron* and that of *Acanthostega* (Fig. 6.11B–D). The pelvic fin skeleton of *Panderichthys* is still unknown, but it seems to have been at least as small relatively as that of *Eusthenopteron.* In this case, the sequence *Eusthenopteron–Panderichthys–Acanthostega* suggests a rapid transition from a fishlike to a tetrapodlike condition, and there are more or less no clues to the intermediate forms.

In fishes such as *Eusthenopteron* and *Panderichthys,* each paired fin

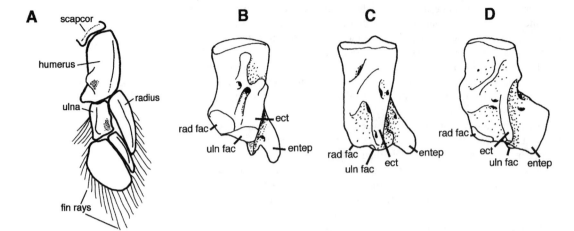

Figure 6.11. (A) Fin skeleton of Panderichthys *(right fin in dorsal view).*
(B–D) Left humeri of Eusthenopteron, Panderichthys, *and* Acanthostega *in dorsal view (from Vorobyeva 2000).*
ect = ectepicondyle;
entep = entepicondyle;
rad = radial facet;
uln = ulnar facet.

was bounded by a fringe of dermal lepidotrichia. Lepidotrichia, being modified scales, are dermal in origin, with their formation controlled by neural crest tissue, so in *Panderichthys,* this tissue was still operating in the fin region of the skeleton. However, in tetrapods, the dermal component of the skeleton is completely absent from the limbs. The jointed elements that form the digits are all endochondral. In other words, the digits are not formed from modified fin-rays but are derived from a different embryonic source. It may be that an animal can have either digits or fin rays but not both, or there may have been a gradual reduction in fin rays alongside an increase in the endoskeletal component that produced digits (Thorogood 1991). The difference may result from differences in timing of the phases of development.

In fins such as those of the lungfish *Neoceratodus,* there are not only fin rays but jointed radials present, some of which are in an equivalent position to digits. It is also true that in lobe-fins generally, the fin rays comprise a smaller proportion of the whole fin than they do in ray-finned fishes. The fin rays in lobe-fins usually form only a narrow fringe around the fleshy lobe. There may be a balance to be drawn between endoskeletal and dermal components.

The most obvious contrast between fishes and tetrapods is the evolution of digits and the associated limb joints, the wrists and ankles and elbows and knees. Some of the features of the earliest tetrapod limbs were described in the section on *Acanthostega,* and the evolution and development of these appendages are considered in the sections below on limbs and digits; their subsequent modifications for terrestriality are discussed in Chapter 10.

In *Panderichthys,* there was a tail fin formed of dermal lepidotrichia running above, below, and around the end of the rather straight tail. However, it has neither dorsal nor anal midline fins. This situation is just as in the earliest tetrapods, *Acanthostega* and *Ichthyostega.* The inference is that the reduction in the activity of neural crest tissue that formed fin rays had begun in *Panderichthys* and was continued among the earliest tetrapods until it was lost altogether in all subsequent forms. The tail fin webbing in these three transitional animals was supported by additional endochondral elements, the supraneural spines or caudal radials. It is not clear whether the supraneural spines disappeared at the same time as the lepidotrichia in the tail. *Ichthyostega* and *Acanthostega* had both, but the appropriate regions are not preserved in the single embolomere specimen with supraneural spines.

Fin Rays and Scales

This sequence is mirrored by the gradual loss of dermal scales from the tetrapod skeleton. *Panderichthys* was scaled all over in a typical fishlike pattern, whereas early tetrapods retained only the belly scales. This ventral armor persisted among tetrapods for much longer, however, and is found even among some reptiles as "gastralia" forming a ventral armor.

When we look at the patterns of skeletal evolution more broadly, it seems that in many different parts of the body, the dermal skeleton was lost at the expense of the endochondral skeleton. This sequence is seen in the evolution of the skull, the shoulder girdle, the limbs, and the axial skeleton, including the scales. The reasons for this must be related to the embryonic development of the animals, but as yet, there are few clues to the underlying causes. Nonetheless, one inference is that reduction of the scale cover and fin webbing in early tetrapods may have been a contingent effect of the decreasing role that neural crest–derived tissue played in the formation of the postcranial skeleton—a kind of domino effect from a developmental accident rather than a direct adaptation to becoming a tetrapod and eventually attaining terrestriality. Subsequent evolution may have been facilitated by such modifications so that tetrapods were eventually able to exploit them in a new environment. Similar effects took place in parallel in the skull among various groups of vertebrates, in that the dermal component of the skull was reduced while the endochondral elements became more structurally important. This kind of change seems to have allowed the opening up and lightening of the skull structure, facilitating skull kinesis and the exploitation of a whole range of previously unavailable morphologies and ecological niches among both tetrapods and ray-finned fishes.

Inferences and Implications: How Did Early Tetrapods Work?

As has been suggested, the best clues about the sequence and timing of changes that came about during the fish–tetrapod transition come from reading the actual fossil evidence. These usually only involve hard parts of the skeleton, with occasional input from trackways, and are aided by the study of the faunal and sedimentological context in which the fossils occur. However, there is another line of evidence that can be used in some cases, and that is the study of present-day animals. This can sometimes furnish ideas about the evolution of soft tissue characters if the characters are treated rigorously, with the foundations laid on phylogeny.

Some of these have been touched on already in looking at skull structure—vision, olfaction, and hearing among sensory systems, and locomotory and breathing mechanisms among others. By looking at modern animals, some gaps in the story may be filled. For example, physiological aspects of early tetrapods such as respiration and excretion may be inferred from modern animals that live in similar conditions, and some insight about locomotory patterns may be gained from looking at both modern tetrapods and some fishes that use "walking" motions underwater. Study of modern animals' sensory systems should provide clues about what the nervous system of early tetrapods might have been capable of—or at least suggest ideas about what questions to ask about modern animals in order to get some useful answers. It is perhaps surprising that not all the work has been done that ideally should be done in this context. There are still more questions than answers.

Sensory Systems

Vision

There were changes to the position and size of the eye socket, so that in general, early tetrapods have larger sockets more closely spaced than in fishes, and they looked as much dorsally as laterally. There was a suggestion that the increase in size may have been connected with retention of a juvenile condition. However, this is not the only change that occurred to the eyes during the transition. Tetrapods whose heads were exposed to the air for any extended periods (whether or not they walked on the land) would have experienced several factors that were far different from life in water. First, the refractive index of water is similar to that of the eyeball, but different from that of air. The refractive index of a substance is a measure of the degree to which light is bent on entering the medium from air. When the eyeball is in water, it experiences very little refraction because the eye is mostly liquid with a similar refractive index to water. When the eye is in air, the light is refracted as it moves from air into the fluid of the eyeball. Second, when it is in air, the eye is exposed to drying conditions. However, it needs to retain its internal fluid and have a smooth surface if it is to function properly, both conditions being compromised by drying. It may also be subject to greater risk of damage from foreign bodies. Third, the transparency of air is much greater than that of water, especially turbid water or water full of organic debris. In air, therefore, vision can come to play a relatively more important part in the animal's appreciation of its surroundings.

These changed conditions are met by solutions that are the hallmarks of modern tetrapods, and it can be inferred that they evolved sometime during the process that produced tetrapods. In some cases, modern amphibians and amniotes have found different solutions, each of which probably arose after these two groups had split. In other cases, modern amphibians show characteristics similar to those of fishes, and many of these can be inferred to be primitive for tetrapods and so likely to have been present in the earliest forms.

Some of these differences lie in the methods of accommodation—that is, on how the lens changes its focal length to focus on images at varying distances from the animal (Fig. 6.11). Much anatomical work on vertebrate vision was published by Walls, and his book is still the standard reference for the subject (Walls 1942). The following observations are derived from his work. In fishes, the lens is hard, incompressible, and spherical. Accommodation is brought about by moving the lens within the eyeball, although not all fishes are capable of this. The lens lies very close to the surface of the eyeball, and the muscles that move it attach round the circumference of the lens. Some aquatic vertebrates such as lampreys and ray-finned fishes move the lens backward in the eyeball to accommodate to distant objects, whereas others, such as chondrichthyans, move it forward. Some tetrapods, including modern amphibians, also use this means in part. Lungfishes apparently have no powers of accommodation, although the condition in *Neoceratodus* is not known. It is thus quite uncertain which condition is primitive for jawed vertebrates, although perhaps the modern amphibians give us the best clue to conditions in early tetrapods.

In amniotes, the lens is slightly flattened and compressible, and accommodation is usually achieved by muscles that change the shape of the lens while it remains stationary in the eyeball. Once in air, some focusing power can be delegated to the thickened corneal layer. As a result of this arrangement, in tetrapods, the lens can lie further back in the eye, which in turn means that it can sit behind the iris, allowing the pupil to be more flexible and to change its diameter. The reason for this is that if the fish's lens were more deeply sunk in the eyeball, without the corneal contribution to refraction, the visual field available to the eye would be very restricted. Once the cornea becomes a significant refracting component, this restriction is lifted.

Changes to the shape of the lens itself in tetrapods are brought about by different means in amphibians, mammals, and all other amniotes. In amphibians, the lens is almost spherical, like that of a fish, and accommodation is brought about as in fishes: by moving the lens. Muscles above and below the lens, the protractor and retractor lentis muscles, are present in frogs and salamanders and are unique to these groups. They still use the fishlike method of accommodation but have specialized it in their own unique way. This distribution of characters can be read to imply that early tetrapods also used the fishlike means of moving the lens to accommodate the eye, but they cannot help to show which muscles they used to do so.

Mammals use a different method from most other amniotes (and snakes are a special case, with a unique arrangement). In mammals, changes to the lens shape are brought about by relaxing the ciliary muscles around the eye, which reduces the tension in the suspensory ligaments supporting the lens. Most other amniotes, by contrast, squeeze the lens at its equator by contractions of the ciliary muscles, which are correspondingly large. If the last common ancestor of mammals and other amniotes lived in the mid–Late Carboniferous, this animal may not have developed either kind of amniote accommodation mechanism but may still have been using the more primitive means of moving the eyeball, as in amphibians.

In amniotes such as lizards and birds, the sclerotic ring of bones lying around the eyeball plays a role in accommodation. Muscles attached on the inner surface help to squeeze the lens, but also the ring counteracts the increase in internal pressure in the eyeball that this squeezing brings about. Each plate may be flat or have its outer surface concave, and this means that the muscle is brought closer to the eyeball (Fig. 6.12). Many early tetrapods also show this ring of plates, but here it is less clear what its role would have been, assuming they used the fishlike method of accommodation. The ring in early tetrapods is not concave but may be flat or slightly domed. Many sarcopterygians also have a similarly constructed ring, and they too presumably still moved the lens rather than squeezing it. It appears that amniotes (except mammals, where it is entirely absent) coopted an existing feature and used it for another purpose.

Other features in which amphibians and amniotes differ can be seen in the sensory cells of the retina. Rods are the cells that respond best to light, dark, and movement, and cones are cells that respond best to color and only work well in good light. Modern amphibians appear to retain some primitive features of their cone cells, but have also evolved some unique retinal cells called green rods. Early tetrapods are likely to have been diurnal and would probably have had both rods and cones in their retinas, but perhaps they would have lacked the specializations of either amphibians or amniotes.

Tetrapods have a unique system of lacrimal glands and ducts that produce liquid to keep the eye moist and drain the fluid away again. Although sometimes it is possible to see the lacrimal duct in fossil material, there is no evidence of it in early tetrapods. Jarvik (1980) suggested that it can be seen in the nasal capsules of osteolepiform fishes, but because there is no nasal capsule in early tetrapods to compare it with, the equation is not

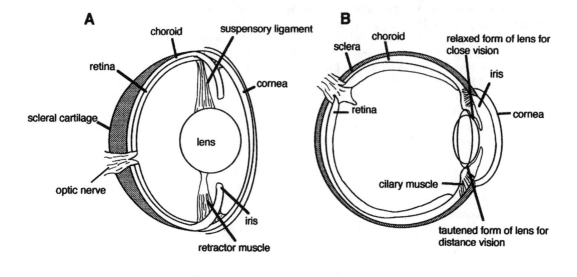
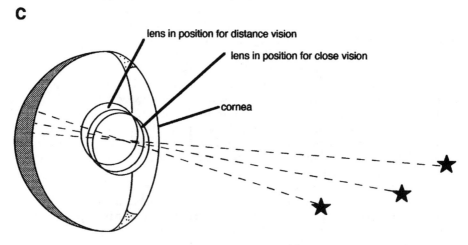

Figure 6.12. Eye accommodation in fish and tetrapods.
(A) Diagram of a fish eye in horizontal section, showing spherical lens set well within the eyeball.
(B) Diagram of an amniote eye, showing compressible lens set close to the outer part of the eyeball, and the shape of the lens shown for close and distance vision.
(C) Diagram of a fish eye showing how the eye moves for distance and close vision.
Adapted from Walls (1942).

easily checked. Tetrapods likewise have eyelids, another unique innovation connected with terrestrial life, and there is a muscle to move the eyelids that only tetrapods have. Bjerring (1987) suggested that *Eusthenopteron* had this muscle, but this depends on the interpretation of a small inpocketing in the braincase. Interpreting muscle insertions is hazardous at the best of times, but in this case, it is particularly fraught with difficulties. Other eye muscles, ones not unique to tetrapods, are equally likely candidates to insert in that small pocket. It is quite likely that it will never be certain when early tetrapods, or their fish relatives, evolved the eyelid or its muscles.

Another eye muscle that is unique to tetrapods pulls the eyeball down into the skull: the retractor bulbi. Some attempts have been made to interpret the position of origin of this muscle in early tetrapods and even in *Eusthenopteron* (Bjerring 1987), but this is made difficult because its position in modern tetrapods is notoriously variable. Again, there are great difficulties in trying to assess whether it was present in early tetrapods.

When artists or museum curators try to make life reconstructions of early tetrapods—and *Acanthostega* is a good example—they often ask, "What kind of eyes did it have?" This is still an unanswerable question and is likely to remain so. The size of the sclerotic ring may tell us something about the size of the eyeball, but not how much of the eye would have been visible in the living animal. It is not even obvious what sort of modern animal it would be best to use as a model. The eyes could have been large and prominent, like those of a flatfish or frog, or they could have appeared small and relatively inconspicuous, like those of a giant salamander or a lungfish. However, the large, prominent kind are associated with very specialized lifestyles and of the two styles, early tetrapods may have been more likely to have had the inconspicuous, unspecialized sort. But this is a guess.

Olfaction

The noses of early tetrapods show both similarities and differences compared with those of fishes. There are similarities in the openings of the nostrils but differences in the arrangement of snout bones and the state of the nasal capsule. Changes to the olfactory system seem to have happened at the transition, to judge from a contrast between living tetrapods and living fishes, and changes to the construction of the nasal capsule may have accompanied these. In fishes, the nose functions purely as an organ of olfaction and is not used in breathing. In modern tetrapods, and probably most fossil forms too, it is used for both. At some point during the fish–tetrapod transition, the change must have occurred. Because the nostrils—internal as well as external—are so similar between *Acanthostega* and *Panderichthys*, it may be safe to assume they were doing the same job in each—namely, olfaction only—although it is not obvious whether either or both were sampling air, water, or both. Probably, however, sampling air chemically preceded taking it in through the nose to breathe. As has been suggested above, *Acanthostega* probably gulped air as lungfishes do, and the guess is that it did not breathe through its nose. This facility arose only later at some unknown stage during the evolution of tetrapods.

In tetrapods, a structure called the accessory olfactory bulb exists in the nasal capsule that is not present in any fishes. This structure, also called Jacobson's organ, is usually associated with detection of airborne chemical signals called pheromones, although that is not its only purpose. It is found in both aquatic and terrestrial forms, although it was once thought to be restricted to terrestrial ones (Eisthen 1992, 1998). It is possible to imagine that at the origin of tetrapods, some major change occurred to the construction of the olfactory organs, with the production of this accessory bulb, and that this coincided with the observed loss of ossification. It may thus also coincide with increasing emersion of the snout. This organ may have been "invented" by the earliest tetrapods.

One more piece of evidence derives from examining the mechanisms by which amphibians and amniotes receive and process olfactory information. There is at least one system, that for detecting sour tastes, that differs fundamentally between modern amphibians and modern amniotes (e.g., Kinnamon and Roper 1987 for the mud puppy and Gilbertson et al. 1992 for the hamster). The implication is that this system was evolved separately in the two groups, but probably some time after the origin of tetrapods as a group.

In contrast to the negative evidence on the sense of olfaction, two senses leave positive traces in the fossil record: the lateral line system and the sense of hearing.

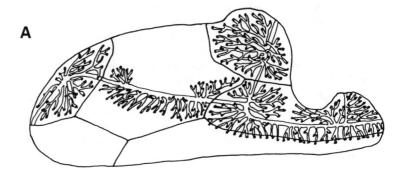

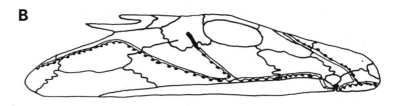

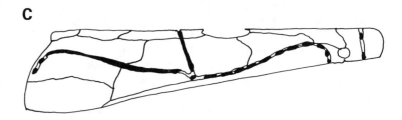

Figure 6.13. Diagram of lateral line expression in different groups.
(A) Cheek of Eusthenopteron showing numerous pores in a network connected to the lateral line canals within the bones (from Jarvik 1980).
(B) Skull roof of Acanthostega in lateral view showing the single row of pores connected to the lateral line canal.
(C) Skull roof of Greererpeton in lateral view showing most of the lateral line canal as open grooves (openings shown in black) (from Smithson 1982).

Lateral Line System

The lateral line system is one that, as mammals and members of the amniote group, humans are not well equipped to understand. Amniotes have lost all trace of the lateral-line system, and people do not even have the language to describe the sensations that it must produce. The system consists of a tracery of tubes or canals running within or on the surface of the dermal bones or skin (Fig. 6.13). The canals or tubes are open to the outside and so the contents are affected by movement and pressure changes in the surrounding water. Inside them lie sensory hair cells or neuromasts, tipped by hairlike cilia, whose movements trigger nervous impulses. As they are moved around by pressure waves from the water, they send signals to the brain. This system can detect pressure changes in water up to about a meter or so away from its owner's body and can inform the brain of the direction as well as the size of the object that has created the pressure changes.

As a rule of thumb, in fossil lobe-finned fishes, the lateral line canals of the head lie within tubes in the dermal bone, opening by a series of pores to the outside. *Eusthenopteron* has a complex ramifying system of tubes with multiple small openings in wide bands on the bone (Jarvik 1980) (Fig. 6.13). *Panderichthys* and *Acanthostega* have a simpler system, and in each, a single series of somewhat larger pores can easily be seen. In most later tetrapods, by contrast, the lateral line canals run in open grooves on the surface of the skull (Fig. 6.13). Although in the more primitive forms of tetrapod some of the canal's length may still be enclosed in the bone, it seems to be the case that most tetrapods show the open condition of the grooves or no grooves at all. What the difference signifies is unknown. It is known that in modern animals, the lateral lines start as superficial structures and gradually sink into the skin or underlying bone as the animal grows. Thus, open grooves gradually become enclosed into tubes. The tetrapods with open grooves may be showing a pedomorphic condition. It is possible to see among modern ray-finned fishes that some have enclosed neuromasts, whereas others have superficial ones in the outer skin, but there seems to be no consistent explanation of this in terms of ecology (Webb 1989). It may relate to the amount of water movement or the water depth that the fish generally lives with, but this is by no means certain. So the differences among the fossil animals can only be guessed at. There may be an ecological signal to be read here rather than a phylogenetic one, but more work needs to be done not only on fossil forms but on living animals.

One thing is quite certain, however, and that is that the amniote lineage lost its lateral line system during the course of evolution. A whole region of the brain and some large nerves that served the lateral line system disappeared in amniotes. It is not clear whether the nerves and brain region were redeployed for other sensory input or whether the structures atrophied (Fritzsch 1989). Neither is it known whether the loss of the lateral line system occurred suddenly at the origin of amniotes or bit by bit during their very early evolution. Modern amphibians do retain superficial lateral line organs (although not grooves) while they are tadpoles, and these are retained in perpetually aquatic forms (Fritzsch 1989). However, those that are predominantly terrestrial as adults lose the lateral line system at metamorphosis. No modern amniote shows any evidence of the lateral line system even in the egg stage, but whether this applied to the very earliest amniotes of all, those of the Late Carboniferous (see Chapter 9), remains a mystery.

In many fishes, the lateral lines continue down the sides of the body borne by a row of specialized scales. No tetrapod scutes are known that carry such lateral line pores, although like modern aquatic amphibians, the lateral lines may have been carried down the body in the skin, so the presence of a trunk lateral line in early tetrapods cannot be ruled out.

Inner Ears

A number of differences can be seen in the soft tissue characters of inner ears in tetrapods as compared with any known fish. These concern the arrangement of pouches and the hair cells they contain. These features, of course, cannot be seen in fossils, so inferences must be based on what can be seen in modern animals. Some of these differences may have been initiated at the same time as the reconfiguration of the otic region of the braincase described above.

The inner ear of all vertebrates is composed of a set of three semicircular canals set at right angles to each other and filled with fluid whose movement informs patches of hair cells of movements in the body. This part of the inner ear is called the pars superior and is primarily concerned with posture control, or "balance," in tetrapods that have gravity to contend with. Situated near the junction of the semicircular canals is a pouch called the utriculus, also part of the pars superior. Beneath this is suspended a further pouch or pouches, the sacculus and lagena, constitut-

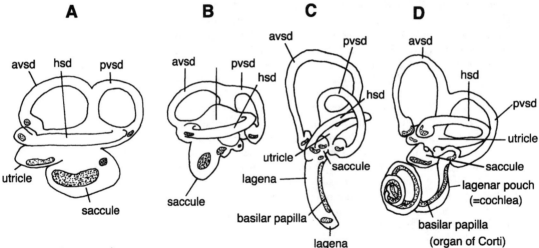

Figure 6.14. Diagram of vestibular systems in different groups.
(A) Lungfish.
(B) Frog.
(C) Bird.
(D) Mammal.
(From Romer and Parsons 1986.)
Avsc = anterior vertical semicircular canal;
hsc = horizontal semicircular canal;
pvsc = posterior vertical semicircular canal.

ing the pars inferior (Fig. 6.14). A sacculus pouch is common to all vertebrates, but a separate lagenar pouch, budded off from the sacculus during development, is found in certain groups, including most tetrapods. Generally the lagena is associated with audition in tetrapods, although in teleost fishes both sacculus and lagena may have auditory functions (Popper and Platt 1993).

One of the differences between modern fishes and tetrapods lies in the structure of the pars inferior and in the contents of the sacs. Another name for these sacs is the otolithic organs. In teleosts and also in the extinct acanthodians, each of the sacs—utriculus, sacculus, and lagena—contains a solid crystalline calcium carbonate otolith, or "ear stone" (Schultze 1990). This rests on the hair cells in the sac and stimulates them when it is moved by gravity or sound waves. Because it has a gravistatic function, it is also called a statolith. Tetrapods and lungfishes lack solid crystalline otoliths (Carlström 1963) but instead have a large number of small crystallites bound together in a gelatinous matrix to form what is usually called a "pasty" statolith. Lungfishes do not have lagenae (Popper and Platt 1996; Retzius 1881), so there are only two of these statoliths. Pasty statoliths are very rarely fossilized, and correspondingly, otolithic material has not been found in early tetrapods. By contrast, only single saccular statoliths have been found in fossil and living coelacanths even though they do have lagenae (Clack 1996).

Fossils of osteolepiforms show that like lungfishes, they appear to have lacked lagenae, as shown in well-preserved material of *Ectosteorhachis* (Romer 1937). If tetrapods are most closely related to lungfishes and osteolepiforms, the earliest tetrapods also lacked this lagenar pouch. Evidence for its existence in tetrapods only appears much later in their evolution.

In addition to differences in the otolithic composition (single crystal versus suspension of small crystals in a pasty matrix) in fishes and tetrapods, there are differences in the patterns in which the hair cells are arranged within the pouches. Although little studied as yet, it seems that, despite the fact that both groups possess lagenae in contrast to lungfishes, coelacanths resemble the more primitive ray-finned fish rather than tetrapods in their pattern of hair cells (Platt 1994).

Lungfishes lack any of the specialized inner ear receptors for processing sound signals that are seen in tetrapods and that are also seen, independently derived, in teleost fishes (modern ray-fins). In tetrapods, the main sound receptor, the basilar papilla, is situated in the lagena. Fritzsch (1987) found a patch of cells in the coelacanth lagena that he considered to be a homolog of the tetrapod basilar papilla—that is, a feature derived in common in tetrapods and coelacanths, rather than as a parallel development (Fritzsch 1992). However, no experiments have been performed on living coelacanths to confirm or refute whether the cells could respond to sound in the same way as in tetrapods. Conclusions of homology would normally be based on more extensive studies of relationship (see Chapter 3).

Another difference between tetrapods and all fishes relates to the way the inner ear pouches are suspended in the inner ear cavity. The fluid inside the inner ear itself is called endolymph, but the system is suspended within a chamber and surrounded by fluid called perilymph. In fishes, this is contained in a simple "periotic" sac, but in tetrapods, the perilymphatic space is organized into a system of tubes and sacs that runs around and about the pars inferior (Fig. 6.15) (see Duellman and Trueb 1986; Fritzsch 1992). There is a main chamber called the perilymphatic cistern, confluent with the periotic sac. They are connected by a tube called the perilymphatic duct. There is no way to know whether early tetrapods had the simple fishlike condition or the more complex one found in modern tetrapods, but it is clear that modern amphibians and modern amniotes have different arrangements from one another (Lombard and Bolt 1979). In amphibians, the duct passes posterior to the sacculolagena complex containing the sensory papillae (see Fig. 6.15A), but in amniotes, it passes anterior to it. These two conditions are mutually exclusive. The implication is that each has been derived independently from a fishlike condition, so that the simple fishlike arrangement might still have been present in the earliest tetrapods. If so, the profound changes to the otic region of the braincase that occurred during the fish–tetrapod transition preceded rather than accompanied those to the inner ear.

The middle ear and the evolution of tetrapod hearing mechanisms are described in more detail in the final chapter. Here, suffice to say that *Acanthostega* does not appear to have evolved some of the refinements that in later tetrapods are seen to be those associated with the reception of airborne sound, nor does it appear to have had the inner ear and brain structures associated with processing and interpretation of the signal. For a model of its hearing capabilities, the other living lobe-fins as well as the more primitive ray-finned fishes seem to provide better models than do modern tetrapods.

Pineal Organ

The skulls of many fishes and tetrapods from the Paleozoic have a foramen situated in the center of the skull table, called the pineal or parietal foramen (Figs. 6.2, 6.6). Some modern amphibians and reptiles have a similar hole, although by no means all. The hole houses the pineal organ, which in mammals is closely allied to the pituitary gland. In most nonmammals, the pineal is a separate organ from the pituitary, and in lizards, it is known to have a role in monitoring day length and in controlling the animal's daily and longer-term rhythms. It has also been shown to be sensitive to polarized light in salamanders, so that the animals can tell the direction of north and south, useful in migrations (Taylor and Adler 1978). In tetrapodomorph fishes, the hole is as a rule rather smaller than that in Devonian or Carboniferous tetrapods, where it can be a conspicuous

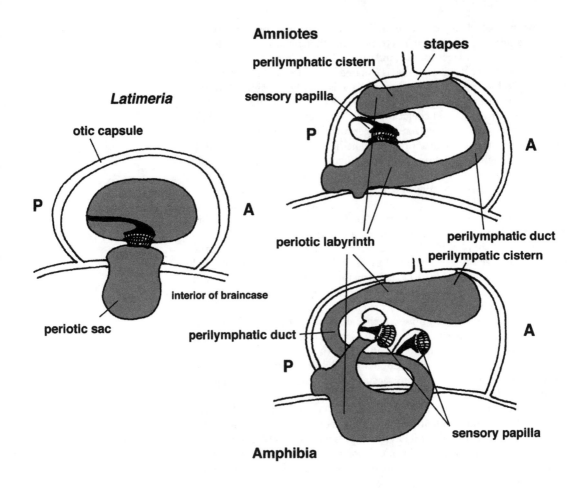

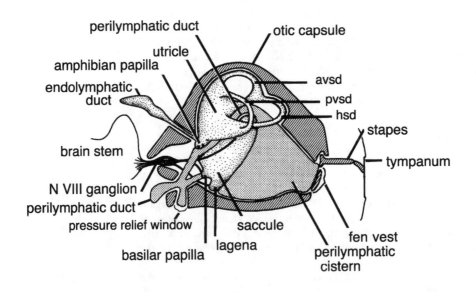

feature. On the other hand, it is lacking entirely in some later tetrapods such as turtles. Even in modern tetrapods such as birds and lizards where there is no foramen, however, the pineal organ can still function to monitor day length. It has been shown that in birds, light can penetrate to and stimulate the pineal either via the eyes, or directly (Barrett and Underwood 1991). Judging from the size of the foramen in early tetrapods, it may have been functioning differently in them than it was in fishes, and its size may reflect the fact that exposure to daylight was more important.

Other Mechanoreceptors

Other receptors are found in the skin and muscles of modern tetrapods, such as stretch receptors and proprioceptors, for telling the brain the position and attitude of the limbs, for example. Fishes and tetrapods have proprioceptors in the paired appendages for informing the brain of what these are touching and their position, but most tetrapods also have receptors in the muscles, known as muscle spindles, that are not present in modern ray-finned fishes (Barker et al. 1974). However, there are simple branched sensory endings, something similar to muscle spindles, apparently found in the fin musculature of rays and lungfishes, as well as in the connective tissue at the base of the walking fins of the gurnard. Presumably there was a type of nerve ending found in the bodies of primitive vertebrates that has been variously adapted and developed in different groups. Muscle spindles are also apparently absent in newts, and it is not known whether caecilians have them or not. In frogs, these spindles are only present in limb and abdominal muscles and are absent from axial muscles. Muscle spindles have been closely studied in mammals and frogs, and in both of these groups, they are highly specialized. Mammals seem to have the most highly developed system of muscle receptor types of all tetrapods; in frogs, the muscle receptors are particularly well developed to control the leg muscles, important in the jumping abilities of the animals. The spindles in amphibians and amniotes apart from mammals are supplied with two types of nerve endings, known as "grape endings" and "plate endings," that may therefore be assumed to be the common inheritance of all tetrapods that have then been modified and specialized in mammals (Barker et al. 1974). Possibly these two types were developed early in tetrapod history at the time when legs were acquiring an increasingly important role in locomotion.

Much research into sensory systems and physiology in the past was dedicated to the basics of finding out how things worked and to the demands of medical application. This means that work has concentrated on mammals and on a few other easily available laboratory animals such as frogs and chickens. More recently, a broader range of animals has been studied, but the perspective of phylogeny and evolution brings new questions that show how rewarding it could be to investigate other animals such as nonteleost ray-finned fishes, lungfishes, and an even wider range of tetrapods.

Unfortunately, some of the most phylogenetically and evolutionarily interesting animals are very difficult or impossible to rear in laboratory conditions, the coelacanth being a prime example. However, a breeding program for the Australian lungfish has begun and promises some exciting and revealing results (Joss and Joss 1995). A possible research program for the future might ask what sensory equipment an early tetrapod might have had, judging from what is present in their closest relatives, more "primitive" fishes, and what is most widely distributed among different tetrapod groups.

Figure 6.15. (opposite page) (Top) Diagram of perilymphatic systems in Latimeria, *amphibians, and amniotes (from Fritzsch 1992).*
(Bottom) Diagram to show how the perilymphatic system relates to the inner ear chambers in a frog (from Duellman and Trueb 1986).

Physiology: Breathing, Respiration, Excretion, Reproduction

Breathing and Respiration

One problem that early tetrapods might have faced was that of desiccation. If the climate was humid, this was not such a great problem, but there may have been a conflict of pressures to be met in any case with respect to the scaly armor of these pioneers. To make more efficient progress over land, it might have been an advantage to reduce weight by losing the body scales. On the other hand, to avoid the drying effects of air, it might have been advantageous to have kept them. The earliest known tetrapods had lost their dorsal scales but kept their belly scales, although several groups of later tetrapods produced dorsal scales again. Belly scales might have protected the undersides from abrasion or assisted in keeping the body's shape during movement, but loss of the dorsal scales would have meant that the most frequently emersed part of the body was most exposed to drying out. An alternative suggestion for the loss of dorsal scales was that when wholly or partially out of the water, this part of the body could be engaged in cutaneous gas exchange, or skin breathing.

In the past, it has been stated that early tetrapods had skins that were too thick and heavily ossified for any cutaneous gas exchange to have taken place through it. Not only that, but it was suggested that, being large, the surface-to-volume ratio of these animals was too poor to allow efficient use of cutaneous gas exchange. These ideas were put forward around 1970 (Gans 1970; Romer 1972a) but have been accepted by many recent physiologists trying to work out how early tetrapods breathed (Ultsch 1987). It is time to look again at the possibilities (e.g., Janis and Keller 2001), because some of the statements made about early tetrapods a few decades ago were based on several assumptions that have led to sources of error. First, they were based on theoretical properties of gas and skin rather than on experiment and observation of living forms, and second, they were based on a view of early tetrapod structure that was limited and has been changed by recent discoveries. Third, the idea that the composition of the atmosphere might have been different from today had not then been put forward (see Chapter 4). Finally, there seems no real reason why early tetrapods should not have used all possible means of breathing, and they might have used lung as well as skin breathing, employing not just the outer body surface but that of the buccal cavity.

One of the main objections to the idea of skin breathing in early tetrapods was their possession of gastralia over the belly and the evidence of dorsal bony scales. However, the gastralia, as well as the dermal bones of the skull and shoulder girdles of early tetrapods, were dermal in origin and so had blood vessels running through them to the epidermis that lay external to bones and scales. Pores in the bones and vascular spaces throughout the bone are evidence of this (Bystrow 1947). Grooves for blood vessels are sometimes visible as well, but even without this, the structure and tissue relations of the dermal bone imply that epidermal tissue lay external to the bone. This means that capillaries carried blood close to the surface, where it could have received oxygen. The thickness of the epidermis is unknown, of course, but in the case of heavily ornamented dermal bone, it may have followed the ornament pattern quite closely, and where the ornament was reduced, skin folds may have hung from it, as suggested for *Crassigyrinus* (Chapter 7). It is known that the palate of *Ichthyostega* bore large blood vessels running over it (Jarvik 1996).

The scales were not similar to the cosmine-covered ones of primitive osteolepiform fishes in that they were not covered in enameloid of any kind, and in this sense, they were not unlike the thinner scales of other tetrapodomorphs such as *Eusthenopteron*. By contrast, the skin of an early tetrapod was not like that of a modern amniote either. Amniotes have incorporated a keratin protein into their epidermis; in most amniotes, this is β-keratin, whereas mammals have α-keratin in addition. This forms epidermal scales, fur, and feathers. Dinosaurs, crocodiles, or lizards with scaly skins, as well as mammals and birds, are effectively waterproofed. Despite this, however, studies on recent amniotes (Feder and Burggren 1985) have shown that even something as armored as a turtle exchanges 30% of its CO_2 and 20% of its O_2 through its skin (some of this in fact occurs through the cloaca and may not be comparable with the situation in other tetrapods). Similarly, lizards and snakes also perform a substantial part of their gas exchange through the skin, using the thin skin between the scales to do so (Feder and Burggren 1985).

At the other end of the spectrum, the heavily enameled scales of the fish *Calamoichthys* do not prevent it from using its skin for about 30% of its gas exchange, although most of this might be involved with superficial servicing of the skin and scales. The large modern amphibian known as the hellbender (*Cryptobranchus*), a permanently aquatic form, also uses its skin for up to 90% of its gas exchange (Feder and Burggren 1985).

The higher oxygen levels that pertained in the Devonian and Carboniferous (see Fig. 4.1) must have enabled a much better uptake of that gas through the skin as well as providing a readier source for breathing via the lungs because of the higher gradient between the air and the body. It seems highly unlikely that early tetrapods failed to exploit every possible means of using this vital resource, however theoretically inefficient. A raised oxygen level in the air has the secondary consequence of raised oxygen levels in the water as well, making oxygen uptake through the skin a more feasible option for aquatic forms.

The elevated oxygen levels pertained just at the period when early tetrapods were beginning to exploit the land and must have had other profound effects on the physiology of these creatures. It means that for every breath, however taken, the amount of oxygen taken in would be greater than for modern animals. If early tetrapods mainly used buccal pumping to ventilate lungs, it was likely to have been less cost effective and efficient on land than it was in the water, partly because the work has then to be done against the influence of gravity. However, the problem would have been reduced because of the increased oxygen that each breath would collect (Graham et al. 1997). Furthermore, the metabolic rate of the animals is of course unknown, but may not have been as high as that of modern tetrapods.

Elevated oxygen levels might have facilitated land dwelling by making more energy generally available and by allowing recovery from anoxia to be more rapid, thereby extending the animal's aerobic capacity (Graham et al. 1997). To move about on land requires expenditure of more energy than moving in water, but the ready availability of oxygen in the air could have meant that relatively inefficient locomotion was feasible in the early stages of terrestrial evolution. Another effect of increased O_2 might have been that the eggs that the animals laid could have been larger than they otherwise would have been, which cuts down on water loss, facilitating the evolution of amniotes.

Because the higher oxygen content of the air meant fewer breaths had to be taken for a given amount of oxygen, a corollary of this is that less water is lost through breathing. The humidity levels in the forests of the time were also presumably high enough to reduce the problem of desiccation.

Carbon dioxide metabolism might have posed more of a problem for early tetrapods than oxygen uptake. With the higher levels of CO_2 in the air in the Devonian and early Carboniferous, diffusion from body surfaces may have been more difficult than in today's atmosphere. However, if animals maintained a high level of CO_2 in the blood, the gradient between body and air might still have allowed diffusion to take away the CO_2, provided the skin could stay moist. CO_2 is as easily soluble in air as it is in water. Modern air-breathing fish provide a possible model here because they can tolerate quite a high level of CO_2 in the blood, more comparable with amniotes than modern amphibians or non-air-breathing fishes. They often live in water that has a very high CO_2 content as well as a low oxygen content (Randall et al. 1981). It has been suggested that early tetrapods, because they lived in vegetation-choked environments, might have used a mode of CO_2 metabolism similar to these fish and thus have been preadapted to make the transition to land (Ultsch 1987). Most modern fishes get rid of excess CO_2 from the gills, although few studies have been done to find whether or what proportion is excreted through the skin as distinct from gills. Certainly no fish is known to use lungs to excrete CO_2, nor do modern amphibians very much. Their lungs are responsible for very little CO_2 excretion. Accumulation of CO_2 in the body is more often a limiting factor in vertebrate physiology than lack of oxygen, and the ability to deal with CO_2 is very important in the stabilization of body fluid acidity. The ability of lungs to cope with this function seems to have been fully accomplished in the amniote lineage only, although again, it must remain unknown at what point in the evolution of the group it evolved. Early tetrapods with gills such as *Acanthostega* may well have continued to use them for CO_2 loss. As the CO_2 levels fell during the late Devonian and early Carboniferous, it may have become less and less of a problem.

Nothing is known about the circulatory systems of the earliest tetrapods. In modern amphibians, much of the system is devoted to supplying the skin with blood for gas exchange, and modern amphibian skin is rich with blood vessels for this purpose, as well as glands to keep it moist. Their skin is also much thinner than that of any amniote. Amphibians appear to have specialized in cutaneous gas exchange, elaborating a primitive mechanism into something more sophisticated. Early tetrapods were probably much less specialized in this respect and used skin breathing as only part of their repertoire. Salamanders, with the most fishlike circulatory system among modern tetrapods, and lungfish may provide clues about the circulatory system of early tetrapods (Fig. 6.16). This is despite the fact that one large group of salamanders has lost its lungs altogether, relying entirely on cutaneous gas exchange. However, these animals are all small creatures, and it is not likely that early tetrapods went to this extreme.

The circulation of a lungfish, a salamander, or indeed some amniotes employs a three-chambered heart with two auricles and a single ventricle. The system is unlike that of a mammal, in which the ventricle is also divided, to separate the lung circulation from that of the body. By this means, in mammals, oxygenated blood is completely isolated from deoxygenated blood, and this is generally supposed to be more efficient than the system amphibians use. However, because the single ventricle of an amphibian receives oxygenated blood not only from the lungs but also from the skin, the system is not as inefficient as it appears at first sight (Fig. 6.16).

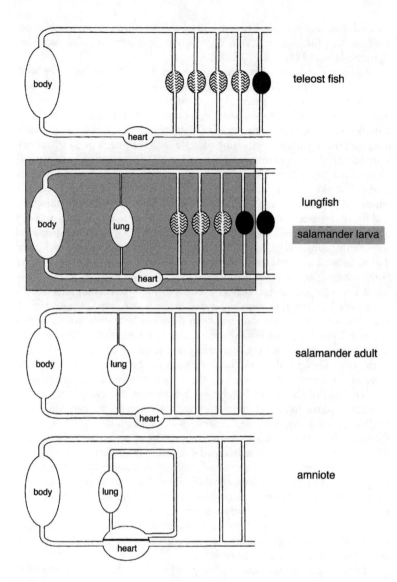

Figure 6.16. Diagrams of the circulatory systems of a teleost fish, a lungfish, salamander larva and adult, and an amniote. The functional gills are shown with wavy hatch marks; nonfunctional gills are black. In the teleost, there are four functional gills and a simple circulation from the heart, through the gills to the body, and back to the heart. In lungfish, there are three functional gills and two nonfunctional gills. In some, the blood from the body and the lung are partially separated (not shown). In the salamander larva, three functional gills have external filaments. These are resorbed in the adult, and gas exchange occurs in the lungs, skin, or both. In amniotes, the blood to the lungs and body are kept separate so as not to mix oxygenated and deoxygenated blood. Early tetrapods may have had a circulatory pattern like that of a lungfish or salamander.

If early tetrapods also had a three-chambered heart and used skin breathing as well as lungs—and bearing in mind the higher oxygen levels—the system might well have worked in a way that was good enough.

More clues to the physiology of early tetrapods have come from study of air breathing in lungfishes and the so-called primitive ray-finned fishes such as *Polypterus, Lepisosteus,* and *Amia.* It is possible to get some ideas from them about the means of respiration in the earliest tetrapods and the circumstances under which air breathing took over from gill breathing. Because these cladistically primitive ray-finned fishes as well as lungfishes breathe air with lungs, lungs are usually assumed to have been present as a primitive feature in all bony fishes (see Chapter 2). *Lepisosteus* and *Amia,* for example, use mainly gills in low-temperature waters, but when the temperature rises, the percentage of oxygen taken up by the lungs increases dramatically, from about 25% to up to 75%. Early tetrapods probably evolved in warm waters. Even at higher temperatures, however, in these modern fish, the gills and skin remain the main site for CO_2 excretion (Randall et al. 1981).

In lungfishes, the dependency on oxygen from the air increases with age and size of the individual. Small, young animals seem to get all their requirements from the water, although where oxygen is in poor supply, they will often congregate at the water's surface (Randall et al. 1981). Lungfish do have some capacity for CO_2 exchange by the lungs, and they use this capacity during estivation. They also have hearts and circulatory systems very like those of amphibians (Fig. 6.15), and they have an epiglottis used to separate air and food passageways. In these respects, they probably make good models for early tetrapod soft anatomy. In practice, it appears that in addition to a function in gas exchange, the lungs of lungfish, and for that matter *Cryptobranchus,* operated also as a buoyancy aid, not unlike that of the swim bladder in some fishes (R. G. Boutilier, personal communication). Perhaps all these Devonian bony vertebrates, including the fishes such as *Eusthenopteron,* had an all-purpose gas bladder, useful both as a lung and a swim bladder.

Early tetrapods almost certainly used a form of buccal pumping to ventilate their lungs, rather than using ribs in the way many amniote groups do. This is another idea that has been disputed by earlier workers, but again, it depends on views of early tetrapods that are oversimplified or have recently been proved wrong. Part of the argument concerns the structure of the ribs. This problem will be dealt with further in Chapter 10, when more of the animals that come into the equation have been described.

The loss of the operculogular system by even the earliest tetrapods may relate to the buccal pumping mechanism, allowing a larger and more flexible gape for this purpose. Other groups of air gulpers—for example, the lungfish—have similarly reduced or lost the operculogular series, and this seems to be related to the changed emphasis from lateral movements of the cheek to ventral movements of the buccal floor. The braincase of lungfish is fused to the palate, another factor that may correlate with air gulping. In many groups of later tetrapods the palate and braincase likewise become solidly attached to each other; that is, they had autostylic jaw suspension, like lungfishes. The flattened broad heads of early tetrapods may have provided a greater area for movements of the buccal floor, as well as a large surface area for gas exchange.

Lungfishes, salamanders, and frogs all use a form of lung ventilation called mixed-air buccal pumping. It appears to be primitive for tetrapods and does not involve the use of ribs. Comparing this mechanism in salamanders and lungfishes reveals some differences in the way that it is brought about and shows another character by which tetrapods differ from other vertebrates. Tetrapods have an extra layer of muscle in the body wall, the transverse abdominal muscle, used in expiration (Brainerd et al. 1993). Contraction of this muscle helps force air out of the lungs, and it is found in amniotes as well as amphibians. It will probably never be known when in history this extra muscle evolved or whether it would have been present in early tetrapods such as *Acanthostega.*

Mechanisms of inhalation vary between amphibians and amniotes. In amphibians, inhalation happens by passive recoil of the lungs and abdomen when the muscles are relaxed, but in amniotes, another suite of muscles is employed. It has been suggested that ventilation of the lungs in air-breathing tetrapods evolved in three stages. The first stage would have used pure buccal pumping, the second stage used muscles only for exhalation, and the third stage, reached only in amniotes, used muscles for both

Figure 6.17. (opposite page) (Top) Diagram of the body wall muscles of a salamander to show which occur in which vertebrate groups.
(Bottom) Cladogram with ventilation mechanisms superimposed. From Brainerd et al. (1993).

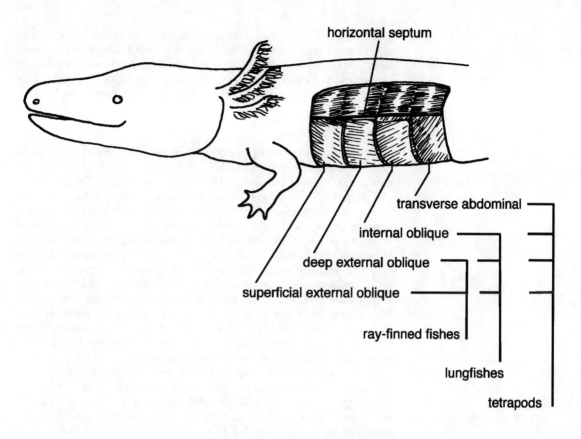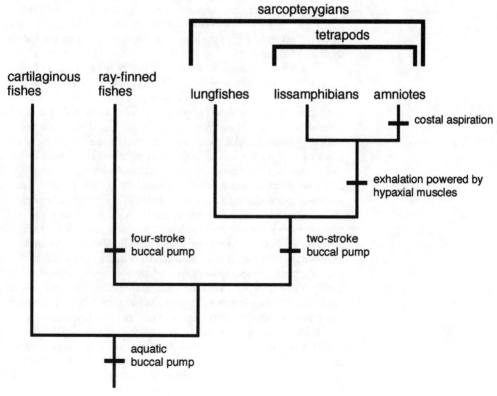

exhalation and inhalation. This is expressed in the cladogram in Figure 6.17 (Brainerd et al. 1993). In some amniote groups, the inhalatory muscles were helped by the ribs, but different amniote groups use their ribs in different ways, or in some cases, such as turtles, not at all. The implication is that the specialized use of ribs in breathing evolved separately in different amniote groups.

Nitrogen Excretion

Understanding the physiology of nitrogenous excretion, like respiration, among early tetrapods can be done best on the basis of what can be seen among modern animals. Neither function is directly reflected in skeletal anatomy. There are big differences between modern fishes and tetrapods in the metabolism of nitrogenous waste. In tetrapods, nitrogenous waste is almost exclusively processed by the kidney and voided either as urine or a uric acid paste. In fishes, this is not the kidney's main function, although in a few fishes, it is responsible for a small proportion of nitrogen excretion. Its main function in fishes is to maintain the salt balance of the blood, and nitrogenous waste is processed and excreted mainly by the gills, with a small proportion excreted through the skin. In most modern ray-finned fishes, ammonia is the main excretory product, but some can produce urea instead (McKenzie et al. 1999; Walsh 1997), especially those exposed to air on a regular basis or those that live in water high in CO_2; some even use quite different methods of dealing with ammonia and convert it to glutamine on exposure to air (Jow et al. 1999). Among those that do produce urea, the kidney can play a small part in nitrogen excretion via an active transport system, although this is always done when the fish is in water and not on land, as far as is known. Lungfishes produce urea especially while estivating, but it is not known where it is excreted. In modern amphibians, nitrogenous waste is excreted as ammonia via the gills (if present) and skin while the animals are in water (e.g., as tadpoles), but they convert products to urea and excrete them through the kidneys when on land (Wright 1995).

Gills were presumably the primitive site for excretion of nitrogenous waste. As for the excretory products, although the production of ammonia is easy for an aquatic animal because it is very soluble in water, production of urea is also probably a capacity that was found widely among ancient vertebrates. The enzymes for doing this are found in invertebrates, chondrichthyans, and coelacanths, as well as some teleosts (Wright 1995). No work on this matter has been done on primitive ray-finned fishes such as *Polypterus* or *Lepisosteus*. For early tetrapods living in water high in CO_2 and that were to some degree exposed to the air or that breathed air routinely, production of urea was probably the preferred option.

During early tetrapod evolution, the kidney acquired the capacity to process the urea and void it, although it was probably a rather gradual process. At first, as in modern air-breathing fish, the animals probably accumulated urea until they returned to water, where it was lost, some via the gills, some via the kidney, and possibly some through the skin (Walsh 1997). Early tetrapods such as *Acanthostega* that retained gills may have used them for nitrogenous excretion as in fishes, but quite possibly, *Acanthostega* may have already begun to use its kidney in addition.

It has been suggested that if early tetrapods such as *Ichthyostega* and *Tulerpeton* had really lost their gills, their ancestors must have been more terrestrial than these descendants were (Janis and Farmer 1999). The basis of this argument is that no modern fish, even an air-breathing one, uses its kidney to excrete nitrogen on land. It retains gills for this purpose as well as for excreting CO_2 on its return to the water. Thus, gills are considered too valuable to lose in an aquatic animal. The argument runs that an early tetrapod that had lost it gills completely would only have done so if the kidney was already being used for its entire nitrogen excretion, and that this would only happen on land. However, because some modern fishes do use kidneys for part of their nitrogen excretion, it may be that air-breathing early tetrapods used the kidneys to some extent in this way. They may have adopted this role gradually along with reduction of reliance on gills for breathing, but while the animals were still aquatic. In any case, the evidence for lack of gills in early tetrapods is based entirely on negative evidence. Gill bars are cartilaginous in lungfishes and may have been so in early tetrapods such as *Ichthyostega* and *Tulerpeton,* for which there is no evidence of gill bars.

Reproduction

It is usually assumed that early tetrapods had what is called an "amphibious" lifestyle. That means that they would have laid their eggs in water, the eggs would have hatched into aquatic larvae that would probably have had external gills, and then at some stage the larvae would have metamorphosed into the adult form. This is the kind of lifestyle seen today in many frogs and salamanders, and it is also seen in a diverse range of fishes. The lungfishes *Protopterus* and *Lepidosiren* as well as the ray-finned fish *Polypterus* employ this strategy. Until the rise of cladistic or phylogenetic methods of classifying animals, the modern Lissamphibians were defined as amphibians because they showed this lifestyle, and early tetrapods were also called amphibians because they too were assumed to have used it. If it is believed that the amphibian style of life history was present in the ancestors of tetrapods and that early tetrapods inherited it, of course that can no longer be used to define a group "Amphibia," because it is more generally distributed than just among amphibians. That is one problem, but not the only one.

For most of the early tetrapods, there is no evidence concerning reproductive strategies, so the amphibious lifestyle is only an assumption. Another problem is that reproductive strategies among modern animals is intimately connected to ecology. An animal may make more investment in production of large numbers of eggs or young that then receive little care, or it may make more investment in parental care of a small number of young (Gould 1977). Which strategy is used is governed by a combination of factors involving how stable the environment might be and the animal's place in the food web. These are sometimes difficult to disentangle, but they have meant that frogs, for example, use the whole range of possible options, from the production of large quantities of frog spawn to production of a single offspring cared for until maturity. Sometimes these both occur in the same genus of frog, so reproductive strategy has little connection with phylogeny. For this reason, it may be a mistake to think of early tetrapods as being like frogs that produce frog spawn in their reproduction strategies.

The most primitive frogs, most salamanders, and all caecilians use internal fertilization. The coelacanths and the chondrichthyans do likewise, so there is no good reason to suppose that early tetrapods might not have done the same. Indeed, coelacanths and many sharks are ovoviviparous, keeping the young inside the body until they are miniature adults. Again, some early tetrapods might have done so too. It is perhaps easier to envisage the development of an amniote egg from such precursors than from a frog spawn–producing animal (Skulan 2000).

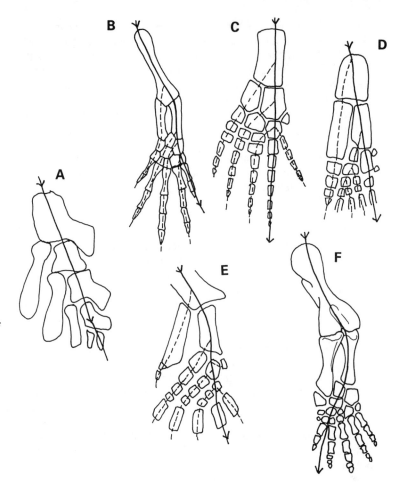

Figure 6.18. Diagram to illustrate historical ideas of the position of the metapterygial axis (shown as an arrow).
(A) Eusthenopteron fin for reference.
(B) Gegenbaur (1878).
(C) Watson (1913).
(D) Gregory et al. (1923) and Gregory and Raven (1941).
(E) Steiner (1934); Jarvik (1980).
(F) Westoll (1943b).

The only way in which these ideas could be tested would be if Devonian tetrapods were discovered in some deposit showing exceptional preservation. Such deposits in the past have yielded the Burgess Shale fauna of Early Cambrian soft-bodied animals. The feathered Jurassic protobird *Archaeopteryx* and the coelacanth *Undina* came from the Solnhofen Limestone. In the latter specimen, babies were preserved inside the mother. Perhaps one day a Devonian tetrapod may be found that also preserves babies inside an adult's body.

Limbs, Digits, and Their Development

Limbs with their fingers and toes and the act of walking are among the most conspicuous attributes of tetrapods, and the combination has given the group its name. In the second part of this chapter, some of the similarities and differences between fins and limbs were explored; the next part examines what is known about how some of those differences came about by using evidence from fossils and living animals, integrating paleontology with embryology and developmental genetics.

There are basic similarities between the fins of a lobe-finned fish and the limbs of a tetrapod in the means of attachment to the body by a single element (bone or cartilage), and in having a pair of further elements at-

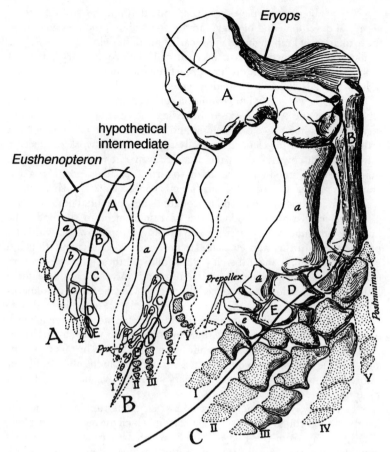

Figure 6.19. Diagram of the fin skeleton of Eusthenopteron, *the pectoral limb of* Eryops, *and a proposed "hypothetical ancestor" between them. From Gregory and Raven (1941). The heavy line denotes the proposed course of the metapterygial axis. Letters on the individual elements denote equivalent (homologous) structures. Roman numerals denote the digits. Note that* Eryops *has been reconstructed with the canonical five digits in its hand, plus theoretical elements called the prepollex and postminimus, whereas more recent work suggests that it had only four digits, as in all temnospondyls where the hand is known in detail.*

tached to that. However, further down the limb and the fin, the differences become more and more obvious, so that it is difficult to see the relationship between the two. In the past, there have been various conflicting ideas put forward to try to reconcile them (Fig. 6.18).

The fin of a lungfish or an osteolepiform is built with a main central axis from which elements branch off. This axis is called the metapterygial axis, meaning "middle of the fin." In the decades from about the 1870s through to the 1950s, people attempted to discover where the equivalent of the metapterygial axis ran in the tetrapod limb (Gegenbaur 1878; Gregory and Raven 1941; Gregory et al. 1923; Jarvik 1980; Steiner 1934; Watson 1913; Westoll 1943b). Because one evolved from the other, the axis should in theory be still recognizable. Because the axis runs more or less centrally in the fin, most people looked for it more or less in the middle of the limb, so that this hypothetical structure would run out either along the middle digit or between two digits, with the digits branching off it in a pyramidal form. It was also thought that the early limb was equipped with a set number of elements or fundamental parts, called a canonical set, that ought to equate with a similar number of elements in the fish fin. Various hypothetical frameworks were derived to try to match them, but none of them completely succeeded, partly because it depended on which fish fin and which tetrapod limb was used as a model (Figs. 6.18, 6.19).

The problems were never resolved, and attention shifted to the techniques of developmental biology; interest focused on processes governing the formation of limb elements in the early limb buds of living forms. These studies were not particularly concerned with the evolutionary origins of limbs from fins but in how the limb grew from the embryo to the adult and in how the cells "knew" where they were, what to do, and when to do it.

Problems with early limb evolution surfaced again in the mid-1980s, when Shubin and Alberch (1986) looked again at the very early development of limbs in a range of amniotes and amphibians to find out in what order and pattern the elements actually appeared. This work produced a suite of key discoveries about the order of events. It was already known that in the early limb bud, bones begin to form as cartilage precursors that appear by aggregation of cells from the embryonic cell matrix. This process is called focal condensation, and it produces a new potential limb element. The first condensation produces the nearest element to the body, the humerus or femur. A little later, the end of this element grows outward and produces two further new elements, in processes called segmentation and bifurcation. Shubin and Alberch (1986) followed these processes in very early embryos from a range of modern tetrapods. After the initial segmentation and bifurcation has produced the elements for the femur or humerus and the next two elements, the process is repeated once more, but only one of the two elements bifurcates further. The side that stops producing new elements forms the radius or the tibia and one or two more small, dependent elements. The side that continues forms the carpus or tarsus on one side, and the ulna or fibula on the other. Springing from the ulnar or fibular bifurcation, a series of further elements arise that segment but do not bifurcate (Fig. 6.20).

Essentially, what results is a chain of elements running around the hand or foot through the wrist or ankle, from which the digits spring along the side away from the body. This chain can be envisaged as the old metapterygial axis, which runs not along or between digits, but forward, bending around the wrist or ankle. The last digit to form is the thumb or big toe. There is some variation among the range of tetrapods studied and between forelimbs and hindlimbs, but the overall pattern is similar in all. The digits lie in a position equivalent to the postaxial radials of the fish fin, whereas most bones of the carpus or tarsus are preaxial. Very few lobe-finned vertebrates form postaxial radials, so the appearance of digits is an unusual development.

It appears that these processes of focal condensation, segmentation, and bifurcation are the same in both fish fins and tetrapod limbs and can explain the similarities in those elements closest to the body in both. Further out, more differences are seen—not only between fishes and tetrapods, but among lobe-fins generally. The more distal parts of the limb or fin are those that form later in growth of the animal, and they may have reached their specialized condition most recently in evolution (see Fig. 2.13).

When the multidigited limbs of Devonian tetrapods were discovered, old ideas about the metapterygial axis and the canonical set of elements could not explain them. However, the more recent discoveries describing the curved axis could accommodate them easily. All that has to happen is that the axis keeps on segmenting and bifurcating longer than it now does in tetrapods (Fig. 6.20). In theory, any number of digits is possible. What is less easy to determine is whether there is any strict correspondence between any one of the digits of the polydactylous tetrapods and any of the digits of a pentadactyl animal. It may be that in the early stages of digit

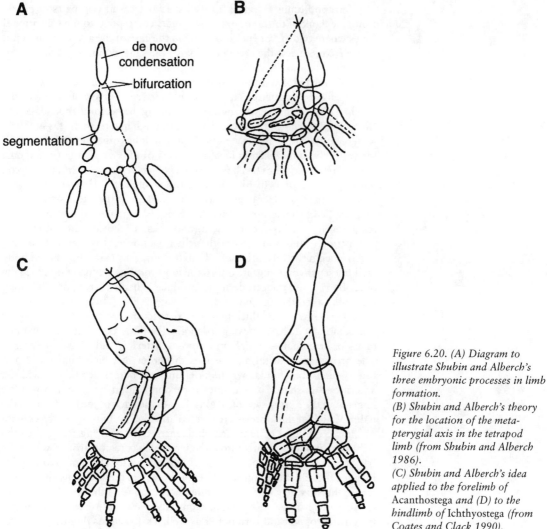

Figure 6.20. (A) Diagram to illustrate Shubin and Alberch's three embryonic processes in limb formation.
(B) Shubin and Alberch's theory for the location of the metapterygial axis in the tetrapod limb (from Shubin and Alberch 1986).
(C) Shubin and Alberch's idea applied to the forelimb of Acanthostega *and (D) to the hindlimb of* Ichthyostega *(from Coates and Clack 1990).*

evolution, genetic instructions were simply issued to "make digits," rather than to "make particular digits." This will be discussed further below.

In recent years, the genetic and developmental control of limb formation (and fin formation by extension) has been the subject of an enormous growth of study, interest, and concomitant understanding. Some of the developmental findings reflect those that are seen in the fossil record, as well as generally among living vertebrates. More details can be found in recent reviews such as those by Shubin et al. (1997) and Capdevila and Belmonte (2000) or in textbooks of developmental biology such as that by Gilbert (2000). Although only a few modern animals have been studied in detail, the conclusions that can be drawn have wide implications for all vertebrates. The interplay of *Hox* genes, morphogens, and transcription factors (proteins that moderate what cells do) (see Chapter 2) are now beginning to throw light on the evolution of tetrapod limbs.

Tetrapod limbs begin to form in special areas at two pairs of points along the flank of the embryo. These points correspond to the transitional zones where two *Hox* genes interact—in the case of the amniote forelimb, where *Hoxc6* meets the action of *Hoxc8*, the same point at which cervical vertebrae are modified into thoracic ones. These points correspond to different vertebral numbers in mammals and birds, but the *Hox* genes influencing the formation are the same. *Hoxc6* also specifies where the brachial plexus (the nerve ganglion controlling the pectoral appendage; see Chapter 10) forms in the zebra fish, so although in tetrapods it is involved with regionalization in the vertebral column, it has other functions as well. Interestingly, some teleosts have more sets of *Hox* genes than tetrapods. For example, the zebra fish has two copies each of *Hoxa, Hoxb,* and *Hoxc,* but only one *Hoxd*. In forming a tetrapod limb, the *Hox* genes are expressed in three distinct phases, with the third phase the one in which digits are specified. The action of the *Hox* genes is regulated in tetrapods by a morphogen such as retinoic acid, and these chemicals can initiate the production of limbs at places other than the normal two pairs of points.

To begin the formation of a limb (or fin), cells from the general body wall (mesenchyme) migrate out to the limb-forming region and induce the outer body layer, the ectoderm, to produce a distinct string of cells along the flank called the apical ectodermal ridge (AER). This provides a major signaling center from which messages pass to and fro to the limb-forming cells, among which are messages defining the three axes that the limb must take up: proximal → distal; anterior → posterior; and dorsal → ventral. The proximal → distal axis defines the order in which humerus/femur, radius–ulna/tibia–fibula, wrist/ankle and fingers/toes are laid down; the anterior → posterior axis defines for example that from thumb to little finger; and the dorsal → ventral axis defines the back of the hand versus the palm. Which of these bones are formed at which point is under the control of the second level of *Hox* gene influence. Cells growing internal to the AER push it outward, first into a small pimplelike region around which the AER runs horizontally, and eventually elongating it as the limb grows. This is called the limb bud. Figure 6.21 explains where these regions appear and where the *Hox* genes have their influence. It appears that the factors controlling axis formation are similar not only among all vertebrates, but between arthropods and vertebrates as well. This suggests that they were established long ago in the history of evolution, in the common ancestor that gave rise to both vertebrates and arthropods. It represents a very deep level of similarity, or homology.

In most vertebrate embryos, forelimbs begin to develop before hindlimbs. Here is a reflection not only of what is seen in the fins of fish but of what is seen in the evolution of fins in general among vertebrates—fore fins are usually larger than hind fins in tetrapodomorph fish, and fore fins are thought to have evolved before hind in early vertebrates (Coates and Cohn 1999).

In the proximal parts of the limb, similar mechanisms of *Hox* gene control act in both fish and tetrapods. One of the critical regions in initiating limb development sits just at the back of the limb bud, where it joins the body wall. It is called the zone of polarizing activity, or ZPA. Here the anterior–posterior axis is defined. What initiates it seems to be the action of a gene called *sonic hedgehog,* the cutely named equivalent of the *hedgehog* gene in the fruit fly *Drosophila* (geneticists have as much fun naming genes as zoologists do naming animals). This gene, under the influence of *Hoxb8,* establishes the ZPA and the anterior–posterior axis in fish fins and tetrapod limbs.

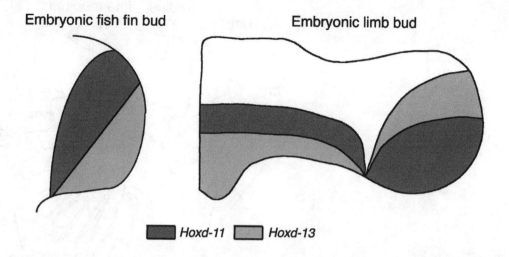

Toward the more distal parts of the limb, controlling the formation of fingers and toes, some *Hox* genes come into play in ways that have no equivalent function among fishes. These are the distally operating sequences of *Hoxa* and *Hoxd*, the groups (paralogues) known as 9–13 in each gene. It is not yet clear how these genes specify the different components of the digits, but it is clear that this region of the tetrapod limb is at least in some respects a unique development and novel compared with the fin of a ray-finned fish. One of the suggestive features is that *Hox* groups 12 and 13 at the ends of *Hoxa* and *Hoxd* show a neat parallel with what is known of the fossil record. These groups, specifying the digits, do not operate in a simple proximal-to-distal way. Rather, their influence follows a path that sweeps from posterior to anterior across the limb bud. It recalls the digital arch—indeed, the metapterygial axis—as suggested by the embryological studies of Shubin and Alberch (1986) (Fig. 6.20). Furthermore, rather than acting in the normal way with groups 9–13 coming into action in sequence, the action of *Hoxd* is reversed here, with 13 coming into action first, then on through to 9. This again suggests that this most terminal part of the tetrapod limb is a novel development, at least insofar as the genes specify a bend along the axis and the identity of particular digits. If so, the appearance of this complex can be dated with some confidence to the Late Devonian.

No one has yet studied the fin development of lungfish, but in some lungfish genera at least, there are parts of the fin that are comparable to the digits of the tetrapod limb—the postaxial radials of *Neoceratodus* are an example. Work is just beginning on the genetic control of development of *Neoceratodus*, now that the Australian research group has been successful in breeding them (Joss and Joss 1995). Understanding the genetic control of limb development in this lobe-fin will provide invaluable clues to the evolution of tetrapod limbs. For example, an interesting question would be, "What specifies the appearance of postaxial radials along a straight axis versus the appearance of postaxial digits along a curved one?" Regrettably, it is unlikely in the foreseeable future that people will be able to breed the other kind of living lobe-fin that would be equally valuable—the coelacanth *Latimeria*.

Figure 6.21. Diagram of Hox *gene domains in the embryo fish fin and tetrapod limb bud. In the fish,* Hoxd13 *is expressed distally and posteriorly to* Hoxd11 *in the embryonic fin, and the fin itself never elongates. In the tetrapod limb bud, in the proximal end, the domains of* Hoxd11 *and* Hoxd13 *are stretched along the posterior margin of the fin in much the same relationship as they are found in the fish fin bud. However, distally, at the end of the limb where the digits form, there seems to have been a "twist" so that not only are their positions reversed, but their expression domains curl around the end of the limb bud. It is possible that this may be responsible for the way the digits are formed around the "digital arch" in the embryonic limb bud. From Gilbert (2000).*

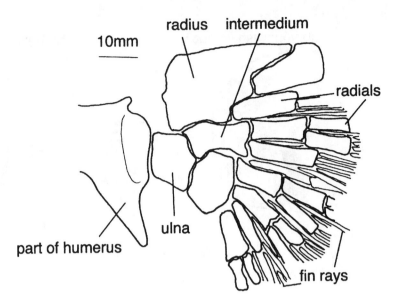

Figure 6.22. Fin skeleton of Sauripteris.

In the rhizodont *Sauripteris*, the arrangement and number of radials is almost equally as complex as the digital arch of a tetrapod (Fig. 6.22). Given the fact that phylogeny suggests this to be a convergent development in the two groups (Daeschler and Shubin 1997), a complex genetic control mechanism can also be inferred to have evolved convergently, although that can never be tested.

By comparing the occurrence of digits in the fossil record with the time at which vertebral specification occurs, it appears that the new regions of *Hox* genes *a* and *d* acquired their digit-forming functions before the precise operation of *Hoxc6* and *Hoxc8* came mutually to specify the cervical–thoracic boundary as found in amniotes. Although in the ray-finned fish *Danio Hoxc6* defines a similar boundary, there does not seem to be an equivalent region in early tetrapods or amphibians, or it occurs in a different place, judging from the position and lack of definition of the cervical–thoracic boundary in these animals. It has also been suggested that *Hoxa* and *Hoxd* came to operate as digit-invoking genes in two separate phases. *Hoxa* might have controlled the appearance of digits, and *Hoxd* came into action later to specify a pentadactyl limb (Shubin et al. 1997). Such a suggestion would fit with the idea that the digits of an early tetrapod such as *Acanthostega* cannot be matched one for one with those of a pentadactyl tetrapod (Coates and Cohn 1998).

One early suggestion in the quest for understanding digit evolution was based on the idea that there were five groups of digit-producing genes on *Hoxd*, and there were five digits: one group per digit might be in operation. This suggestion carried the implication that for a polydactylous animal, there could only be five *types* of digit, and if there were more present, then at least some must be duplicates (Shubin et al. 1997; Tabin 1992). The idea is refuted by *Acanthostega*, in which of the eight digits, none is a duplicate in either size or configuration of any of the others. It remains a puzzle why so few animals have evolved polydactyly as a normal body form, and why, if extra digits are needed by a modern animal, they are almost always formed from coopted material derived from elsewhere (see Gould's [1980] essay "The Panda's Thumb"). (For example, most frogs have only four fingers on the hand, but some, which use their hands for digging, "invent" an extra digit from wrist bones.)

To make the individual digits, not only positions but individual differences among digits must be established, and it is still not clear how this is done, although it does involve complex interplay between the *sonic hedgehog* gene in the ZPA, the action of fibroblast growth factors in the AER, and morphogens. However, it is known that digit formation can only happen by a process of orderly cell death (apoptosis), with cells that lie in the limb bud between where digits form being sacrificed to make the spaces between the digits.

Paleontology may not only be able to pose but also to answer some questions about the development and evolution of limbs. It has been suggested that forelimbs evolved before hindlimbs (Coates and Cohn 1998, 1999), although throughout tetrapod evolution, in nature (as opposed to experiment), changes that happen to one set tend to happen to the other. For example, the reduction in the number of toes in horses and the increase in the number of phalanges in ichthyosaurs (marine dolphinlike amniotes of the Jurassic and Cretaceous periods) happen to both sets. This, however, is not an invariable rule, and functional necessity dictates that birds and bats have only one set of paired limbs modified into wings. The number of toes in early tetrapods may test the hypothesis of similarity between fore- and hindlimbs. So far, it appears that there were the same number of digits on the fore- and hindlimbs of *Acanthostega* and *Tulerpeton*. However, the hand of *Ichthyostega* is unknown. The hypothesis of similarity predicts that it will have seven digits on the forelimb. At the same time, its complement of toes on the hind foot is so unusual that it is hard to see how it could be combined with the format of the forelimb, which is so distinct from the hind in structure, proportions and presumably, function. The most recent developmental work has, by contrast with the hypothesis of similarity, suggested that although fore- and hindlimbs are specified by similar transcription factors, these factors, although closely related, are not identical (Coates and Cohn 1998).

The hypothesis of similarity also predicts that digits will appear on fore- and hindlimbs at the same time, so for example, no animal would have had, say, digits on the forelimb while still retaining fin rays on the hindlimb. Fossils would theoretically be able to refute this hypothesis. Because of the apparent evolution of forelimbs before hindlimbs, a corollary to this idea would be that it would be still more unlikely for an animal ever to have retained fin rays on the forelimb while having acquired digits on the hindlimb. Something that appears to have happened in the evolution of limbs is the suppression of the dermal fin rays found in the lobe-fins and other fishes. No limb has yet been found with both fin rays and digits, although *Acanthostega* and *Ichthyostega* retain fin rays in the tail. The rhizodont *Sauripteris* retains fin rays at the same time as producing an enlarged endoskeletal fin skeleton, so the appearance of both together might not be impossible in a very early tetrapod.

Fossils can also refute hypotheses of "forbidden morphologies." That is, sometimes it appears that the way some developmental pathways work precludes the existence of certain morphologies. One such applies to the configuration of digits on the hand or foot. A theoretically forbidden morphology is to have the shortest digit in the middle of the set, rather than at one end or the other (Holder 1983; Thomson 1988). The hind foot of *Ichthyostega,* with its tiny digit as number three from the front, refutes this idea. However, it is also refuted by some living animals—some seals have small digits at the center of the hind flipper, with long ones forming each

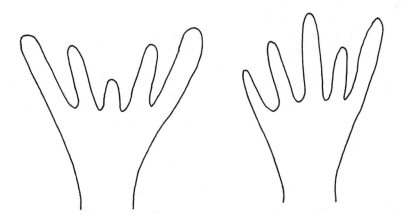

Figure 6.23. Theoretically forbidden morphologies according to earlier hypotheses. Compare with the hind foot of Ichthyostega (Fig. 5.13) and of the modern elephant seal (Fig. 5.15).

edge (Fig. 6.23). Part of the problem of posing such theoretical ideas is that they are based on the study of such a limited sample of experimental animals. Zebra fish, mice, chickens, and the frog *Xenopus* are the most commonly studied laboratory animals in developmental biology, and these animals all belong to species that are highly specialized members of their vertebrate groups. This means they may not always be representative of tetrapods as a whole, and they certainly cannot account for many of the extreme specializations that some tetrapods show.

The variable number of digits seen among the Devonian forms does fit another observation of developmental biology. Early in the evolution of a number of systems, variation seems to be the rule, followed by stabilization only later, as the genetic mechanism controlling it tightens up. This is seen in the number of segments found in arthropod bodies, for example, or the number of gill slits in the earliest or most primitive vertebrates, or the number of vertebrae in a tetrapod group. Stabilization may be forced by some functional necessity. In frogs, birds, and mammals, for example, the number of vertebrae is very strictly controlled, in response to the needs of their respective locomotory requirements. This may be true also of insect bodies. *Hox* genes may well be implicated in this tightening-up process.

Although it is still unknown why the tetrapod limb stabilized at five digits, two factors may play a part. The limb bud is a physically restricted area, so it may only be possible to make either larger numbers of smaller digits or a smaller number of larger digits, in combination with the material needed to make wrists and ankles. The amount of material available in relation to the demands of walking may predicate a smaller number of larger digits if a functional wrist or ankle is to be formed. Alternatively, perhaps five digits is the number that fits not only most economically but also most stably round the wrist or ankle for taking the weight of the animal.

Until it is clearer whether the configuration of pentadactyly evolved once or more often, the answer to the question of "why five" will almost certainly remain elusive. Some suggestions have placed the six-digited *Tulerpeton* on the stem lineage of amniotes (Coates 1996; Lebedev and Coates 1995), and this suggestion implies at least a double origin for the pentadactyl hindlimb. These same analyses suggested that an "amphibian" lineage only ever had a four-digit manus, implying that the four- versus five-digit manus are alternative, mutually exclusive configurations. While early

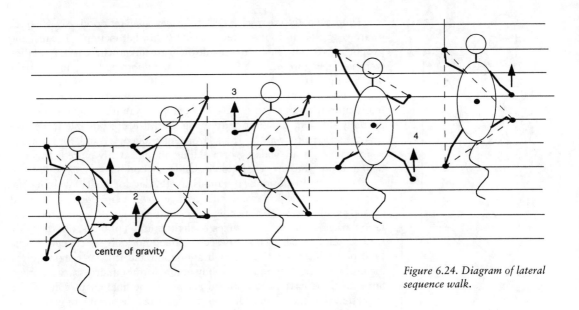

Figure 6.24. Diagram of lateral sequence walk.

tetrapod phylogeny remains controversial and uncertain, such questions as these must likewise remain the subject of debate.

Locomotion

During the fish–tetrapod transition, along with the changes to the bone and muscle structure, the mode of limb movement must also have changed. A fish swims by moving its body in a sinusoidal wave that passes down the body and pushes against the water. The fins, attached to the sides of the body, are generally moved in coordination with the bending so that diagonally opposite pairs of fins move more or less together. A few fish that "walk" on the bottom of the sea use this kind of movement to perform something like a walking trot (Pridmore 1995). Early tetrapods that moved primarily in water would probably have used this kind of movement to begin with, and it would probably require little if any rewiring of the neural pathways of the brain and limbs to control it. Modern salamanders use this movement in water today.

There is a debate over which method of locomotion would have been used by early tetrapods moving out of the water (Edwards 1989). If the body were to be lifted off the ground, the animal would have needed three legs on the ground at any one time if it were not to fall over. To move one leg at a time in an orderly fashion is called the lateral sequence walk (Fig. 6.24). This may have required considerably more complex control by the brain than that produced by simple axial bending to move opposite corners at the same time, and so may not have been achieved immediately (Pridmore 1995). Early tetrapods may therefore have been belly crawlers on land, at first using limbs only for purchase against the ground and for steering and braking, much as fish fins are used. Limbs used as support structures probably evolved only later. When moving quickly on land, modern salamanders first change from the lateral sequence walk to the walking trot, but eventually abandon use of limbs and move by wriggling the body. Some have suggested that the walking trot may have been used by early tetrapods with substantial tails, such as *Acanthostega*, in which the tail could have functioned as a fifth leg.

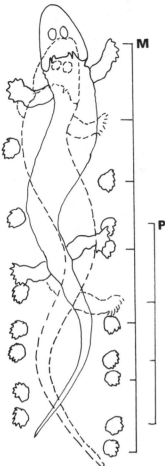

Figure 6.25. Diagram to show how an Acanthostega-like tetrapod might produce tracks like those found at the Genoa River.

Unfortunately, the trackways left by Devonian tetrapods (see Chapter 4) are very difficult to interpret (Clack 1997a). For example, because the proportions of the animals that made them are unknown, so is the gait they were using. Figure 6.25 shows how an *Acanthostega*-like animal might have produced tracks like those of the Genoa River sequence if its body was partly supported by water. However, it is not clear whether the tracks were made entirely on the land surface or whether the animals might have been partially submerged. Without knowing that, it is not possible to be sure that the apparently walking tracks from Australia or Ireland, for example, were made by truly walking or paddling animals. Only more fossils will help interpret the early evolution of land locomotion.

During the evolution of tetrapods and the development of walking, the emphasis in muscle power and bulk changed from preponderance of axial or body wall muscles to appendicular or limb muscles. (People tend to eat limb muscles of mammals and leg or flight muscles of birds, but the body wall muscles of fish.) This was accompanied not only by changes to the limb bones but also to the vertebral column and ribs, as they acquired means by which to support the body and constrain unnecessary twisting. Some of these have been outlined above. In the final chapter, it will be suggested that these same changes also affected the means by which the animals breathed in air and by which they fed on land, with domino effects to hearing resulting. Because the animals must have continued to work as a coordinated whole, each of the morphological changes must have occurred in concert with the others. The suggestion here is that they began with breathing, which affected feeding, which in turn affected locomotion and hearing.

Summary

Skeletal characters that changed over the transition can be summarized as follows (some are summarized in Fig. 6.1):

From *Eusthenopteron* to *Panderichthys*:
Losses: median fins except for the tail; mosaic of frontal series of bones; cranial mobility across skull roof; reduction in relative size of opercular region
Gains: lengthening of snout; enlargement of eyes and change of location to lie more dorsally

From *Panderichthys* to *Acanthostega*:
Losses: opercular series, supracleithral series (except anocleithrum); dermal fin rays on paired appendages, dorsal scales; mosaic of nasal bones (except for median rostrals and anterior tectals); reduction of cleithrum
Gains: consolidated braincase with suture between anterior and posterior units; fenestra vestibuli enclosing head of stapes; large, bulbous basipterygoid processes; ribs attaching to neural arch and centrum; limbs with digits; enlarged scapulocoracoid; large dermal interclavicle; large pelvic girdle, sacral rib; enlarged femur, tibia, and fibula, mosaic of ankle bones

From *Acanthostega* to *Greererpeton* (see Chapter 10 for further details):
Losses: dermal fin rays from tail; reduction in number of digits per limb to no more than five; lateral lines mainly in tubes opening by single pores
Gains: occipital joint; elaborated zygapophyses; further elongation of ribs; radius/ulna and tibia/fibula form weight-bearing joint

Seven
Emerging into the Carboniferous: The First Phase

The Carboniferous World

It was the Carboniferous period that really saw the advent of tetrapods onto the land and their radiation into a wealth of body forms, niches, and families, among them fully terrestrial forms that were the forerunners of modern amniotes. Regrettably, the early parts of the period have a very poor fossil record, not only for tetrapods but also for plants and invertebrates. By the time tetrapod fossils become at all common, during the later parts of the Carboniferous, they had already acquired many of the characteristics that mark them as terrestrial tetrapods, which makes understanding this sequence of events very difficult. To place the tetrapods in context, it is necessary to survey what is known of their contemporaries in the plant and invertebrate world, because these paved the way for the tetrapods to follow.

The Carboniferous period is usually divided into two phases, called the Early and Late Carboniferous in Europe and the Mississippian and Pennsylvanian in North America. Although the boundaries between these divisions have not been considered equivalent on each side of the Atlantic, this may be an effect of incomplete stratigraphic correlation. In turn, each of the two parts of the Carboniferous is divided again into stages, with the Early Carboniferous consisting of an earlier Tournaisian and a later Viséan division. Among vertebrate paleontologists, the names widely used for the

		STAGE	LOCALITY	ANIMALS
Late Carboniferous	PENNSYLVANIAN / Namurian A	Aportian		
		Chokerian		
		Amsbergian	Greer	colosteid, anthracosaur
		Pendleian	Burghlee Loanhead Cowdenbeath	*Eoherpeton, Proterogyrinus, Crassigyrinus*
Early Carboniferous	MISSISSIPPIAN / Viséan	Brigantian	Gilmerton East Kirkton	*Crassigyrinus, Loxomma*, colosteid anthracosaurs, temnospondyls, microsaurs, *Eucritta, Westlothiana*
		Asbian	Delta Burdiehouse, Pitcorthie Broxburn Pumpherston Cheese Bay	*Whatcheeria*, microsaur, colosteid adelogyrinids, colosteid *Casineria*
		Holkerian	Wardie Middle Paddock	*Lethiscus* (aistopod) Australia: tetrapod skull and limb fragments
		Arundian		
		Chadian		
	Tournaisian	Coureyan	Dumbarton Horton Bluff	Undescribed articulated skeleton humeri, pectoral girdle fragments
			Devonian-Carboniferous boundary	

Late Carboniferous are Namurian for the earliest part, Westphalian for the middle, and Stephanian for the latest. Because of more detailed work by sedimentologists and stratigraphers, these names are no longer universally accepted because they are imprecise. For example, the earliest part of the Namurian is now considered to be part of the Early Carboniferous, and parts of the Stephanian may fall within the Permian. The older names are nonetheless convenient to use for the study of tetrapods because they seem by and large to coincide with changes to the faunas. As far as plants are concerned, the Namurian stage flora shows more similarities to the Early Carboniferous than to the Westphalian and Stephanian. Figure 7.1 sets out a simplified stratigraphical table of the Early Carboniferous and indicates the localities at which some of the animals described in this chapter are found.

Biogeography and Paleoecology: Coal

Much of the Carboniferous was a time of mountain-building activity. During this period, a huge chain of mountains was progressively pushed up, and the time was marked by periods of increased volcanic activity. The mountains stretched all across Laurussia, dividing it into two, and the mountain range is possibly responsible for somewhat different, although comparable, faunas and floras in the two parts. These mountains, as they exist today, stretch all along the eastern United States—the Appalachians—across Europe and into North Africa—the Atlas and Alpine Mountains—and on into Asia, forming the early stages of the Himalayas. This sequence of Carboniferous earth movements is called the Allegheny or Variscan orogeny. Throughout the Early Carboniferous, Gondwana was drifting southward away from the equator, and the regions now making up Siberia, Kazakhstan, and North China were isolated. No tetrapod remains from this period have been found there, although it is not necessarily because there were none living there at the time. They may simply not have been found yet.

Throughout the Carboniferous, the climate became gradually more humid and warmer, at least in equatorial and tropical latitudes. During the Early Carboniferous in the United Kingdom, for example, at least the early and mid-Viséan is thought to have been seasonal with periods of severe water stress for plants and animals (Falcon-Lang 1999). This can be read in sequences of sediments and also in the type of growth rings found in some of the larger plants. Glaciation still affected the southern continents, and the deposits from localities in their middle latitudes from this time show an impoverished flora and fauna. However, all across the lowlands around the northern shores of Iapetus, an increasingly dense and speciose forest flourished, stretching across Laurussia from what is now the central United States to the Urals in Russia. This forest has left its legacy in the form of coal deposits, and has also provided almost the entire record of Carboniferous tetrapods (Milner 1993a).

The trees and other plants found in the Late Carboniferous forest flourished because of the warm, humid climate that the whole of the equatorial region experienced at that time. They show no growth rings in their wood (DiMichele and Hook 1992). In the Carboniferous swamp forest, there were no seasons to affect plant growth in this way. Trees flourished, and when individuals died, their remains fell into the swamp water made anoxic by all the fallen vegetation. So instead of decaying totally, as they do in modern rainforests, they accumulated first into peat, which was then turned into coal.

Throughout the Carboniferous, changes to the gas composition of the air continued the trends begun in the mid-Devonian. The changes appear to

Figure 7.1. (Opposite) Stratigraphic table of the earlier parts of the Carboniferous with localities and their faunas indicated.

coincide with the vast expansion of the coal forest vegetation (Graham et al. 1997). A large proportion of the carbon material that was used by the plants was apparently not recycled but became locked in swamp deposits as peat and later fixed as coal. The result was that by the end of the Carboniferous, CO_2 levels had dropped to about the same levels as today—that is, about 0.03%—whereas oxygen content had risen to about 30% (the modern level is about 21%) (see Figure 4.1). The effects must have influenced in their turn the breathing and gas exchange capabilities of all animals, not just air-breathing ones, because the atmospheric gas content must also have had a rebound effect on the water. It is notable that the current increase in CO_2 in the atmosphere results in part from the release of this ancient source of stored carbon from 300 million years ago, as fossil fuel is burned. Why this carbon was stored in such quantities during the Carboniferous, rather than being recycled as decayed matter as it would be today, remains unclear.

The Carboniferous period gets its name from the fact that most of the world's coal deposits derive from rocks of this age. Vast quantities of carbon are tied up in the remains of plants that lived during this time interval and that became the source of so much of the present-day world's fuel. A look at how, where, and why these coal deposits arose helps us understand what happened to the tetrapods of the time, which in turn helps us understand the Carboniferous world in general.

One striking phenomenon seen in the formation of coal deposits is a regular repetition or cyclicity of layers. Like the Devonian red-beds described in Chapter 4, this may partly be a result of Milankovich cycles. It also seems to correlate with periodically increased glaciation of the southern continents, which in turn affected the global sea level.

The trees in the Carboniferous grew in a clayey, often iron-rich soil, and with their root remains, this layer is called an underclay, or seat-earth. This layer always lies under the coal seam, which is formed by the decay of layer upon layer of plant material. Although the land surface is exposed, the lowlands and marshy regions can support such extensive plant growth. Gradually, however, as the sea level rises as a result of the melting of the ice cap, the sea floods the land surface and the forest dies. This may happen in a fairly restricted local region, or it may be a more extensive, large-scale event. Finally, however, the coal forest sediments are covered by marine deposits, first of shale from shallow waters and then of limestone from more fully marine conditions. Then as the glaciation resumes, the sea levels fall again, and the sediments become less marine in character, with nonmarine limestones and shales eventually giving way to terrestrially deposited sandstone. Eventually the land surfaces are recolonized by plants, and the whole sequence starts again. The series of deposits so formed are called cyclothems and are characteristic of coal forest deposits; some localities show up to 100 of them in a sequence.

Until recently, it was assumed that the coal swamps formed exclusively in freshwaters, partly because among modern trees, very few are well adapted to tolerate salt water. However, more recent investigations of some coal deposits have revealed tidal influences and faunas characteristic of marginal marine conditions (Schultze and Maples 1992; Schultze et al. 1994; Schultze 1995). This is a similar problem to that of interpreting Devonian paleoenvironments explored in Chapter 4.

Early Carboniferous Plants and Invertebrates

Many great changes occurred to the flora between the Late Devonian and the later parts of the Early Carboniferous. During this period, *Archae-*

opteris and *Rhacophyton* became extinct, and during the Tournaisian stage, most of the forest cover of the Devonian disappeared. For about 5 million years in the Tournaisian, the flora consisted mainly of low, shrubby pteridosperms only about 2 m high.

New kinds of plants evolved during this period, including true seed plants, the gymnosperms related to today's conifers. Although swamps were still dominated by the lycopsids, these plants were essentially conservative. New species colonized coastal swamps and lagoons, and *Calamites*, related to today's *Equisetum* or horsetails, grew along stream margins, and ferns and pteridosperms grew in better-drained places (DiMichele and Hook 1992). Much information about the Early Carboniferous floras has come from localities preserved under volcanic sediments (e.g., Rolfe et al. 1990, 1994), and Chapter 8 shows how this kind of preservation has thrown light on the early evolution of tetrapods. Other sources of information include that from fossilized charcoal (known as fusain), which becomes much more common in the fossil record from the Early Carboniferous. Which kinds of plant are preserved as charcoal in contrast to those which are not can give quite detailed information about plant ecology and climate. It appears that the Early Carboniferous in Laurussia was monsoonal and that fires were a regular occurrence; in some instances, even the regularity of forest fires can be estimated, suggesting periods of between 105 and 1085 years between events (Falcon-Lang 2001).

The structure of Viséan stage forests was much like that of the preceding Tournaisian, although the constituent species were different. By this time, the forests showed a complex mosaic of habitats and communities, as they do today, with some plants adapted for wetlands, some for better-drained places, and some for uplands. However, in the Carboniferous, ecology and taxonomy were linked, with related groups of plants specializing in some habitats and other groups in other habitats. Nothing like this separation is seen today (DiMichele and Hook 1992). In modern floras, the same plant group has many different species in different habitats.

By the end of the Namurian, there is evidence of more distinct and varied interactions between plants and animals. Truly herbivorous arthropods had evolved, and winged insects flew in the trees. Plants evolved devices to combat herbivorous arthropods, such as hard coatings to their seeds, and the arthropods, mainly insects, would have evolved ways to get around them, such as stronger mouthparts. However, the vertebrates would have had little impact on the plants until toward the end of the Carboniferous. No vertebrate evolved the ability to deal with plant food until the Stephanian (DiMichele and Hook 1992).

Tetrapods of the Early Carboniferous: Experimenting with Shape and Form

Among the forests, rivers threaded their way, forming pools and channels in which the water-dwelling tetrapods and fishes lived. Sometimes the channels became blocked off from the main stream, as in oxbow lakes today, and these blocked channel deposits are often those that produce the best record of fossil tetrapods (Hook and Ferm 1988). Other localities result from shallow-water pools, and yet others preserve animals from deeper water environments.

In any discussion of Carboniferous tetrapods, something to bear in mind is that almost all the information comes from a very few, but very exceptional, sites representing such coal-swamp pools and channels. Almost nothing is known of more upland environments, and only a little is known of the more terrestrially adapted tetrapods of the time. Many of the fossils of coal-swamp tetrapods derive from coal deposits worked in deep mines in the 19th century by coal miners who dug coal by hand and noticed fossils as they dug. Many of the best are from such old, deep mines, no longer available to paleontologists, such as those of the Low Main Seam mined at Newsham in Northumberland, England, and those of the Scottish localities of Gilmerton, Loanhead, and Borough Lee. Today, most coal fossils go unnoticed because coal is dug out by machines, unless fossils are spotted by enthusiasts searching open-cast sites. Some of the more recent discoveries of coal fossils have been of this sort, including the Namurian site of the Dora bone bed in Cowdenbeath, Scotland (Andrews et al. 1977), and a single specimen from the Coal Measures of Wigan, Lancashire, England (Milner and Lindsay 1998).

After the end of the Devonian, the fossil record of tetrapods becomes very sparse, and until very recently, only a few fragments of limb bones had been found from rocks representing the succeeding 30 million years. The significance of this gap in the record was noted during the early 1950s by paleontologist A. S. Romer, who pointed out that much of the evolution of terrestriality probably occurred during this period for which there is almost no information. It has been called "Romer's Gap" in his honor (Coates and Clack 1995). The interval covers the whole of the Tournaisian and most of the Viséan. Since Romer's time, the record of tetrapods from the Viséan has been increased by some spectacular finds, so that now, several tetrapod genera represent the top part of this period. Most recently, a new specimen of an articulated tetrapod has been discovered and dated as Late Tournaisian, falling just in the middle of Romer's Gap. Perhaps, as with the Devonian tetrapods in the early 1990s, the logjam is beginning to break and more and more Tournaisian and early Viséan tetrapods will soon be discovered.

The interrelationships of early tetrapods constitute a problem that for the time being is unsatisfactorily resolved. There seems to be no obvious connection between the tetrapods of the Late Devonian and those of the Early Carboniferous. The Late Devonian forms share a suite of features that are at present regarded as primitive for all tetrapods, with the only possible exception to this being *Tulerpeton*. However, once over the Devonian–Carboniferous boundary, a completely different complement of taxa is found. It may be that most of the Devonian forms became extinct at the end of the Devonian, as suggested in Chapter 4. By the time they are found in any numbers in the fossil record, Carboniferous tetrapods had reradiated into a wide range of morphologies and ecologies.

Some of the genera that are coming out of the Early Carboniferous challenge the present understanding of tetrapod evolution in that they do not fit readily into any scheme of classification based on the forms appearing later during the Late Carboniferous. They hint at a diversity of form and function among these earliest tetrapods representing ecologies and lifestyles no longer found on Earth today. For many of them, the part they played in the evolution of tetrapods can only be guessed at. How they are related to each other or to more recent forms, or even whether or not they were descended from more terrestrial ancestors, remains to be discovered.

When the tetrapods that lived throughout the early parts of the Carboniferous period, the Tournaisian, Viséan, and early part of the Namurian are compared, it seems there was a different complement of genera from those that lived during the Westphalian. Only one genus (*Loxomma*; see Chapter 9) crosses this boundary, and indeed, there is another noticeable

Figure 7.2. Horton Bluff in Nova Scotia, Canada, from which some Tournaisian tetrapod elements have been recovered. Photograph by J.A.C.

gap in the tetrapod record from the middle part of the Namurian to the early part of the Westphalian (Smithson 1985a). The next section describes these animals roughly in their order of appearance in the fossil record to try to convey some of the strangeness that stretches the imagination in trying to make sense of them.

Horton Bluff Tetrapods

Until recently, the only evidence of tetrapods from the earliest phase of the Carboniferous, the Tournaisian, consisted of a few isolated elements from a locality known as Horton Bluff in Nova Scotia, Canada (Fig. 7.2). These have recently been described for the first time and consist of limb and girdle fragments (Clack and Carroll 2000). At least two kinds of humeri are represented, one resembling that of an animal such as *Tulerpeton* or possibly an early anthracosaur such as *Eoherpeton* (Chapter 9). Another is more like that of a juvenile colosteid. There is a partial shoulder girdle preserving the lower part of the scapulocoracoid, and this somewhat crescent-shaped bone shows one feature in which it resembles that of *Ichthyostega*. Partial interclavicles from Horton Bluff also resemble those of anthracosaurs in being kite-shaped. At a nearby locality, there is also a length of trackway made by a large tetrapod of about the size that the largest of the Horton Bluff specimens would predict (Sargeant 1988). Even by the Tournaisian, a range of tetrapod morphologies was present.

Casineria kiddi

A single small specimen from a coastal locality not far from Edinburgh, Scotland, represents one of the most tantalizing finds in recent years from the Early Carboniferous (Paton et al. 1999). Found on the coast at Cheese Bay, its name is derived in part from the Latin word for "cheese." It consists of a nodule in part and counterpart showing a tetrapod skeleton with well-preserved vertebrae, ribs, and forelimbs, less well preserved hindlimbs (water-worn and only on half of the specimen) and measuring only a few centimeters from neck to pelvis (Fig. 7.3). The vertebral column is solidly ossified and is of the form called gastrocentrous, having large pleurocentra. It has long, slender, curving ribs and oval-shaped gastralia. Most remarkable, however, is the preservation of one of its forelimbs. This shows that

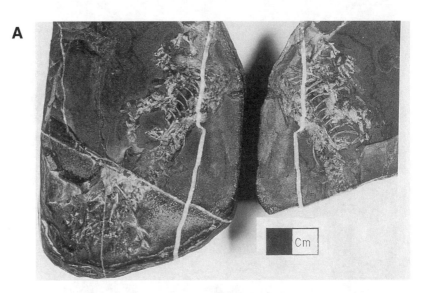

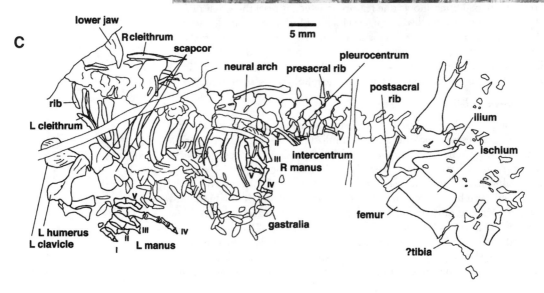

Figure 7.3. (A) Part and counterpart blocks of the specimen of Casineria kiddi. (B) Enlargement of the region around the hand (NMS.G1993.54.1). Photograph by S.M.F. (C) Specimen drawing of Casineria, taken from part and counterpart (from Paton et al. 1999).

the humerus is much more slender than that of any other Early Carboniferous tetrapod, with an obvious shaft and with the two ends set at different angles to each other (known as torsion). The radius and ulna are also slender, with an olecranon process on the ulna (although this is not easily visible in the figures). These features alone strongly suggest a fully terrestrial animal. However, in addition, the manus has five slender digits, with the fourth substantially longer than the others, and on the underside, the individual phalanges show tiny V-shaped grooves that would have housed ligaments to the next phalanx. The last phalanx on each digit (the ungual) is noticeably curved, and the whole arrangement suggests a hand capable of grasping. No other Early Carboniferous tetrapod shows digits like this, but they are similar to those found in Late Carboniferous early amniotes. *Casineria*, however, comes from a deposit dated as early Viséan!

Perhaps the most striking thing about *Casineria*, in view of its date, is its small size. Most Early Carboniferous tetrapods so far known are large, with skull lengths of 100–300 mm, and apart from an undescribed microsaur from East Kirkton (see Chapter 8), *Casineria* is by far the smallest. It has been suggested that the origin of amniotes is connected with an evolutionary step involving small size (see Chapter 10), and here is a specimen that accords well with this theory.

Unfortunately, the head of *Casineria* is not preserved, which makes it difficult to determine its relationships to other tetrapods. In some respects, it resembles the more terrestrial anthracosaurs such as *Gephyrostegus* (see Chapter 9). Despite the lack of a skull, a phylogenetic analysis of a range of early tetrapods that included several amniotes placed *Casineria* among the early amniotes of the Late Carboniferous (Paton et al. 1999). If this is its true position, it suggests that the origin of amniotes might have been much earlier than otherwise suspected, and furthermore that fully terrestrial tetrapods had evolved very soon after they had emerged from the water in the Late Devonian (Ruta et al. 2002).

Aïstopods

Until the recent discovery of a Tournaisian tetrapod (see below), the earliest articulated post-Devonian tetrapod was a creature called *Lethiscus*, a member of a group known chiefly from the Coal Measures, the aïstopods (Fig. 7.4) (Carroll et al. 1998). The name *Lethiscus* comes from the name of the River Lethe, the river of forgetfulness, one of the rivers in ancient Greek mythology that flowed into Hades. *Lethiscus* comes from Viséan rocks found at a locality near Edinburgh in a sequence called the Wardie Shales. This marine sequence of shales and ironstone nodules is more famous for its many fine fossil fishes, including ray-fins and sharks.

Although it is more primitive in some ways than its later relatives, *Lethiscus* is nonetheless highly specialized and scarcely recognizable as a tetrapod at all. Aïstopods were snakelike creatures that had lost all trace of limbs and girdles, and *Lethiscus* had increased the number of vertebrae to at least 80 (the specimen is incomplete). Aïstopod skulls were also very strange, in some cases reduced to a series of struts and bars or a mosaic of small plates. These will be dealt with in more detail in Chapter 9.

In *Lethiscus*, the skull, which is about 60 mm long, shows several of the characters that distinguish aïstopods. Its eyes are large and far forward on the skull, and many of the bones of the cheek have been lost, their places instead represented by open spaces. The ribs, which evolved into unique K-shaped structures in later aïstopods, were more or less straight in *Lethiscus*.

Lethiscus is acknowledged as a tetrapod even though it had no limbs because it shows features of the skeleton that are thought to characterize

Figure 7.4. Skeletal reconstruction of Lethiscus stocki.

tetrapods. These include a skull in which a large part of the orbit margin is formed from the jugal bone and in which the snout is formed by the expected series of paired bones, rather than the mosaic found in fish. The well-ossified vertebrae and ribs provide further evidence of its tetrapod nature.

The existence of such a specialized animal so early in the fossil record of tetrapods shows that a great deal of evolutionary diversification and adaptation must have taken place in the time since tetrapods first evolved. Presumably *Lethiscus* had ancestors with legs and digits, although it is impossible to say how many digits this ancestor might have had. The Devonian tetrapods are known to have shown variation in the numbers of digits, but with the discovery of *Casineria,* it is clear that by the mid-Viséan, at least some tetrapods had evolved the more usual five. *Lethiscus,* with none at all, provides a puzzling contrast.

How *Lethiscus* lived is a mystery. *Lethiscus* is the only tetrapod to have come from this much-collected site. Its occurrence is thought to have been "erratic," meaning that its presence at Wardie was the result of accidental transport of the carcass from elsewhere, and its original habitat might have been either aquatic or terrestrial.

Adelogyrinids

Another group of limbless animals is found during the Viséan and early parts of the Namurian. These are the adelogyrinids (Fig. 7.5). Several

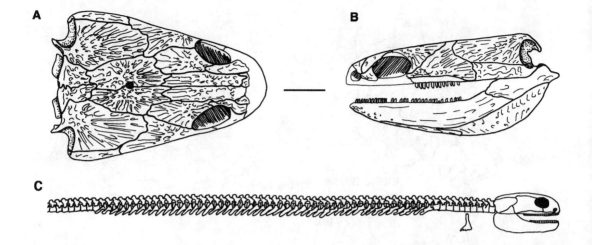

Figure 7.5. Adelogyrinids. (A, B) Skull of *Adelospondylus watsoni* in dorsal and lateral views. (C) Skeletal reconstruction of an adelogyrinid.
Scale bar for skulls = 10 mm.
From Andrews and Carroll (1991).

specimens have been found in different localities, all in Scotland, and all seem associated with fish faunas: they are considered to have been mainly aquatic. There are four named genera, of which *Adelogyrinus* was the first to be named, but because each is incompletely preserved, there is a question about whether they are all valid. They could all belong to the same taxon, preserved in different ways (Andrews and Carroll 1991).

Like the aïstopods, the adelogyrinids have elongated bodies and highly modified skulls with the eyes placed far forward. Unlike the aïstopods, they retain a substantial dermal shoulder girdle, and although no limb elements have been recognized, such an extensive dermal girdle suggests that the limbs were present but that the bones remained as cartilage and so were not preserved. The vertebrae were holospondylous—single complete cylinders—but the ribs were not modified like those of aïstopods but instead were more like those of a conventional primitive tetrapod.

Unlike the aïstopod skull, that of the adelogyrinids was completely roofed with ornamented dermal bones, although the complement was different from any other tetrapods and reduced by comparison with more primitive tetrapods. At the back, there was a rounded embayment, or temporal notch, occupying the rear margin of the cheek.

They appear to have been capable of a large gape. The teeth were sharply chisel-shaped and arranged in close order like a hacksaw blade. These might possibly have been used as a filter-feeding device through which water could be expelled once the mouth was closed, and several specimens are found in strata that preserve numerous small crustaceans called ostracods. The teeth show no evidence of wear and are unlikely to have been used either for eating plants or subduing active prey. The hyoid apparatus was extremely well ossified and relatively large, although it does not show the grooving found in that of *Acanthostega* and more likely contributed to a mechanism of suction feeding rather than breathing. This would be consistent with the tooth morphology and large gape and implies that the animals were aquatic.

Although adelogyrinids share several derived characters with other animals usually included among the "lepospondyls" (see Chapter 9), such as vertebral construction, and although they have elongate bodies with small or absent limbs, in other respects they appear to be the most primitive members of this assemblage.

Emerging into the Carboniferous • 201

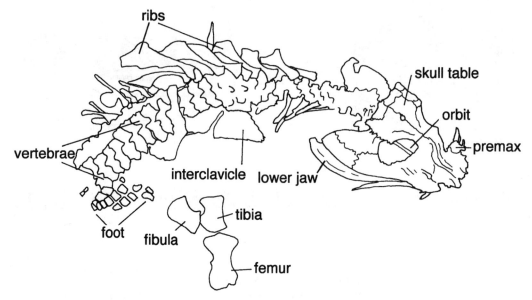

Figure 7.6. Outline drawing of one of the specimens of Whatcheeria.

Whatcheeria deltae

One of the recently discovered and described tetrapod species from the Early Carboniferous is *Whatcheeria deltae*. This creature, unlike almost all its near contemporaries, comes not from Scotland but from the United States. It was found near the town of What Cheer, near Delta, Iowa, in an abandoned limestone quarry (Bolt et al. 1988; Bolt and Lombard 2000; Lombard and Bolt 1995). The sediments in which it was found are thought to represent lake, river, and swamp environments in a coastal lowland setting, and it is possible that on occasion the water in the swamp was brackish as the sea made incursions into the area. The locality also yielded colosteids (see below) and other fragmentary tetrapods, as well as fish and invertebrates. *Whatcheeria* itself was the most common tetrapod found at the site, and several articulated and nearly complete skeletons were recovered.

The skull of this animal was about 200–250 mm long and armed with an array of large, recurved teeth (Fig. 7.6). The profile was high rather than flattened, with deep cheeks and large orbits facing more sideways than upward. The midline of the skull was strengthened by buttresses running along its length. The back part of the cheek extended far backward beyond the back of the skull table, which itself was relatively short, and a steep angle was formed between the skull table and cheekbones. This kind of shape and arrangement of bones is seen again in *Crassigyrinus* and the new Tournaisian animal (below) as well as in embolomeres (Chapter 9), and at first, both *Whatcheeria* and *Crassigyrinus* were taken to be related to embolomeres. This idea has yet to be fully tested.

The skull of *Whatcheeria* retains a number of primitive characters, such as the preopercular bone, and some characters of the lower jaw, which are lost in later tetrapods. Unfortunately, little is known of the palate.

The postcranial skeleton of *Whatcheeria* is well preserved in most respects and shows an animal of up to about a meter long. Notable are the expanded ribs found in the thoracic region of the trunk, somewhat reminiscent of those of *Ichthyostega,* and flattened, paddlelike limbs, reminis-

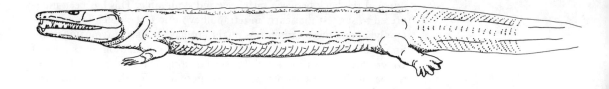

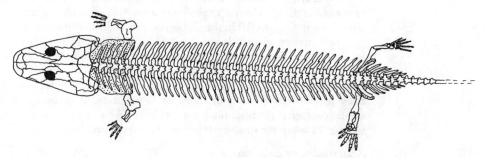

cent of both *Ichthyostega* and *Acanthostega*. The fossils show that at least five digits were present on the limbs, but none of the specimens is complete, and it is possible that more were present.

Figure 7.7. Drawings of Greererpeton. *(Top)* Life reconstruction. *(Bottom)* Skeletal reconstruction viewed from above (from Godfrey 1989a).

Colosteids

One group of Early Carboniferous tetrapods seems to be more commonly and more widely represented than any of the others. These are the colosteids. They are known from the United Kingdom and North America.

A couple of North American genera are well known from nearly complete skeletons: *Colosteus* (after which the group is named) and *Greererpeton* (Fig. 7.7) (Carroll 1980; Godfrey 1989a, 1989b; Hook 1983; Romer 1972b; Smithson 1982). The two genera are very similar, with *Greererpeton* the more completely known. The genus *Greererpeton* was named after the quarry where the animals were first found, in West Virginia, in rocks that date from around the Viséan–Namurian boundary. Several well-preserved and nearly complete skeletons from this quarry lay beside one another in the deposit, suggesting that the animals had died more or less simultaneously, perhaps in a drying riverbed. This locality has also yielded several other well-preserved genera, indicating that a rich fauna was living at the site (Holmes 1984).

The skulls of most specimens measure from between 150 and 180 mm, but because of the abundance and size range of specimens, not only the basic anatomy of *Greererpeton* but also the growth stages have been studied. As in many animals today, in the smaller, younger forms, the eyes were relatively larger, although as the skull grew, the snout did not get progressively relatively longer, as it usually does in other genera.

The animals were flattened with shallow skulls, and it is clear that they were fully aquatic. They had short limbs with a robust ventral covering of scutes and also a dorsal covering of rounded scales. The length of the tail is not known, but it does seem to have been deep and flattened from side to side. The skull outline was a smooth U shape, the arms of which diverged a little toward the rear. The large orbits looked dorsally, although the size of the eyeball is not known, nor whether they would have been flat inside the skull or bulged out above it.

Unlike many early tetrapods, the skull of *Greererpeton* did not have a notch at the back between the cheek and skull table. This has been interpreted to suggest that early in tetrapod evolution, the notch found commonly among these early forms might not have housed a tympanic membrane for hearing, as was at one time assumed. In addition, the stapes of *Greererpeton* was robust and flattened, resting against the dorsal part of the palate. This discovery gave rise to alternative theories about the early role of the stapes in tetrapods, suggesting that it originally had not much to do with hearing but rather acted as a bracing member between the braincase and the palate (Carroll 1980; Smithson 1982). This idea is explained more fully in Chapter 10 in the section on the evolution of hearing.

The modern giant salamander *Andrias,* which lives in present-day Japan, gives an idea of colosteid ecology. This creature lives in streams and rivers 2–8 m deep, hiding in rock crevices. The larger, older animals tend to favor the deeper water. *Greererpeton* seems to have had similar limb and body proportions to *Andrias,* and the distribution of specimens fits a similar ecological distribution. It was probably likewise a lurking predator feeding on whatever small vertebrates or invertebrates it could catch.

Acherontiscus caledoniae

This small animal is represented by a single specimen from Scotland and is named, as are several other serpentlike lepospondyls, after one of the rivers of Hades, tributaries of the River Styx. There are no data on locality or collector, and the dating remains in doubt. However, spore analysis performed in 1969 suggested that it was late Viséan or early Namurian (Carroll 1969; Carroll et al. 1998). The bone in the specimen is poorly preserved, so the interpretation of the morphology is also in doubt. The skull appears to be that of a notchless, small-eyed, adelogyrinidlike form, and it has an elongate vertebral column with reduced or absent limbs, as in those animals. The vertebral column, however, is anomalous in that the centra consists of two more or less disc-shaped parts, somewhat like the later embolomeres (see Chapter 9). Informal suggestions include the idea that it might be a juvenile colosteid. It is another of those enigmatic forms that suggest a much greater diversity of early tetrapods in the Early Carboniferous than those that are currently known.

Antlerpeton clarkii

The single specimen of *Antlerpeton* was found in a most unlikely place for an Early Carboniferous tetrapod. It comes from a talus slope in the hilly desert of central Nevada—the Antler Highlands after which it is named—and is the only early tetrapod to have been found in that part of the world (Thomson et al. 1998). The area has been little explored, and recent expeditions have failed to relocate the spot at which it was found. It appears to have been an erratic find in that it was totally isolated, although somewhere as yet undiscovered there may be other such treasures waiting to be found. Because it was found as a loose block, it cannot be dated with certainty, but it appears to have come from Early to mid-Carboniferous beds. The sediments include some marine deposits, indicating the possible close proximity to the sea of this animal's habitat. The matrix in which the specimen was preserved is also unusual for a Carboniferous tetrapod, being an intractable red sandstone, quite different from the coal, shale, or ironstone in which they are typically found. The specimen was therefore prepared as a natural mold from which latex peels were made.

There is no skull preserved, nor much of the limbs or girdles, although parts of the femora and pelvis appear to be present. The remains chiefly consist of vertebral centra and ribs preserved in an unusual ventral view that makes them difficult to interpret, and they were described as being unlike those of any other known early tetrapod. It remains possible that they were rhachitomous. This tantalizing specimen is another that hints at greater diversity than paleontologists are yet aware of.

Crassigyrinus scoticus

One of the most peculiar and least understood creatures from the early part of the Carboniferous is *Crassigyrinus scoticus*. Fossils of this creature have been known since the first specimens of it were discovered in the 1850s, but they were not immediately recognized for what they were. This accounts for the name: *Crassigyrinus* means "shallow wriggler," and it was given because only half the skull was known and it was considered to have been a very flat-headed animal—very far from the truth as it is now understood! In recent years, more material has been found so that most of its anatomy can be described (Clack 1998b; Panchen 1985; Panchen and Smithson 1990; Watson 1929). Despite this, it remains a puzzling animal.

Crassigyrinus is known exclusively from Scotland and from only a couple of sites, one a deep coal mine at Gilmerton near Edinburgh, and the other the Dora bone bed from an open-cast mine north of the Firth of Forth at Cowdenbeath. These localities represent times around the Early to Late Carboniferous boundary.

The head of *Crassigyrinus* was about 350 mm long, narrow, and deep, with a short snout and long rear cheek region (Fig. 7.8). The skull bones were ornamented around the mouth with deeply incised pits and grooves, but elsewhere, the ornamentation was shallow and irregular. The eye socket was rhomboidal in shape, and the deep skull ensured that the eyes looked laterally rather than dorsally. At the front, above the nose, lay what appears as a large hole in the snout, apparently connected to a hole in the palate, although it is not known whether these passages were separated by soft tissue in life. Running back from the hole in the snout was a groove, which seems to have become shallower toward the top of the head until it disappeared. The buttresses flanking the groove may have strengthened the snout. There were also buttresses on the inside of the skull running down the front rim of the orbit. Massive teeth, especially on the palate, may have required extra strengthening of the snout that bore them. The lower jaw received some of these teeth into special holes when the jaws closed. The length of the cheek means that the lower jaws hinged very far back on the skull, allowing them extra swing as they closed like a trap.

Between the short skull table and the cheek there was a notch, which has been interpreted as housing a spiracle (Carroll, in Panchen 1985). The nostril was large, but a breathing hole on top of this high skull might have been useful to an aquatic animal.

Crassigyrinus was certainly aquatic. It had a long, narrow body, probably nearly 2 m in length (including the tail), which matched the skull in narrowness and tapered gradually toward the rear. A deep tail has been reconstructed for it from a few postsacral vertebrae. While its hindlimbs were relatively small for the size of its body, its forelimbs were extremely reduced. The humerus or forearm bone was only as long as the long dimension of its orbit. From the limbs and body, *Crassigyrinus* is thought to have been an actively swimming, permanently aquatic animal, and the skull and dentition are those of a fearsome predator (Fig. 7.9).

Crassigyrinus is puzzling in having not only many curious adaptations of its skull and limbs but because at the same time, it shows a number of very primitive features. In particular, the pattern of bones in the palate is

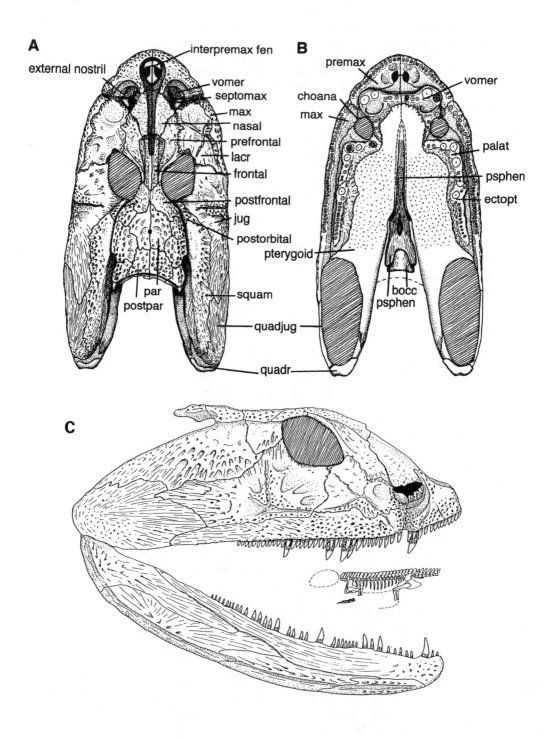

Figure 7.8. Skull reconstructions of Crassigyrinus. *(A) Dorsal view. (B) Ventral view. (C) Lateral view shown to the same scale as a reconstruction of* Casineria. *From Clack (1998b).*

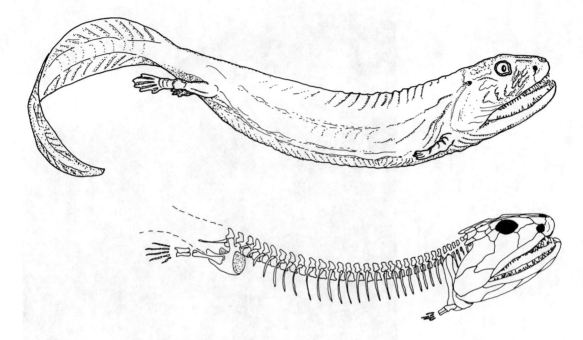

Figure 7.9. Drawings of Crassigyrinus. *(Top) Life restoration. (Bottom) Skeletal restoration.*

one of the most fishlike known among early tetrapods. The vertebrae were not very well ossified, and the neural arches had only poorly developed articulations joining them together. Paleontologists have debated whether this combination of characters indicates that *Crassigyrinus* was a secondarily aquatic animal adapted from a previously more terrestrial form, or whether in fact it is a long-surviving relict of the earliest tetrapods that had not yet become fully terrestrial. The most recent analyses suggest that indeed it is a secondarily aquatic form and that the fishlike features of its palate have been reacquired (Clack 1998b). This is because computer-generated phylogenies have placed it closer to anthracosaurs, baphetids, and *Whatcheeria*, none of which have these fishlike tooth arrangements.

Early Carboniferous Tetrapods in Gondwana

Until very recently, finds of Carboniferous tetrapods had been confined to Euramerica, despite the presence of Devonian tetrapods in Australia. In 1996, finds of tetrapods from a site in Queensland, Australia, proved that they existed well outside the Laurussian continent, in Gondwana (Thulborn et al. 1996). The finds are dated as the equivalent of the early–mid-Viséan and are among the earliest of tetrapod fossils found anywhere. The material consists of isolated, often broken bones that are difficult to identify as belonging to particular groups. Possible colosteids, temnospondyls, and anthracosaurs were identified at first, but more recent study has suggested that these identifications might not be correct. Until more complete material is discovered, their exact relationships will remain obscure, but biogeographically, they are no less important as showing that tetrapods did inhabit Gondwana at this time.

An Articulated Tetrapod Specimen from the Tournaisian

A single specimen of a tetrapod has been discovered from late Tournaisian deposits of western Scotland, and it is the only Tournaisian tetra-

Figure 7.10. The site at Auchenreoch Glen near Maryland Farm, close to Dumbarton in Scotland, where the Tournaisian tetrapod was found. The specimen was found in situ in a concretion. Photograph by J.A.C.

pod so far to be preserved as an articulated specimen (Clack and Finney 1997). The sequence in which it was found consists of shales, clays, and limestones that originated in marginal marine conditions (Fig. 7.10). The specimen, Hunterian Museum Glasgow specimen GLAHM 100815, is an almost intact skeleton, lacking only the tail, some parts of each limb, the top of the head, and most of the lower jaws (Fig. 7.11). Work is still in progress on this specimen, which at this time does not even have a name. However, some things are already clear. Its skull is very like that of the Viséan *Whatcheeria*, with only a few detailed differences. Its stapes is well preserved and looks like a larger version of that of *Acanthostega*, not surprising in a proportionately larger animal. However, its postcranial skeleton is different from that of *Whatcheeria* in many respects, so it represents a different genus.

The humerus resembles most closely that of the recently discovered *Baphetes* specimen, marked in particular by a prominent, spikelike process

Figure 7.11. The specimen of the Tournaisian tetrapod (GLAHM 100815) when preparation was complete. Scale in centimeters. Photograph by S.M.F.

for the latissimus dorsi muscle that attaches to the shoulder girdle and raises the forearm. Two forelimb digits only are preserved, those of the right hand, but these two are rather different from each other. One is short and stubby, and the other is more slender and relatively tiny. This tiny digit appears to be a more anterior one than the other, suggesting the tiny digits at the leading edge of the hindlimb of *Ichthyostega*. Another feature suggestive of *Ichthyostega* is the form of the ribs. They have triangular flanges on the shafts, as indeed do some of those of *Whatcheeria*, but in the case of the new tetrapod, there is also a curious pronglike process separate from and dorsal to the triangular flange. The hindlimbs of the new animal are quite robust and do not show the flattening of the tibia, fibula, and phalanges characteristic of *Whatcheeria*. The inference from this is that the new animal was less aquatically adapted than *Whatcheeria*. Five digits are preserved on the foot of the new animal, which may be the full complement, although this cannot be taken for granted (Fig. 7.12).

Only a full analysis of all its characters will help place this new animal in its context, but once that is done, it may help resolve the relationships of some of the other curious creatures of the Early Carboniferous, whose interrelationships are considered in more detail at the end of Chapter 9.

Early Carboniferous Tetrapods: An Overview

By the mid-Carboniferous, tetrapods had produced a wide range of body forms, skull shapes, and sizes of animals that far exceeded those of the Devonian, so far as is known. From tiny, mouse-sized animals such as *Casineria* to large, predatory forms such as *Crassigyrinus* to the limbless aïstopods, tetrapods broke away from the conservatism of their Devonian forerunners. This range also exceeds what is seen in the lobe-finned fish of the Late Devonian, although by the Late Carboniferous, the rhizodonts had produced some of the largest bony fish that have ever evolved, and the ray-finned fishes were filling the water with small forms unparalleled among the lobe-fins. The diversity of form, size, proportion, and structure produced by this early radiation of tetrapods is matched only later by the diversification of amniotes in the Early Triassic and by the evolution of teleosts among the ray-finned fishes of the Late Jurassic and Early Cretaceous. It is as if some kind of threshold had been crossed that allowed the animals to colonize the new territory and exploit the variety of niches that

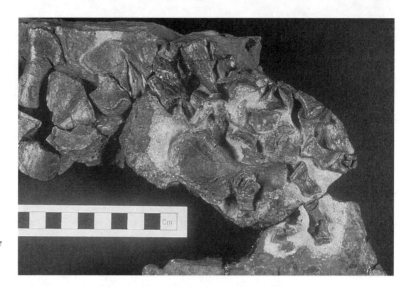

Figure 7.12. The hindlimb of Tournaisian tetrapod showing the bases of five digits. Photograph by S.M.F.

were being newly provided among the plants and invertebrates. More complex forest structure and the subsequent diversification of terrestrial invertebrates, including molluscs and insects, may have triggered this diversification, or the ecological niches may have become exploitable by tetrapods as a result of some structural or developmental innovation.

A suggestion for this apparent developmental breakthrough is provided by the observation that the Tournaisian saw the advent of cooler and drier conditions operating throughout the period. By the early Namurian, the climate had once more become monsoonal. The cooler, drier conditions of the Tournaisian period could have led to two outcomes, both of which may be involved in the lack of a fossil record and the appearance of many groups of small tetrapods during that time. First, the conditions may not have favored preservation except in a few rare circumstances. After an extinction event, the animals might have been scarce in any case. But another possibility links the climate to developmental changes among tetrapods. Under environmental stress, many animals undergo a shortening of their developmental periods, in order to reduce the risk factors of desiccation or predation in harsh conditions. This is a strategy seen today, especially, for example, among frogs that habitually live in temporary pools. By comparison with their relatives in more permanent water bodies, the frogs of ephemeral pools have larvae that are generally faster developing and adults that are rather smaller (Gould 1977). The process is known as progenetic dwarfing. It can result in modifications to skeletal and skull structure and proportions, including dermal skull roof bone reduction, reduction of the skull size relative to the braincase and ear regions, and reduction of endochondral ossifications. These are in many ways consistent with the differences seen among the first Carboniferous tetrapods compared with their Devonian relatives. The climatic stress of the Tournaisian period might have been a contributory factor in such changes. It is also possible that such processes might have played a part in changes to methods of breathing and feeding that had profound consequences for skull architecture, as suggested in the final chapter.

Many of the important questions about early tetrapod evolution and diversification—as well as about how they eventually became more fully terrestrial—can only be answered by discoveries of more fossils to represent this crucial period. Now that the picture of Devonian tetrapods is beginning to fill out and the Late Viséan is becoming better known, the discoveries with most impact are going to come from the Early Viséan and Tournaisian periods. That will depend on renewed collecting in these areas. Almost all the known Viséan tetrapods have been found in Scotland, and most of these are from long-known and now inaccessible localities. The Scottish Midland Valley has historically been one of the most productive and important areas for early tetrapod paleontology. However, this in itself makes the sample biased. Much remains to be done in the Early Carboniferous of other parts of the world, and recent finds in Australia and North America emphasize the possibilities. A recently discovered locality in Ohio has unearthed more Viséan tetrapods that will make a considerable contribution to knowledge of this period. Collecting from Early Carboniferous localities is probably one of the most potentially rewarding paleontological investigations that could be undertaken at this time.

Eight
East Kirkton and the Roots of the Modern Family Tree

Background to the East Kirkton Locality

A small former mining town called Bathgate, on the outskirts of Edinburgh, has recently been made famous in the paleontological world for being the location of a window through which to view an extraordinary episode in evolutionary history. At the edge of a housing estate lies a quarry where in the 19th century a rock called the East Kirkton Limestone was dug out. It had some curious qualities that made it attractive as a building stone and hard-wearing for making the local farm walls.

It is composed of thinly alternating bands of dark carbonaceous limestone, pale silica, and hardened gray volcanic ash called tuff, and often the bands are speckled with small white nodules or twisted and distorted into intriguing curls and waves. In the 1830s, fossil collectors also found some unusual specimens, which are now recognized as the carapaces of eurypterids, or sea scorpions, as well as many plant remains. The quarry was closed in about 1844, and although geologists visited occasionally afterward, it was eventually forgotten and became grown over.

In about 1984, fossil collector Stan Wood decided to investigate the East Kirkton Limestone again, first by investigating the now disused and neglected farm walls. It was here, after some years of searching, that he found the remains of tetrapods (Wood et al. 1985). Among these were some

Figure 8.1. East Kirkton Quarry site once clearance was complete and a section had been excavated through the sequence, with the author standing at the level of unit 82, where most of the best tetrapod specimens have come from. Photograph by R.N.G.C.

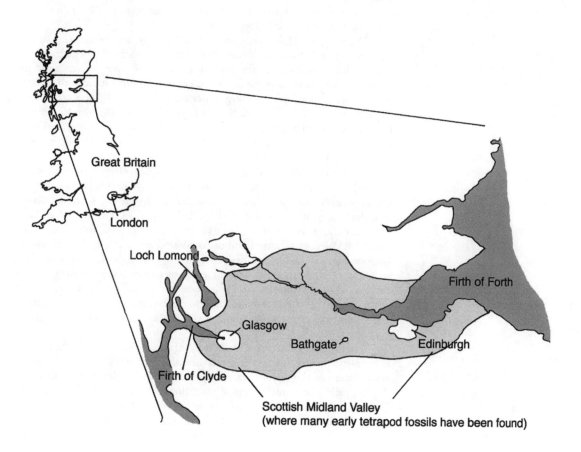

Figure 8.2. Main map shows the British Isles; inset shows the Scottish Midland valley and the position of Bathgate.

completely articulated skeletons representing new species. However, that was only the first of a series of discoveries revealing unusual and important aspects of the fossils of East Kirkton. The quarry was subsequently relocated and opened up, with a section dug to expose the stratigraphic sequence (Fig. 8.1). Figure 8.2 shows a map of Scotland and the location of Bathgate.

The East Kirkton Limestone lies within a sequence known as the Bathgate Hills Volcanic Formation, which gives a clue to its significance. This sequence dates from the later part of the Early Carboniferous, the Viséan, making it one of the earliest known sites for tetrapod fossils, after those from the Late Devonian. It predates most of those that yielded the animals described in Chapter 7. The site subsequently yielded not only many new taxa of tetrapod but of invertebrates and plants. Many of these proved to be of exceptional interest because they were the earliest known representatives of their groups to be demonstrably terrestrial. In short, East Kirkton had yielded the earliest certainly terrestrial assemblage of animals, including tetrapods, yet discovered. The circumstances allowing the preservation of these animals have themselves proved extraordinary because the site itself represents remarkable physical and ecological conditions.

Over the years that followed Stan Wood's discovery, much work has been carried out on the sedimentology, stratigraphy, geophysics, geochemistry, paleoecology, and paleontology of the site, culminating in an international conference in 1992. Accounts from a wide range of experts were published in 1994 by the Royal Society of Edinburgh (Rolfe et al. 1994). The rest of this chapter is devoted to describing some of the findings and their significance for tetrapod evolution.

The Bathgate Hills to the south of Edinburgh are the remains of old volcanic uplands that rose during the latter part of the Early Carboniferous, spilling basalts and other volcanic sediments over the surrounding region. The uplands and the active volcanoes lay not far from the coast, where rivers flowed out, forming estuaries, lagoons, and deltas and where thick forests grew on the resulting rich soil. The East Kirkton Limestone formed as the bed of a predominantly freshwater lake, filling gradually by sediments derived from the surrounding country as well as by occasional volcanic outflows.

The nearby volcanoes had a number of different effects on the animals, plants, and sediments of the East Kirkton lake. The first was that after an eruption, sediments of volcanic ash settled and were then washed down into the lake forming thick deposits of tuff (Fig. 8.3). The hot ash or lava appears to have set fire to the forests on occasion, as volcanic eruptions do today, and pieces of charred and broken wood are found in the tuff layers. Poisonous gases and fires killed many nearby animals, so that as well as wood, the flowing ash picked up their carcasses, broke them up, and carried the bones into the lake. It is notable that the isolated bones found in the tuff layers are by far the largest of the tetrapod specimens to have been found at East Kirkton, suggesting that the animals did not live in the lake itself.

Another, less direct effect of the vulcanism was the production of mineral-rich water in the lake. Both the concentration of minerals and the temperature of the water fluctuated, perhaps seasonally or annually, and this had an effect on the fauna and flora inhabiting the lake waters. Calcium, silica, magnesium, iron, manganese, and sulfur are all found in relatively high quantities in some of the layers. Some layers were formed almost entirely of calcite, whereas others were more highly carbonaceous. When the water became saturated with minerals, these precipitated out of

Figure 8.3. Detail of the tuff (fossilized volcanic ash) layers from which some isolated tetrapod bones have been recovered. Photograph by J.A.C.

the water, clinging to any available surface. Common in the sediments are small nodules of calcite, which apparently formed by accretion around single-celled algae, ostracods (small ovoid crustaceans), plant debris, or other minute cores, forming the characteristic whitish speckles in the rock. It is thought that sometimes these accretions formed blankets that covered the surface or floor of the lake, choking whatever was underneath. Occasionally, these accretions of calcite formed larger structures that distorted the over- and underlying layers to form waves and ripples in the sediments (Fig. 8.4). Mineral-rich water occasionally penetrated plant stems, replacing or impregnating the tissue cell by cell so that the fossils are able to reveal the stem structure in exceptional detail (Scott et al. 1994). Along with plant debris and broken fronds, fossilized charcoal, called fusain, is frequently found in some levels. This is taken to indicate the activity of forest fires, perhaps caused by volcanic eruptions.

Many of the tetrapod specimens found at East Kirkton are complete or almost complete skeletons, but these are exclusively of relatively small animals, no more than about 400 mm long. This suggests that the animals were not transported very far and died within close proximity to the water, if not in the water itself. Because the skeletal structure of most of these animals suggests terrestrial habits, one idea is that they fell into the water by accident, perhaps attempting to escape from the volcanic eruptions or influx of poisonous gases. Other possibilities are that the lake itself occa-

Figure 8.4. Sample of rock showing distorted layers and calcite bands in East Kirkton sediments. Scale bar = 10 mm. Photograph by S.M.F.

sionally produced upwellings of poisonous gas, which overcame the creatures living nearby. Another suggestion is that the creatures ventured out onto algal mats that grew along the lakeshores, until they reached the thinning edges over the water, whereupon they fell through the mats and drowned. Certainly, in many cases, bone structure is poorly preserved, and the animals may have been "cooked" before preservation.

The closest modern analogies to the East Kirkton of Viséan times might be the Yellowstone National Park in the United States and the Virunga National Park in the Ruwenzori Mountains at the headwaters of the River Nile in the Congo Republic. Here, volcanic activity produces places where hot streams can literally boil unlucky animals in seconds, and others where pockets of carbon dioxide, accumulating in the night, can asphyxiate unwary creatures that stray into them, followed by predators that attempt to scavenge their corpses. The bodies of both lie undisturbed and decay without being disarticulated.

The sediments of the East Kirkton limestone are about 15 m thick, and a series of bore holes and careful excavations allowed the sequence of beds to be recorded. This survey showed some of the changes that occurred to the lake flora and fauna during its existence, demonstrating that toward the end of deposition in the lake, it became confluent with other nearby water bodies. The evidence also suggests a change from a drier to a wetter environment toward the top of the sequence, with fish remains found for the first time in the topmost levels. It is not clear what length of time the sediments represent, except that it was relatively short, geologically. A few hundred to a few thousand years seems most likely.

Plants and Invertebrates of East Kirkton

Plants

Like the animals, the flora of East Kirkton shows a change in character through different parts of the sequence. Early on, the flora is dominated by treelike early relatives of modern conifers. These had large leaves, and their wood was dense and thus preserved well. These plants enjoyed well-drained soils, which suggests that early in the life of the lake, conditions were warm and dry. From the fact that none of the remains seems to have been transported very far, it is inferred that they grew close to the lake. One genus of this kind is called *Stanwoodia*, for obvious reasons. These kinds

Figure 8.5. Two plants from East Kirkton (NMS specimens). (Top) Sphenopteridium crassum, *one of the most common pteridosperms at East Kirkton. (Bottom)* Spathulopteris obovata, *a pteridosperm. Scale bars = 10 mm. Photographs by N.M.S.*

of plants and conditions are associated with the base of the sequence from which most of the best vertebrate specimens have also come (Fig. 8.5) (Scott et al. 1994).

Later in the history of the lake, represented by the East Kirkton limestone itself, so-called seed ferns and true ferns are found. At the top of the sequence, as the lake finally became silted up, there is a further change to the vegetation. Lycopods (club mosses) become the most common component, and these plants suggest damper soils. This may mean that the climate became wetter or that the land was beginning to sink again.

Invertebrates

Much interest in the East Kirkton site has been generated by the invertebrate fauna. This includes not only the eurypterids (Jeram and Selden 1994), but scorpions (Jeram 1994), myriapods (Shear 1994), and a single specimen of a harvestman.

Figure 8.6. Pulmonoscorpius kirktonensis, *one of the giant scorpions from East Kirkton. About 700 mm long (NMS G1987.7.136). Photograph by N.M.S.*

Scorpions

Probably the most important thing about the scorpions of East Kirkton has been their state of preservation. This has allowed much more detailed information to be gained than from any other previous site, and it has meant that not only the East Kirkton scorpions but also some of those from other sites can be interpreted more accurately. Another factor has been the large number of specimens recovered.

The East Kirkton scorpions are often preserved as isolated pieces of cuticle, the resilient skin in which the animals were covered. Like all arthropods, scorpions shed their skins periodically, and for some reason, at this site, many of these shed skins have been found. Instead of being preserved as compressed and blackened stains in the rock, which is usual for fossil scorpions, here they have been preserved almost in their original three-dimensional state, often with what appears to be their original brown or yellow color (Fig. 8.6). Because of the difference in the chemical nature between the rock and the scorpion cuticle, it has been possible to dissolve away the rock to leave behind fragments or larger pieces of scorpion material. Details of the claws, eyes, and bristles can be made out. Most of the specimens discovered appear to belong to a single species, called *Pulmonoscorpius kirktonensis*. Several stages of growth are represented, from juveniles about 17 mm long up to one that is 700 mm long. A few other scorpion species are also present, represented by rarer material.

Pulmonoscorpius has proved important by establishing that it possessed an air-breathing mechanism called a book lung. This discovery shows that *Pulmonoscorpius* definitely breathed air, adding to the evidence that East Kirkton preserves a terrestrial fauna. *Pulmonoscorpius* also had large eyes, suggesting that vision played a significant role in its life. Modern scorpions have reduced eyes and are predominantly nocturnal, whereas this Paleozoic form seems to have been active during the day (Jeram 1994).

It is not clear what *Pulmonoscorpius* would have eaten, but perhaps insects formed part of its prey. No insect remains have been found at East Kirkton, although winged insects are known from the Namurian of Germany. With the characteristic scorpion sting at the end of its tail, a large *Pulmonoscorpius* was probably equipped to deal with small tetrapods that might have crossed its path.

Eurypterids

Eurypterids, sometimes known as sea scorpions, were not really scorpions at all. They were arthropods closely related to horseshoe crabs. Although the early forms found in the Silurian were marine, those of East Kirkton were certainly not so, and like the scorpions, they have been shown to be largely terrestrial.

Three genera of eurypterid have been found at East Kirkton, the most common being *Hibbertopterus*. Specimens of *Hibbertopterus* have been found preserved most commonly in the tuff layers, where they occur as isolated fragments or more complete pieces of carapace with some remaining three-dimensional structure. Like the scorpions, sometimes the carapace can be dissolved from the rock to reveal clear detail. The head of *Hibbertopterus* was rounded in outline and slightly domed, with small eyes near the center (Fig. 8.7). The largest carapace found measures 650 mm across, making it one of the largest ever found. The anterior appendages that fixed to the underside of the carapace lack any evidence of specialization for swimming or predation, and for juvenile animals, it has been suggested that they raked through soft sediment with their anterior appendages. As the animals grew larger, their mode of feeding may have changed to a form of sweep feeding using comblike structures developed on their limbs (Jeram and Selden 1994).

No very small eurypterids are represented at East Kirkton, suggesting that they did not breed there. Where there is much broken-up plant material, an abundance of broken-up cuticle fragments is also found; both kinds of debris were probably transported from elsewhere. Large hibbertopterids may have fed on small invertebrates at the lake edges, perhaps as a seasonal occurrence, and then moved elsewhere to breed.

Myriapods, Harvestmen, and Ostracods

Although not particularly well preserved, a few specimens of myriapods have been found at East Kirkton. These represent the earliest Carboniferous fossils of millipedelike animals, although representatives are known from the Silurian and Devonian. One of the East Kirkton specimens shows the earliest evidence for the kind of pores that modern millipedes possess to secrete noxious chemicals, effective in repelling predators. The substances that they produce are known to be effective against amphibians, and the tetrapods of East Kirkton would have attempted to feed on the millipedes of the day (Shear 1994) (Fig. 8.8).

One specimen of an opilionid—harvestman—has been found at East Kirkton. It has only four remaining legs preserved, but then, many living examples lose one or more legs and function adequately, so it might have lost the other four while it was alive rather than after it had died (Fig. 8.9). Like modern opilionids, it had a single part to the body, and each joint of the legs bore a slight swelling. These and the long, very thin legs make it recognizable as an opilionid rather than a spider, to which harvestmen are closely related. True spiders are known from the Devonian, but this Kirkton specimen is the earliest known opilionid. Unfortunately, the specimen shows no detail other than those that identify it as a harvestman.

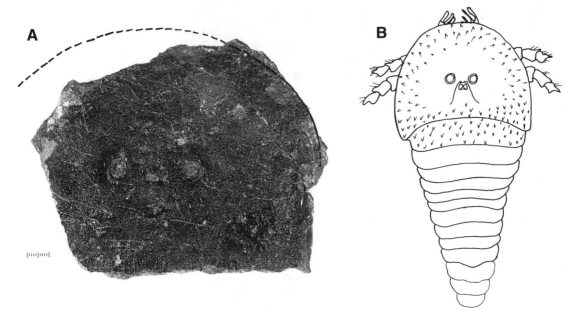

Figure 8.7. (above)
(A) Hibbertopterus scouleri *head shield from East Kirkton. The dotted line shows the restored outline of the head shield. Note the eyes and the ornamented surface of the cuticle in this specimen (UMZC specimen). Scale bar = 10 mm. Photograph by S.M.F.*
(B) Life restoration of Hibbertopterus. Hibbertopterus grew up to about 650 mm across.

Figure 8.8. (right) Unnamed myriapod specimen from East Kirkton (length of animal, about 6–8 mm) (NMS G1992.21.1p). Photograph by N.M.S.

Many layers of the East Kirkton Limestone are crowded with the remains of small, oval-shaped organisms that look like tiny bivalve mollusc shells. However, they are not molluscs (whose shells are absent from the deposits) but small crustaceans called ostracods. Their shells are hinged like those of bivalves, but inside lived a little creature with feathery feeding arms like those of a tiny barnacle. Ostracod shells often have surface sculpture on the outside, and in well-preserved samples, they can be identified as to species by means of the details of this ornament. However,

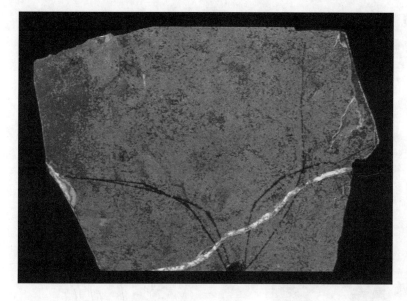

Figure 8.9. Opilionid (a harvestman) from East Kirkton. Specimen from the Hunterian Museum, Glasgow, Scotland. Photograph by W.D.I.R.

those from East Kirkton are not sufficiently well preserved to do this. The abundance of ostracods in the lake speaks of a rich food source for both invertebrates and tetrapods and suggests that at some periods at least, the water was rich in even smaller organisms, such as unicellular animals, on which the ostracods would have fed.

One group of invertebrate that has not been found at East Kirkton are insects. Although six-legged insect relatives are known from the mid-Devonian, flying insects have not yet been found before the Namurian. Their absence from the East Kirkton deposits is surprising given the type of preservation there, and it may mean that flying insects had indeed yet to evolve.

This brief survey of the variety of plants and invertebrates found at East Kirkton provides a view of a habitat rich in a diversity of niches for animals to live in and a glimpse of what it must have been like for the tetrapods of the time. The invertebrates, as well as the vertebrates themselves, no doubt formed complex food chains, with the tetrapods as well as the larger invertebrates feeding in the forests surrounding the lake.

Vertebrates of East Kirkton

The vertebrate fauna of East Kirkton has proved most remarkable for its collection of tetrapods, which seems to represent the earliest undoubtedly terrestrial assemblage of species. Although many diverse tetrapods must have evolved before the period spotlighted at East Kirkton, there is no clear record of them. Furthermore, groups whose descendants come to play a significant role in the subsequent evolution of tetrapods are found for the first time in the rocks of the East Kirkton Limestone. There are others whose relationships remain obscure and that seem to have left no descendants.

In addition to the tetrapods, several kinds of fish are found in the sediments toward the top of the sequence. They include acanthodians, ray-finned fishes, and chondrichthyans, but they add little to the story of the origin and radiation of tetrapods, and they will not be dealt with further.

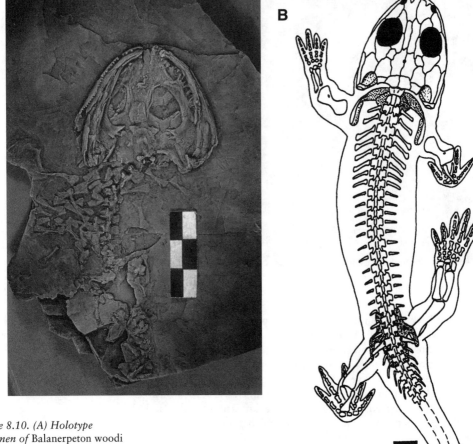

Figure 8.10. (A) Holotype specimen of Balanerpeton woodi *(Hunterian Museum specimen GLAHM V2051). Photograph by A.R.M.*
(B) Drawing of Balanerpeton *(from Milner and Sequeira 1994).*

The best specimens of tetrapods, including some quite small ones, come from a black shale horizon known as Unit 82, and this layer is also notable for producing the filter-feeding eurypterids. The fine grain of the rock and the high carbon content, together with the good preservation of animals, indicate that the water was deep and quiet enough for fine sediment to accumulate and for an undisturbed oxygen-depleted layer to develop at the bottom of the lake. This allowed the animals' bodies to be covered over before they could rot, which would not have been possible if the water had been disturbed and oxygenated.

Temnospondyls

The most common tetrapod by far to be found at East Kirkton is a creature called *Balanerpeton* (Milner and Sequeira 1994). Its name means "crawler from the hot spring or bath." Most specimens of this creature are quite small, with skulls ranging from 25 to 48 mm long. Although no complete specimens are known, if extrapolations are made from known specimens, body lengths of 175 to about 440 mm can be estimated (Figs. 8.10, 8.11). It had a rather flattened head with a rounded snout, and large circular orbits suggest eyes looking dorsally.

Figure 8.11. A juvenile specimen of Balanerpeton. *The skull is a little disrupted and much of the vertebral column is missing. Three of the four limbs are represented (UMZC T1313). Scale bar = 10 mm. Photograph by S.M.F.*

Balanerpeton belongs to a group called the temnospondyls. Temnospondyls became very common and diverse during the Paleozoic and early Mesozoic; their story is continued in the next chapter. However, at this point, it is appropriate to introduce the group and some of its significant features.

One uniting character of temnospondyls is seen very clearly in *Balanerpeton,* and that is found in the palate. All the animals described in this book up to this point have had what are called closed palates—that is to say, they are formed from a complete bony sheet, mostly contributed by the pterygoid. In all temnospondyls, by contrast, there are large openings or vacuities where the parts of the pterygoids lying near the midline of the skull have been embayed (this is illustrated in detail in the next chapter; cf., e.g., Fig. 9.7 and Fig. 9.10; the feature can be seen in Fig. 8.10). In early temnospondyls such as *Balanerpeton,* the margins of the vacuities are formed mainly by the pterygoids and vomers, but in some later ones, the vacuities enlarged even more so that the palatines were involved as well. In yet others, the ectopterygoids disappeared altogether. Fossils of some Carboniferous animals show that a mosaic of denticle-bearing bony platelets lay in skin that stretched over the vacuities, and it is assumed that in other temnospondyls, skin covered the vacuities in life.

The form of these palatal vacuities reflects a similar condition in frogs and salamanders today. In frogs, the flexible skin of the vacuities allows the eyeballs to be pulled down into the mouth cavity to help the animal to swallow food. The same process is also used in breathing as the animal swallows air. It could be that in temnospondyls, the vacuities were used in the same way. Certainly the possession of these vacuities is one character that is thought to unite temnospondyls with frogs and salamanders.

Where the palate and braincase joined, at the basal articulation, *Balanerpeton* shows the beginning of a trend seen among all later temnospondyls, which is for the basal articulation to become a fixed, immobile joint. In early tetrapods, there appears to have been a mobile, cartilage-capped (synovial) joint at this junction, which persists among some groups and can still be seen in some modern reptiles. Although the joint was not sutured in *Balanerpeton,* the style of junction suggests it was not movable (see Fig. 8.10). One suggestion is that the joint no longer needed to move once the palatal vacuities were formed and allowed an alternative region of flexibility in the palate.

The bones of the palate itself, like those of most other early tetrapods, were covered with shagreen, a field of small denticles embedded in the skin. Where other early tetrapods have a row of teeth, only a pair of somewhat enlarged teeth were found on the vomers, palatines, and ectopterygoids. This seems to be true of most early temnospondyls. The parasphenoid had a very narrow process at the front, but at the back, where it lay under the braincase, it was quite broad and bore a further patch of denticles. Again, this seems characteristic of early members of the group.

As with most temnospondyls, there was a rounded otic notch at the back of each cheek that might have supported a tympanic membrane. The stapes was a small bone, and although only part of it is preserved in any specimen, there is enough to show that it was probably slender and rodlike. The footplate fitted into the braincase, and a hole penetrated the base of the bone. In modern tetrapods, a small branch of the stapedial artery runs through such a hole, and it probably did so in these fossil creatures too. The combination of the embayment and the rodlike shape of the stapes suggests that *Balanerpeton,* like many other temnospondyls, had a reasonably good hearing system.

The teeth in the premaxilla and maxilla were numerous, small, and undifferentiated, although those on the dentary seem to have been somewhat larger. *Balanerpeton* does not seem to have been much of a predator and probably fed on small invertebrates. Indeed, some specimens have the remains of myriapods preserved in the position where the stomach would have been.

The limbs were quite short but sturdy, with the hindlimbs somewhat longer than the forelimbs (Fig. 8.10B). The humerus was robust, waisted in the middle, and expanded at each end, but with little true shaft. There are no well-preserved wrist bones, although there is some evidence that they were ossified. There were four fingers on the hand and five toes on the feet. The femur, like the humerus, was robust with expanded ends, although the ankle was more substantially ossified than the wrist. The pelvic girdle bore a single posteriorly directed dorsal process. The ribs were short, with little differentiation among them along the column, except for in the cervical region, where they are slightly expanded at the ends. It appears that a substantial tail was present, although its length is unknown. As with many early tetrapods, *Balanerpeton* had a covering of dermal scales over its belly, known as scutes or gastralia.

Almost as interesting as what is preserved of *Balanerpeton* is what is not preserved. For example, there is no evidence of any lateral line system or any branchial arches. This, along with the well-ossified ankle and limb bones, and the possible presence of a tympanic ear, suggests quite strongly that *Balanerpeton* was mainly a terrestrial creature.

Some large ribs of a second type of temnospondyl were also recognized at East Kirkton, although none of the rest of the animal has yet been found. This is a pity, because knowledge of early temnospondyls may help throw light on the origins of the modern frogs and salamanders. It is widely believed that temnospondyls gave rise to these two groups of modern amphibians. Because East Kirkton provides the first evidence of temnospondyls in the fossil record, it could be said that the evolution of these modern groups started there.

Anthracosaurs and Their Kin

Although by far the majority of tetrapod specimens (about 50) from East Kirkton are of *Balanerpeton*, there are about 20 specimens that show other kinds. Among these is a great diversity of forms, each represented by only a few fossils. It is curious that the temnospondyls show such a limited diversity despite a large number of specimens, whereas other forms show a much greater diversity despite only a few fossils. Perhaps the temnospondyls were more local inhabitants than some of the less common forms. Of these other forms, the anthracosaurs are the most numerous.

Anthracosaurs are a group of fossil tetrapods that are best known from specimens dating from the Late Carboniferous. In the relatively recent past, they were considered important from the point of view of the origin of amniotes, but although anthracosaurs are still thought to be more closely related to amniotes than they are to temnospondyls, the relationship may not be as close as was once thought. Anthracosaurs and their relationships will be discussed in the next chapter.

At East Kirkton, there are two named species of anthracosaur, which differ from each other in many respects. These are *Silvanerpeton miripedes* (Clack 1994b) and *Eldeceeon rolfei* (Smithson 1994).

Silvanerpeton miripedes

The name *Silvanerpeton* derives from the Roman god of the woods, Silvanus. It was created to honor its discoverer, Stan Wood, and the fact that the East Kirkton quarry was a wood when Stan rediscovered the site. The specific name *miripedes* ("wonderful feet") refers to the fact that the holotype specimen is preserved with both hindfeet intact and in beautiful detail, showing an extra toe bone in the fifth toe that is unique to anthracosaurs (Fig. 8.12).

Most of the *Silvanerpeton* specimens are quite small, showing an animal about 200 mm from snout to pelvis, with a skull about 44 mm long. The length of the tail, as with so many East Kirkton specimens, is unknown. Its body was covered ventrally by an armor of interlocking scutes, narrower than those of *Balanerpeton*, and unlike *Balanerpeton*, they were plain rather than striated. This feature is sometimes helpful in diagnosing very incomplete specimens as one or the other. Its lozenge-shaped interclavicle was more elongate than that of *Balanerpeton*, and again, this is a useful distinguishing character. One of the most conspicuous differences between *Silvanerpeton* and *Balanerpeton* lies in the ribs. In *Silvanerpeton*, the ribs were elongate and curved and must have reached at least halfway round the body, even quite far back along the abdomen. Where *Balaner-*

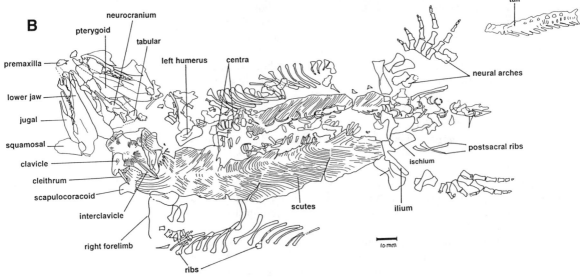

Figure 8.12. (A) Silvanerpeton miripedes *holotype specimen (UMZC T1317). Photograph from UMZC archives. (B) Interpretive drawing. From Clack (1994b).*

peton had four, *Silvanerpeton* had five fingers on the hand, in common with most other early tetrapods except the Devonian forms.

The key to recognizing an anthracosaur really lies in the skull, and in particular the relationship of the bones of the skull table to one another. In the skull table, rather than having the primitive pattern in which the postparietal stretches across the skull table to meet the supratemporal (e.g., *Acanthostega* and *Ichthyostega*), in anthracosaurs an alternative and mutually exclusive pattern is found in which there is a contact between the tabular and parietal bones (see Chapter 8 for more details). *Silvanerpeton* shows this feature, which marks it as belonging to the anthracosaurs. Unfortunately, some of the other key features are either lacking or are not preserved sufficiently well to be sure about. Therefore, the position of *Silvanerpeton* with respect to other anthracosaurs is a bit uncertain. Some of the features that *Silvanerpeton* does share with anthracosaurs are apparently primitive features that might be expected in many early tetrapods and that are seen, for example, in *Acanthostega*. These include a closed palate, without any of the large vacuities that characterize temnospondyls (see Fig. 9.17 for an anthracosaur palate). At the point where the palate and braincase meet, the braincase sends out a prominent process that fits into a socket on the palate in what appears to have been a synovial joint. This contrasts with *Balanerpeton,* where this junction is considered to have been immobile. A moveable joint here may be left over from the condition found in osteolepiform fishes, although in tetrapods even as early as *Acanthostega,* the articulation had specialized to a degree.

Where the skull table joins the cheek, anthracosaurs have a conspicuous joint known as a kinetic line. This abrupt junction has been considered to have allowed a limited amount of flexibility between the two skull units, reflecting that which was found in osteolepiforms at the same point. It has been thought to be related to the presence of the mobile palatal articulation. *Silvanerpeton* shares this character with many later anthracosaurs.

The back of the cheek sloped diagonally backward and downward from the corner of the skull table, in a manner that at one time would have been considered to have formed an otic notch, as in *Balanerpeton*. For reasons that are addressed in the last chapter, this interpretation now seems unlikely.

Eldeceeon rolfei

This animal is only known from three specimens, but they show a very different animal from *Silvanerpeton,* despite being about the same size and apparently related (Fig. 8.13). *Eldeceeon*'s name comes from the letters LDC, which stand for the Livingston Development Corporation, who contributed to the purchase of the specimen for the Royal Museum of Scotland. It has a much more robust skeleton than *Silvanerpeton,* with much more heavily ossified vertebrae and limbs. Like *Silvanerpeton,* it has elongate scutes, but its interclavicle, rather than being lozenge-shaped, has a narrow posterior process. Other differences between the two include a shorter vertebral column anterior to the pelvis and a ribcage that seems only to have surrounded the anterior part of the trunk, much in the same way as it does in mammals, but in contrast to the more complete ribcage of reptiles such as lizards or crocodiles. Alternatively, the more posterior ribs may simply have been lost from the specimen. The ribcage shows no evidence of shorter ribs toward its posterior end, which one would expect normally.

The most conspicuous difference lies in the hindlimbs. In *Silvanerpeton,* they are relatively short, only a little over one-third of the length of the skull, whereas in *Eldeceeon,* they are nearly half the length of the skull. The ankle is sturdily ossified, and the feet appear very large for the size of the body. Although in some respects *Silvanerpeton* looks more like an aquatic form, or perhaps one for which only subadult and incompletely ossified individuals are preserved, *Eldeceeon* appears to have been unquestionably terrestrial. Its relationships are unclear because many details of the skull roof are poorly known, but some features, such as the closed palate and kinetic line and features of the vertebrae and pelvic girdle, sug-

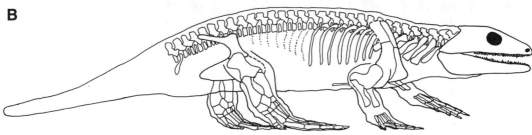

Figure 8.13. (A) Holotype specimen of Eldeceeon rolfei *(NMS G1986.39.1p). Photograph by N.M.S. (B) Drawing of skeletal reconstruction of* Eldeceeon. *(There are no ribs preserved posterior to those illustrated.) From Smithson (1994). Scale bar = 30 mm.*

gest that it belongs with the anthracosaurs. For neither *Eldeceeon* nor *Silvanerpeton* is there evidence of what they might have eaten because the teeth of the jaws are unspecialized and those of the palate are unknown.

Eucritta melanolimnetes

One of the most recently described species from East Kirkton is one that produced great puzzlement among the paleontologists who first looked at it. Those who worked on anthracosaurs shook their heads and thought, no, it is not an anthracosaur, it must be a temnospondyl—look at the shape of its head and its large otic notch. Those who worked on temnospondyls also shook their heads and said no, it is not a temnospondyl, it must be an anthracosaur—look at its closed palate, its lozenge-shaped interclavicle, its double-pronged ilium. For this animal showed a mixture of characters normally associated with each of these two separate groups, but not normally found together. It was given the name *Eucritta*, based on the American vernacular word "critter," meaning "creature," and the specific name *melanolimnetes* meaning "from the black lagoon," a reference to the nature of the East Kirkton locality (Clack 1998a).

Eventually, five specimens were recognized, of different sizes in a growth sequence. The specimen chosen as the holotype was a medium-sized example, with some well-preserved postcranial material, although no vertebral elements (Figs. 8.14, 8.15). The skulls showed subtle but consistent differences from *Balanerpeton,* despite their superficial similarities, although the rounded shape, large open notch, large orbits, and short snout were originally deceptive. The closed, denticulated palate and the long, stylus-shaped parasphenoid were not those of a temnospondyl. Significantly, the largest and best-preserved skull roof showed that the orbits had slight but clear anterior extensions, also unlike *Balanerpeton.*

The animal had quite a short body, with robust hindlimbs and large feet, not unlike *Eldeceeon* in proportion, but its ribs were shorter and straighter, more like those of a temnospondyl than an anthracosaur. The forelimbs were badly preserved, so it was not possible to say whether it had four or five digits on the hand. Cladistic analysis has suggested that *Eucritta* may belong with a later Carboniferous group called the baphetids (see Chapter 9).

Westlothiana lizziae

Among the many specimens from East Kirkton that were found by Stan Wood was one that caused particular excitement. When it was first discovered and studied, only one specimen was known, and preliminary results suggested that this animal was the earliest known reptile. It received the name "Lizzy the lizard" from Stan. The reason for the excitement was that previous to this discovery, the earliest known reptile was a creature called *Hylonomus* (see Chapter 9) from a locality called Joggins in Nova Scotia, in rocks about 15 million years younger than those of East Kirkton. An amniote from East Kirkton therefore would put back the known origin of this important group by this considerable amount of time.

The specimen was purchased by the Royal Museum of Scotland with help from West Lothian District Council, the district in which East Kirkton lies and whose name was thus incorporated into the generic name of the animal, *Westlothiana*. It was given the specific name *lizziae,* after Stan Wood's nickname for it. Unfortunately, the specimen was not well preserved in some key areas that would truly establish the creature's identity as an amniote. However, a little while later, a few more specimens came to light, allowing more thorough study of this species.

What all the information showed when put together was a small, rather long-bodied animal with short legs (Figs. 8.16, 8.17) (Smithson et al. 1994). The vertebrae are robust, with the centrum of each vertebra fused to its neural arch. This is one of the features that first made *Westlothiana* look like an amniote and is one generally associated with terrestrial forms. Although the limbs are rather short, the ankles are well ossified, and the bones of the limbs are themselves gracile, rather than short and stubby. It was the form of the skull that helped form the first impression of an amniote, however. The back of the skull shows no otic notch (see Chapter 10), and there are fewer or smaller bones in the temporal region than in almost any other contemporary nonamniote group. The tooth row extends far forward in front of the eye but not very far behind it, and the eyes are large (Fig. 8.17).

The additional specimens showed some regions that were not visible in the first one. They showed that a feature such as a particular flange on the palate, diagnostic of early amniotes, was in fact not present. Most amniotes have two enlarged bones in the ankle, the astragalus and calcaneum, which were also absent in *Westlothiana*. The temporal series of bones was difficult to interpret, although they seemed most probably to be amniotelike, whereas other potentially diagnostic characters, such as the form of the occipital condyle, were not visible in any specimen.

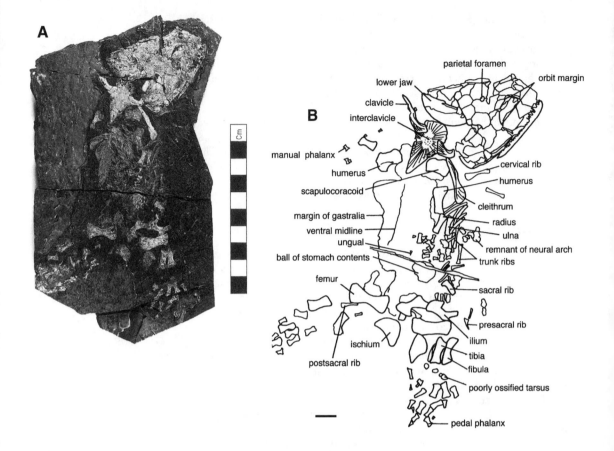

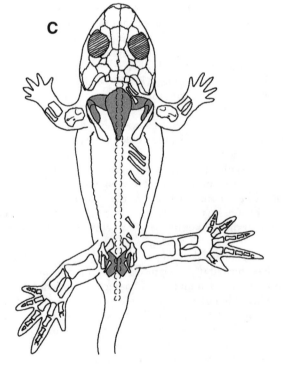

Figure 8.14. Holotype specimen of Eucritta melanolimnetes *(UMZC T1347). (A) Photograph by S.M.F. (B) Interpretive drawing. (C) Skeletal reconstruction of* Eucritta. *The inner surfaces of the girdles that are visible in dorsal view have been shaded. Modified from Clack (2001a).*

Figure 8.15. An isolated skull of Eucritta (NMS 1992.14) showing its similarity to Balanerpeton. This specimen has split into part and counterpart, with the skull roof bones mainly on one side (left), and the natural mould of the internal surface on the other (right). In this photograph, the part specimen has been placed alongside the counterpart. Photograph by S.M.F.

Figure 8.16. Westlothiana lizziae. Specimen in part and counterpart (NMS G1990.72.1). Scale bar = 10 mm. Photograph by N.M.S.

An Early Terrestrial Fauna • 231

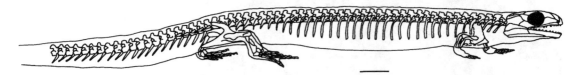

Figure 8.17. Skeletal restoration of Westlothiana. *From Smithson et al. (1994).*

When included with other amniotes and nonamniotes in a computer program (PAUP) that analyzes data to produce phylogenetic trees, the position of *Westlothiana* became a little clearer. Rather than being a true amniote, it slotted into the bottom of the family tree of the group containing amniotes; in other words, it belongs to the stem-group of amniotes. This group also contains the anthracosaurs, although they slot in below *Westlothiana*.

If this analysis is correct—and the difficulties of studying this animal mean there is some doubt about this—it is nonetheless an important find. It means that at East Kirkton, representing a period about 333 million years ago, members of both the lineage that gave ultimate rise to modern amphibians (*Balanerpeton*) and that which gave rise to amniotes (*Westlothiana*) were already present. The implications of this are that these two lineages must have separated before this time, and most likely well before it. The subsequent discovery of *Casineria* supports this view.

Despite the attractiveness of this hypothesis, there is an alternative that must be considered, and further study may resolve the problem in its favor. This hypothesis argues that *Westlothiana* is not a stem amniote at all but belongs to a separate group, the microsaurs (see Chapter 9). Many of its features are consistent with its being a microsaur, and all the features that could resolve the issue one way or the other are either poorly preserved or not visible on the known specimens. Only further specimens could really prove the case.

Other Tetrapods

There are at least three other taxa of tetrapods from East Kirkton, one of which is still undescribed. One of them is a member of the very specialized group called the aïstopods. Chapter 7 introduced *Lethiscus,* which is the earliest known member of the group, but at East Kirkton, the genus *Ophiderpeton* is represented by two or three very poorly preserved specimens (Milner 1994). This genus is also found in rocks of the Late Carboniferous of Ireland and shows some features in common with other, later aïstopods that *Lethiscus* does not show. There is some degree of specialization of the teeth among aïstopods, with *Ophiderpeton* showing broad, conical, recurved, and widely spaced teeth, in contrast to later genera in which the teeth are narrow, straight-sided, and closely packed in the jaw. However, the different niches that these peculiar creatures occupied are matters for speculation. It is thought that with their snakelike bodies and lack of limbs, they were terrestrial rather than aquatic.

The microsaurs may still be represented at East Kirkton, even if *Westlothiana* is not one. These creatures became quite common during the Carboniferous (see Chapter 9), and by the Permian, they were highly specialized. A single specimen from East Kirkton is possibly their earliest known member and the only one recognized in the United Kingdom. The single specimen is very poorly preserved, and the main diagnostically microsaur feature is the form of the gastralia. These show the characteris-

Figure 8.18. Reconstruction of the East Kirkton landscape and fauna. The scene shows two Balanerpeton *in the foreground, with* Hibbertopterus *at the left and a harvestman and a scorpion close by on the rocks.* Silvanerpeton *and* Eldeceeon *are seen in the background. Illustration by J.A.C.*

tic fine striations found in many microsaurs but in no other groups. This specimen is unusual in having a deepened swimming tail, with webbing above and below the tail vertebrae formed by soft tissue, not by bony processes.

Overview

The single locality of East Kirkton, with its unusual preservation, gives invaluable insight into the early diversification of tetrapods, as well as other terrestrial organisms. It provides minimum dates for the origin or separation of major tetrapod groups that can be compared with results from molecular analyses of living forms. It gives a glimpse of some of the range of niches into which tetrapods had evolved by this early stage, as well as some of the interactions that went on between vertebrates and invertebrates. Whichever analysis of tetrapod interrelationships is accepted (see Chapter 9), the lineages leading to modern amniotes and modern amphibians had already split from each other by the time represented by the East Kirkton locality. Although excavations have finished for the present time, much of the tetrapod-bearing horizon remains to be explored, and it is to be hoped that the locality has not ceased to yield unknown creatures and will continue to provide important surprises for early tetrapod paleontology.

Figure 8.18 shows a reconstruction of East Kirkton as it might have looked in the Early Carboniferous.

Nine
The Late Carboniferous: Expanding Horizons

Late Carboniferous/Early Permian Biogeography and Paleoecology

During the Late Carboniferous, the continents, which had slowly moved southward through the Devonian and Early Carboniferous, changed direction and began to rotate so that Gondwana and Euramerica gradually collided, initiating the formation of the supercontinent Pangaea. The world's vegetation had differentiated into continental regions so that, for example, the Gondwana flora became quite distinct from those of Euramerica and of what are now China and Siberia. At this time, Euramerica, positioned in the tropics, was covered by a vast swamp forest, whereas to the north and south of it, evaporite deposits speak of arid climates (Milner 1993a).

The forest has provided an excellent fossil record of both plants and animals. There was a huge increase in diversity not only of plants but also of invertebrates and vertebrates. Partly this can be explained by the quality of the fossil record and the distribution of fossil collectors, but the increase of taxonomic diversity, body forms, and ecologies among the animals parallels but lags behind that among the plants, suggesting that the phenomenon is real, not simply an artifact. In the Late Carboniferous, more orders and classes of plants were contributing more diversity to the flora than at any other time in Earth's history (DiMichele and Hook 1992).

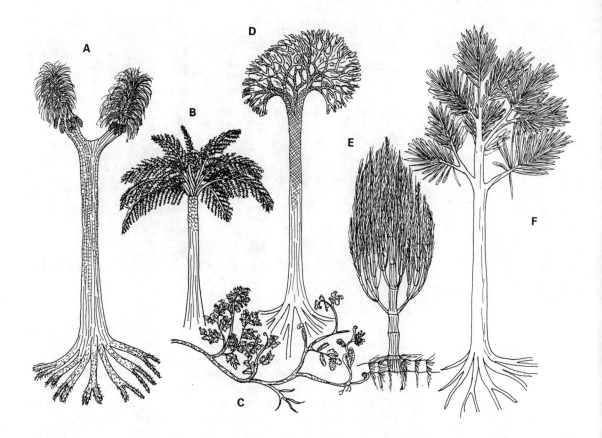

Figure 9.1. Drawings of Late Carboniferous plants.
(A) Sigillaria *sp.*, a lycopsid (clubmoss) with a trunk up to 1 m in diameter.
(B) Psaronius *sp.*, a tree fern related to the modern family Marattiales.
(C) Callistophyton *sp.*, a trailing pteridosperm.
(D) Lepidodendron *sp.*, a lycopsid (clubmoss) up to 54 m high.
(E) Calamites carinatus, a horsetail (Equisetales), with a trunk up to 0.5 m in diameter.
(F) Member of the Cordaites family, a gymnosperm with trunk diameter up to 1 m.

Most of the plants from this time are those of the lowland swamps, but a few localities preserve more upland floras showing adaptations to drier conditions. The dominant plants were still lycopsids in the form of lepidodendrons. These grew into huge trees up to 30 m tall. They had a life history like no modern plants. In their prereproductive phase, they consisted of a single polelike trunk that at a certain point in their growth sprouted a crown of branches bearing the conelike fruiting bodies. The branches consisted of thickened woody tissue, making them unappetizing to arthropod herbivores, but the cones, containing a rich filling of spores, would have been appreciated. Some of them were eaten by the large *Arthropleura* type of millipedes (see Chapter 7 and below). Probably only flying insects could reach the cones borne in the tops of the trees until the cones were shed (DiMichele and Hook 1992). Figure 9.1 shows a selection of Late Carboniferous plants.

The lepidodendroid trunk was formed of a relatively thin but tough outer casing, whereas the inner tissue was soft and open, unlike the structure of woody trees today. When a lepidodendroid died, the inner material rotted faster than the outer, sometimes leaving a tall, hollow cylinder standing upright after the rest of the plant decayed. This construction is important for vertebrate paleontologists because it has led to the occasional and very unusual preservation of some small, terrestrially adapted animals.

At one or two localities, notably South Joggins in Nova Scotia, hollow lepidodendroid trunks stayed upright long enough for the sediment to

Figure 9.2. A rotten palm tree base photographed in Myakka State Park in Florida, showing how the Carboniferous lycopod tree stumps might have appeared at Joggins. Photograph by J.A.C.

accumulate around them until it was level with the top of the trunk cylinder. The hollow cylinder then became a very effective pitfall trap into which passing unwary small tetrapods fell and were unable to escape. Figure 9.2 shows a palm trunk in Florida, decayed into a similar type of pitfall trap, subsequently filled with leaf litter. The unfortunate side effect for paleontologists was that the animals then preyed on and scavenged each other until their remains became almost totally disarticulated and mixed up, causing years of taxonomic confusion for those trying to understand them. Nevertheless, without these flukes of preservation, almost nothing would be known about small, terrestrially adapted tetrapods of the Late Carboniferous.

Pteridosperms were very diverse during the Late Carboniferous, and some reached heights of 6–10 m. Individual leaves could be several meters long, and when they died, the fronds remained on the trees, drooping around the trunk in a skirt that would have provided important refuges for small arthropods or tetrapods (DiMichele and Hook 1992). As with the lepidodendroids, the stems and branches offered tough food for herbivores, but the fruiting bodies, once the protective outer coat had been penetrated, would have provided a rich source of food. As well as trees, scrambling or vinelike habits were common among pteridosperms. True ferns were also part of the flora, low in numbers but high in diversity. Both low-growing, ground-covering forms and tree ferns were present, as well as climbing and vinelike forms. Their leaves were soft and might have been eaten by herbivores, but today's ferns produce toxins to deter this. Perhaps this ability evolved first in the Carboniferous with the first radiation of herbivorous insects.

Sphenopsids, relatives of modern horsetails, provided major ground-covering forms in the sphenophylls, whereas the calamites formed trees with trunks up to half a meter in diameter. The sphenophylls formed thickets with branches linked by hooks and barbs, much like brambles do today. The undergrowth would have been impenetrable to most large animals, but it formed cover for smaller ones.

Cordaites evolved and radiated explosively in the Westphalian. They were quite closely related to modern conifers, and like them, many of the

Figure 9.3. Trail of a giant arthropod found at Joggins. The rock hammer gives scale. Photograph by J.A.C.

Carboniferous forms are associated with drier environments. They had true seeds, sometimes winged and sometimes with fleshy outer coats that could have attracted animals to eat them, a device used by many modern plants to aid dispersal. True conifers also evolved during this stage, and as with modern forms, many had needlelike leaves and were associated with drier environments. Many examples are preserved as fossil charcoal, or fusain, implying that the conifer forests were periodically burned.

Much has been learned about Late Carboniferous arthropods from one or two exceptionally preserved sites. Mazon Creek in Illinois has produced nodules in which soft body parts are preserved, so that as well as information about arthropods, there is information about soft-bodied invertebrates and vertebrates such as lampreys (Shabica and Hay 1997). The diversity of terrestrial invertebrates increased greatly during this period, including the first appearance of land snails. Many of these invertebrates were feeders on detritus, both plant and animal, such as the myriapods—arthropleurid millipedes. The latter grew up to 2 m in length, and their large size suggests that at this stage, they were not sharing their niche with tetrapods. Figure 9.3 shows the trail of an arthropleurid preserved at Joggins in Nova Scotia. As with the East Kirkton fauna, some of the millipedes were equipped with glands that produced toxic substances, and the smaller forms were eaten by tetrapods, as shown by fossilized gut contents in a few specimens.

Much evidence from chewed leaves and bored seeds and the mouthparts of insects themselves suggests that herbivory was well developed in

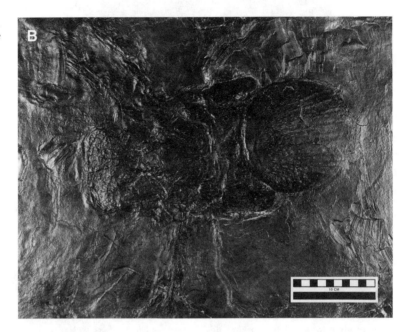

Figure 9.4. Giant fossil spider, Megarachna *from the Late Carboniferous in Argentina. (A) Life reconstruction from the Sedgwick Museum, Cambridge, England, UK. Scale bar = 30 mm. (B) Photograph of a replica in the Sedgwick Museum, Cambridge, England. Photograph by S.M.F.*

many forms. It has been suggested that insect flight and herbivory coevolved, allowing insects to exploit the tops of the trees (DiMichele and Hook 1992). Many Carboniferous insects reached large sizes, the most famous being a primitive form of dragonfly with a 430-mm wingspan. Its flight and landing abilities were probably quite limited. Primitive cockroaches and their relatives were common, and early relatives of grasshoppers and plant-sucking bugs made their first appearance. Many insects were predatory and could have been in competition with tetrapods; some were big enough to have preyed upon small individuals, such as juvenile amphibians (DiMichele and Hook 1992). Spiders and scorpions lurked among the plant debris, and although most spiders were small, one fossil spider had a leg span of 400 mm (Fig. 9.4). Many of the insects, perhaps most, had aquatic larvae, as do their modern descendants. The larvae may have formed food for tetrapods, but again, as today, large predatory forms such as dragonfly nymphs may have preyed upon larval or hatchling tetrapods.

A link has been suggested between the increased oxygen supply in the Late Carboniferous, the concomitant higher air density, and the appearance of winged insects, as well as the subsequent evolution of giant insects and other arthropods toward the end of the period (Dudley 1998). It has been suggested that the denser air had a direct effect facilitating the evolution of flight, that more oxygen in the air allowed the animals to be more active, and that their means of gaining oxygen by their system of minute tubes throughout the body (the tracheal system) was better able to distribute the oxygen under these conditions (Graham et al. 1997). Similarly, this would enable the animals to reach a much larger size than is possible in today's relatively oxygen-poor air. Not only air density is affected when the proportions of these gases change; the viscosity, diffusivity, and heat conductivity also change. Some of these would have had unknown effects on the evolution of the animals of the time. One of the effects of increased oxygen in the atmosphere might well have been to provoke the evolution in animals of mechanisms to prevent or to deal with oxidative damage to tissues.

Another link has been suggested between increased oxygen levels in the Carboniferous and the prevalence of deposits of fusain, evidence of forest fires. Such a high level of oxygen is almost at the point at which organic matter spontaneously combusts, and certainly if a forest fire started, it would spread widely and rapidly (DiMichele and Hook 1992; Graham et al. 1997).

The increasing diversity among tetrapods that began in the Early Carboniferous continued and intensified during the Late Carboniferous, so that by the end of the period, the foundations were laid for the appearance of most major groups, including those that remain today (Milner 1987). However, most of the information comes from a handful of especially productive Westphalian localities, most of which, except for South Joggins and a similar site at Florence, Nova Scotia, preserve predominantly aquatic forms. At Linton, Ohio, and Nýřany in the Czech Republic are found the faunas of relatively shallow lakes. Linton shows the animals preserved in an abandoned river channel (Hook and Baird 1986), although Nýřany seems to have been a shallow pool that existed for perhaps no more than a few hundred years (Milner 1980b). Despite some environmental differences and their geographical separation, the two share many genera in common. Newsham, in Northumberland, United Kingdom, like Linton, was an abandoned river channel, but the fauna preserved there, consisting for the most part of larger animals than are found at Linton, is indicative of deeper water.

A few localities investigated in recent years show clear indications that they experienced transitions from marine to freshwater sequences. For example, three localities in North America, Robinson, Hamilton, and Garnett, in Kansas, form a series showing increasing influence from marine or tidal conditions, yet they have all produced tetrapods and represent coal swamp environments. They demonstrate that coal swamp flora and fauna could tolerate at least some salinity and that such assemblages cannot be assumed to be associated exclusively with freshwater (Schultze 1995; Schultze and Maples 1992; Schultze et al. 1994).

Westphalian and Stephanian tetrapod localities are more or less geographically restricted to Europe and North America. As yet, virtually nothing has been found from these periods in other parts of the world (Fig. 9.5). As far as Gondwana is concerned, the reason for this is that during this part of the Paleozoic, it was undergoing a prolonged period of glaciation—it was simply too cold for the tetrapods to live there (Milner 1993a).

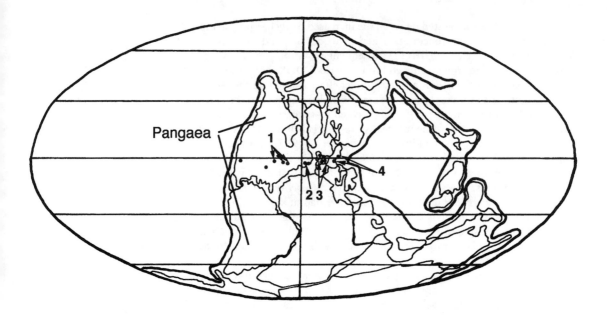

Figure 9.5. Late Carboniferous paleocontinents with major tetrapod sites indicated.
(1) North American sites including Mazon Creek (Ill.) and Linton (Ohio).
(2) Canadian sites including Joggins and Florence (N.S.).
(3) English, Scottish, and Irish sites including Jarrow (Eire), Scottish Midland Valley sites, Newsham, and Yorkshire sites.
(4) Czech and German sites including Nýřany and Tremosna. All are grouped around the Equator, most to the south of it.

However, it is less clear why tetrapods from this time have not been discovered in what were the more easterly parts of the northern Paleozoic continents—today's Asia. The reason for this may be historical and may have more to do with collecting bias. Perhaps the tetrapods are there but are still waiting to be found.

During the Early Permian, which followed the Stephanian stage, the tetrapods reached a kind of phylogenetic stability. About 60 families of tetrapods are known from the Late Carboniferous and Early Permian, and all except a very few occupied the same biogeographical region (Milner 1993a). This was the equatorial European and North American continental landmass. There is some evidence that tetrapods lived beyond the limits of this region by the later Late Carboniferous or early Early Permian. Two tetrapod footprint trails have been found from this period in South America, which would have been well into the Gondwanan continent at that time (Bell and Boyd 1986). Also, one species has been found in what is now North Africa, which similarly represents a more southerly radiation. One little group of tetrapods, the fully aquatic amniotes called mesosaurs, are known from Gondwana in the Early Permian, but they represent almost the only tetrapods from the Early Permian of Gondwana. They are significant in several ways. Not only are they the first amniotes known to have been marine, but their distribution across South America and southern Africa was one of the first pieces of evidence to support the idea that these two continents had once been joined together. There are three genera in the group, and their distribution is known in some detail, with shallow-water forms found around the periphery of what was an extensive seaway, and deeper-water forms found more centrally (Oelofsen 1987) (Fig. 9.6).

From the early Permian, the climate had been slowly changing; by the end of this period, the coal swamps had all disappeared. In the equatorial region of the new supercontinent, they were replaced by red-beds and evaporites, indicating the return of a more seasonal climate. Several tetrapod families disappeared at this time, although at least one new one appeared (Milner 1993a).

The next section of this chapter examines some of the major tetrapod

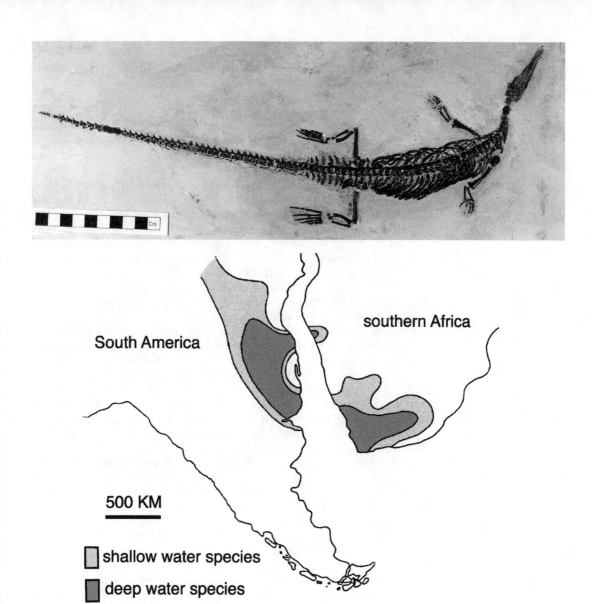

Figure 9.6. Specimen of one of the Brazilian mesosaurs (UMZC T1342) and map showing the distribution of the group during the Late Permian. Photograph by S.M.F.

groups of the Late Carboniferous and Early Permian to show how the roots of the family tree found at East Kirkton had produced a bush of many and varied forms. Some of these are strikingly modern in appearance, whereas others resemble no creature living today.

Tetrapods of the Coal Swamps

Baphetids

Among the most bizarre and enigmatic tetrapods of the Carboniferous period are the baphetids. Although the earliest of them occurs in the Early Carboniferous, the majority are Westphalian. These animals are known from a few skulls and lower jaws, but almost no fossils of their vertebrae or other postcranial remains have been found. At this time, very little is known about their limbs, except for two specimens that show some shoulder girdle elements, one of which also preserves a humerus, a radius, and a few phalanges. Baphetids were among the first of the Carboniferous fossil tetrapods to be found and were first described by William Dawson (1863). They were discovered at a time when Paleozoic tetrapods were only just beginning to be studied, and they helped found ideas about what early tetrapods were like.

Four genera comprise the core of this group: *Loxomma, Baphetes, Megalocephalus,* and *Spathicephalus,* with only one or two species in each (Beaumont 1977; Beaumont and Smithson 1998; Milner and Lindsay 1998). *Eucritta,* described in Chapter 8, may be the most primitive member of this family (Clack 1998a, 2001a). Each of the four core genera has its own peculiarities, and very few unique characters tie them to each other. The one distinctive feature they share is one that is easily recognized and described but much less easy to explain: in each member of the group, the eye socket has a forwardly directed extension, making what has been called a keyhole-shaped orbit.

Loxomma ("slanting eyes") is from rocks of the Scottish Midland Valley dated between the Viséan and the middle of the Coal Measures. In some ways, it is the most primitive genus of the four core genera. The skull was about 250 mm from the tip of its snout to the back of the cheek, with a bluntly rounded snout. It was rather flattened in profile, but the eyes would have been positioned quite high on the head, to allow it to look out of the water while the rest of the animal was submerged. The jaws and palate were armed with a battery of large, curved, and slightly keeled teeth, making the animal a formidable predator.

At the back of the skull, the skull table included a small bone called the intertemporal, which is thought to be a remnant feature left over from the tetrapods' fishy ancestors, although it is not found in the skull roofs of *Acanthostega* or *Ichthyostega*. The intertemporal is lost in most later tetrapods, including most of the other baphetids. The skull table was marked off from the cheek region by a strongly defined embayment. This feature has been called an otic notch, but more recent research suggests that rather than having any connection to hearing, it was more likely to be connected with breathing and to have housed a spiracle (see Chapter 10).

Megalocephalus ("big head") is a later, larger, and more common relative of *Loxomma,* known from the Coal Measures of Scotland, England, and Ireland. It had a longer, although relatively narrower and more pointed head than *Loxomma,* reaching about 350 mm in length, possessing even larger teeth than *Loxomma* and with an altogether more crocodilelike appearance (Fig. 9.7). Although nothing of its skeleton is known apart from the head, it must surely have been an active predator, perhaps occupying a similar niche to that of crocodiles today. It probably fed mainly on fish caught in the swampy waters of the coal measures forest, although it presumably ate other tetrapods. Its legs have never been found, so whether it ventured onto the banks or marshy ground to catch more terrestrial prey remains speculation.

Baphetes was contemporary with *Megalocephalus* but seems to have occupied a different niche. Its head was much broader and flatter, although it possessed numerous large teeth like its relative and was also clearly predatory. Figure 9.8 shows a recently discovered specimen of a baphetid skull that probably represents a new species. It was found as an erratic boulder on the coast of Northumberland, England, where it had fallen out

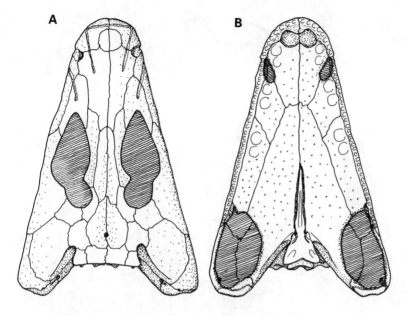

Figure 9.7. Drawing of skull of Megalocephalus *in (A) dorsal and (B) ventral views. Skull is about 300 mm in length. From Beaumont (1977).*

Figure 9.8. Skull of an undescribed baphetid specimen (NEWHM:2000.H845). Photograph by S.M.F.

of the overlying boulder–clay deposit. Presumably it had been picked up during the Pleistocene glaciation from some unknown locality and moved along until it eventually eroded out of the cliff.

The relationships of baphetids one to another have not really been worked out, and it is not clear if the broad-headed *Baphetes* was an ancestor or a descendant of possibly the most bizarre of them all, *Spathicephalus*. Certainly *Spathicephalus* was earlier than *Baphetes*, coming from the Namurian and early Westphalian of Scotland and Nova Scotia, but it appeared later than *Loxomma*, so time alone is not much help in judging relationships.

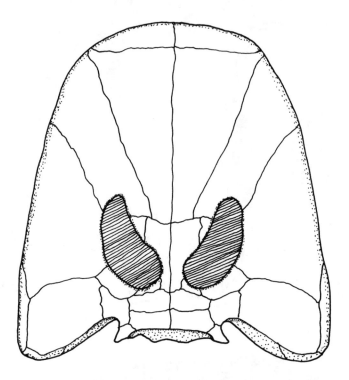

Figure 9.9. The skull of Spathicephalus, *about 220 mm long. From Beaumont and Smithson (1998).*

Spathicephalus ("spade head") had a skull that was almost totally flat and shovel-shaped, and in this genus, the form of the eyeholes was taken to extremes (Fig. 9.9). The eye sockets were very close together in the middle of the skull, with the unique baphetid keyholes giving the face a very strange appearance. It is possible to imagine its eyes raised above the surface of the skull—as, for example, in a flatfish or a frog. It is not clear what *Spathicephalus* may have fed on, but its teeth were unlike those of the other baphetid genera. Rather than being large, lanceolate piercing teeth, those of *Spathicephalus* were small, chisel-shaped, evenly sized, and very numerous. They seem to have formed a comblike structure, which might have been used for filter feeding or for picking up large quantities of small, slippery invertebrates. Unlike almost all other Paleozoic tetrapods, most of the teeth were present in the jaw at the same time, forming a continuous band. In most other contemporary vertebrates, about one-third of the tooth positions were empty or being replaced at any one time. *Spathicephalus* must have had an unusual system of tooth replacement to furnish its specialized dentition, but the mechanism and the purpose are unknown (Beaumont and Smithson 1998).

The function of the anterior extension of the orbit in baphetids has given rise to much debate. Several theories have been suggested, but objections can be found to all of them. One of the suggestions is that the space was occupied by a salt gland, ridding the body of excess salt. The bone formation around the rim, however, suggests that whatever was situated there did not lie outside the skull roof, as would be expected of a salt gland, but was housed inside (Beaumont 1977). In most analyses, the coal swamps in which baphetids lived are considered to have been freshwater, an unlikely habitat for an animal to develop a salt gland. More recently, however, it has seemed less certain that these swamps were purely freshwater, and there is some suggestion of tidal or marine influence, perhaps making the water brackish. If that were so, a salt gland might be a more likely explanation.

Another possibility is that the space was occupied by the mass of a muscle used for closing the lower jaw rapidly and powerfully, the pterygoideus muscle (Beaumont 1977). With the shallow skull characteristic of baphetids, there may have been little room for this muscle, which is thought to have run dorsally along the inside of the palate, to bulge during contraction. Extra room may have allowed the muscle to develop more power for snapping the jaws shut. Alternatively, the muscles may have originated around the rim of the hole, providing a firmer attachment. However, the course of the muscle is conjectural.

A third possibility is that the space was occupied by an electrosensory organ (Bjerring 1986). Such organs are not uncommon in modern fishes that live in murky, vegetation-choked water, and they help the animals detect electrical impulses produced by the muscles of fishes or tetrapods when vision is of limited use. All these hypotheses are, unfortunately, speculative and cannot be scientifically tested.

Although the interrelationships of the baphetids remain to be fully investigated, they have been analyzed in the context of an attempt to decide where *Eucritta* belongs phylogenetically. Putting all the available characters into a computer analysis produced some equivocal results. Five most parsimonious trees were found showing two major topologies. One placed *Eucritta* among the baphetids with either the anthracosaurs (three trees) or with the temnospondyls (two trees).

Variations within the two main topologies consisted only of alternative placements of *Loxomma* and *Baphetes* or *Gephyrostegus* and *Crassigyrinus*. A consensus tree, derived from combining the results of these five, placed *Eucritta* with the baphetids but could not resolve their position with respect to anthracosaurs and temnospondyls. This result not only reflects the mixture of characters found in *Eucritta* itself, but also the labile position that baphetids as a whole have held over the years since their discovery. *Eucritta* is only slightly older than the more conventional baphetid *Loxomma,* but it does appear to be more primitive and suggests that the baphetids arose from more terrestrially adapted forms, perhaps during the Tournaisian stage. If conventional wisdom is correct, and if amniotes share their lineage with anthracosaurs and lissamphibians with temnospondyls, baphetids appear to be basal crown group tetrapods, showing a mosaic of characters later stabilized in each of the two main tetrapod lineages (Clack 2001a).

Caerorhachis bairdi

Originally interpreted as a type of temnospondyl, this is another mid-Carboniferous animal known from a single, patchily preserved specimen (Holmes and Carroll 1977). Limited data suggest that it was basal Namurian in age. The skull, which is preserved as a natural mold, shows only a part of the external surface of the skull roof. Although some is represented by the internal surface, this is more difficult to make out, and it has been interpreted in different ways. The palate is also preserved as a natural mold, but one half of it is almost complete. It has been interpreted as having the characteristic palatal vacuities of temnospondyls, but at the same time, the skull has been described as notchless. Both these characters have been challenged by a more recent assessment that suggests that a shallow notch and small interpterygoid vacuities were present in a relatively larger skull than suggested by Holmes and Carroll (Ruta et al. 2002). The pelvis ap-

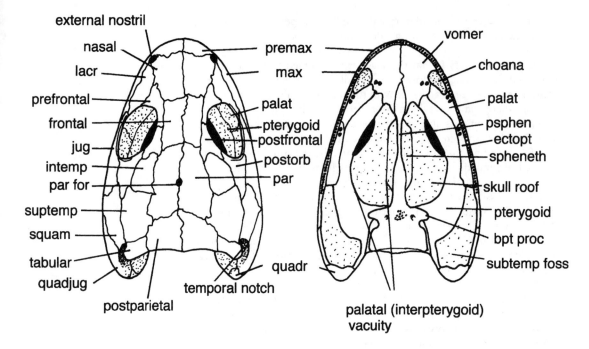

Figure 9.10. Skull of Dendrerpeton *in (left) dorsal and (right) ventral views. Abbreviations as for Figures 2.1 and 2.10, plus* intemp = intertemporal. *From Godfrey et al. (1987).*

pears to have the single dorsal process associated most closely with temnospondyls, although this may be an effect of uneven preservation. Ruta et al. (2002) suggested that the dorsal process may in fact have been present, represented by a broken base. Uncharacteristic of temnospondyls are the vertebrae of *Caerorhachis,* which are gastrocentrous, with large pleurocentra and small ventral intercentra (Ruta et al. 2002). Also uncharacteristic of temnospondyls—and indeed any other known early tetrapod—is the lining of denticulated shagreen that covers not only all of the palate, but the whole of the inner face of the lower jaw. Such a complete cover of shagreen on the lower jaw is distinctive and unique to *Caerorhachis.* Ruta et al. (2002) found that *Caerorhachis,* rather than being a peculiar temnospondyl, more consistently appears among the amniote stem lineage than elsewhere in their analysis.

Temnospondyl Diversity

By the Westphalian, the temnospondyls, whose earliest known members were found in the Viséan of East Kirkton, had radiated into a diversity of forms that paleontologists place into about 10 different families. The temnospondyls were the largest group of fossil amphibians (see Chapter 8 and below for the justification of the term *amphibian* in this context), producing the largest number of species, the largest individuals, and the greatest diversity of body forms extending over the longest time range (Milner 1990, 1993b). One of the best-known genera, and apparently the most primitive, is *Dendrerpeton* (Holmes et al. 1998). Its name means "tree creeper" in honor of the fact that its remains are found in and among the tree stumps of Joggins (Fig. 9.10). Unlike most other small, early temnospondyls with large notches, *Dendrerpeton* appears to have had a rather robust stapes, large for the size of its skull and presumably primitive in this respect (Clack 1983). Like *Balanerpeton, Dendrerpeton* was a relatively small form, apparently predominantly terrestrial, and these two animals suggest that the temnospondyls were originally a terrestrial group. However, later, the temnospondyls produced many aquatic forms, and the last ones of all to survive were entirely so.

The palatal vacuities, seen in *Balanerpeton* (Chapter 8), make one of the most conspicuous distinguishing characters of temnospondyls and it is found throughout the group. Similar vacuities are found elsewhere, so it is not, unfortunately for systematists, exclusive to the group. Other characteristics of temnospondyls include an occipital condyle formed only by the exoccipitals, vertebrae that are usually rhachitomous (later ones lose the pleurocentra altogether), a relatively small, usually diamond-shaped interclavicle, and an ilium with only a single dorsal process. None of these by itself is reliably diagnostic.

It was during the Carboniferous and early Permian that temnospondyls produced their greatest range of body forms and ecology (see Holmes 2000 for a review) (Fig. 9.11). Some, such as the trematosaurs and archegosaurs, were large, ghariallike forms with very long, slender snouts, sometimes over a meter and a half in length—although complete snouts have not been found for all genera and the intact skulls of some could

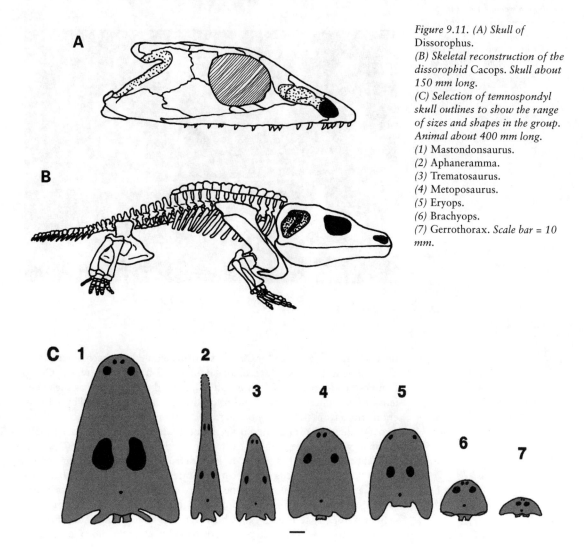

Figure 9.11. (A) Skull of Dissorophus.
(B) Skeletal reconstruction of the dissorophid Cacops. Skull about 150 mm long.
(C) Selection of temnospondyl skull outlines to show the range of sizes and shapes in the group. Animal about 400 mm long.
(1) Mastondonsaurus.
(2) Aphaneramma.
(3) Trematosaurus.
(4) Metoposaurus.
(5) Eryops.
(6) Brachyops.
(7) Gerrothorax. Scale bar = 10 mm.

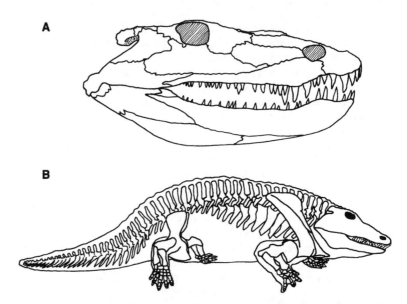

Figure 9.12. (A) Line drawing of the skull of Eryops (from Sawin 1941). Skull about 400 mm long.
(B) Skeletal reconstruction of Eryops. From Romer (1966). Animal about 2 m long.
(C) Skull of Eryops megacephalus (UMZC T250); photograph by S.M.F.

therefore have been even longer. These forms would have been aquatic and their elongate snouts adapted for snatching fish from the water. Other families produced forms with broader, flattened snouts, such as the cochleosaurids, trimerorhachids, and eryopids. Presumably, they fed by a different method from the long-snouted forms, perhaps by gulping. Again, these two families were predominantly aquatic, although *Eryops,* a Permian eryopid, was an exception in that it seems to have been largely terrestrial.

The anatomy of *Eryops* is well known from many specimens preserved in the Permian red-beds of the United States (Fig. 9.12). Before *Balanerpeton* and *Dendrerpeton* became better understood, *Eryops* was often viewed as the embodiment of not just a typical temnospondyl but as a generalized early tetrapod. Now, however, it is seen as rather unrepresentative and to have given a somewhat distorted view of the morphology and ecology of early tetrapods in general and temnospondyls in particular. *Eryops* is nonetheless interesting because recent studies have shown that several species of the genus replace each other in time and space throughout the area and span of the Early Permian of the southern United States. It could be one of the few vertebrates of use in stratigraphy (Milner 1996).

Another well-known family from the Early Permian is the Dissorophidae (Fig. 9.11). This is a family of small- to medium-sized animals with some terrestrial adaptations, including vertebrae strengthened by dermal armor plates along the back; well-ossified limbs, wrists, and ankles; and specialized nasal regions suggesting adaptations for dry or even desert conditions (Dilkes 1993).

The dissorophids are related to another suite of families that are surprisingly different from them in their ecology and body forms. These are the peliontids, micromelerpetontids, and branchiosaurids (Figs. 9.13–9.15). They are grouped with the dissorophids on their anatomy, and the whole collection is known as the dissorophoids (Milner 1990, 1993b; Trueb and Cloutier 1991) (Fig. 9.11). The dissorophoids are of most interest when discussing the possible origin of modern amphibians. These families show a series of features that suggest that pedomorphic processes were important in their evolution. These processes were at work in the evolution of tetrapods from their fish ancestors, and the origin of lissamphibians could be another instance. In these temnospondyl families, many genera, for example, are known to retain the juvenile body form with external gills and fleshy tail fins throughout their lives—some genera are thought never to have metamorphosed, much in the same way that many modern urodeles (newts and salamanders) behave, for example (Fig. 9.14). Some genera retain skulls with very short snouts, very large eye sockets, and with many of the skull bones remaining poorly ossified or absent altogether, and some have larvae that look remarkably like tadpoles (Boy 1974) (Fig. 9.15). Studies have been performed that show that these are the same bones that have been lost by all modern amphibian genera, and indeed some of the branchiosaurs in particular have skulls that look very like those of frogs (Boy and Sues 2000 and references therein). Frogs and salamanders share with temnospondyls the structure of the occipital condyle, formed only from parts of the exoccipital bones, and the palatal vacuities that are so conspicuous in temnospondyls (Milner 1990, 1993b; Trueb and Cloutier 1991), as well as peculiarities of the tooth structure (Bolt 1969, 1991). Frogs also share with some of the dissorophoids unique details of the ear structure (Bolt and Lombard 1985). It has been suggested that lissamphibians could have arisen from dissorophoids by continuation of the pedomorphic processes at work in Paleozoic members of the group (Bolt 1979; Boy and Sues 2000).

Pedomorphic processes are also thought to have been responsible for producing many of the later members of the temnospondyls. Some large forms from the later Permian were short-headed, sometimes bizarrely so, and some possibly retained external gills as well as losing the froglike ear of their ancestors (Milner 1990, 1993b).

After the end of the Permian, most families of temnospondyls were wiped out, and apparently the main survivors were some of the large aquatic forms, with those from Gondwana providing the source from which the Mesozoic forms reradiated (Milner 1990). The last surviving temnospondyls are now thought to have been Early Cretaceous (Warren et al. 1997), and these were some of the large forms that would have been gulping predators in freshwater, perhaps populating these niches before they were taken over by the newly evolving fish groups.

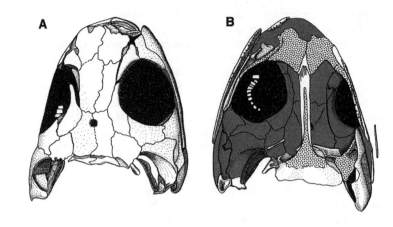

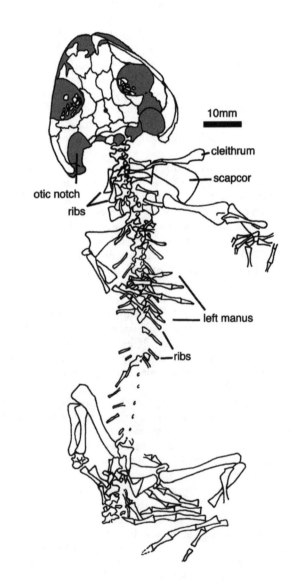

Figure 9.13. Skull of Amphibamus grandiceps (YPM 794) in (A) dorsal and (B) ventral views. Light shaded areas: in A, palatal elements; in B, skull roof bones visible through palatal vacuities. Vacuities through skull shown in black.
(C) Drawing of articulated skeleton of Platyrhinops lyelli from Carroll (1964) and personal observations (AMNH 6841). Shaded areas, matrix and palatal bones.

Figure 9.14. Branchiosaurs.
(A) Micromelerpeton credneri *showing some soft tissue preservation—note the faint close ridging on the animal's left side (UMZC T970).*
(B) Apateon pedestris. *This not only shows body wall and tail outline (as above, note the close ridging round the abdomen), but also tiny strings of ossicles from the gill filaments just behind the skull (UMZC T955). Scale bar = 10 mm. Photographs by S.M.F.*

Anthracosaurs

The anthracosaurs, from their known beginnings in the Viséan, radiated like the temnospondyls into a variety of body forms and ecologies, but their success was in some ways more limited. Their interrelationships have not been fully assessed by use of modern cladistic methods, but from work done in the 1980s, it seems that they radiated from small- to medium-sized more or less terrestrial forms to the large predatory embolomeres, which dominated the waters of the Late Carboniferous coal swamps (see Smithson 2000 for a review). One of the semiterrestrial forms, called *Eoherpeton*, is known from the Namurian of Scotland and this gave the first hint that early anthracosaurs may not necessarily have had the so-called otic notch long associated with all early tetrapods (Smithson 1985b) (Fig. 9.16).

Anthracosaurs are characterized by contact between the tabular and parietal bones in the skull table (see also Chapter 8) in combination with the presence of an intertemporal (a primitive character), a closed palate with small or no vacuities in the midline, and often a skull table that is separated from the cheek plates by a noninterdigitating suture. This has

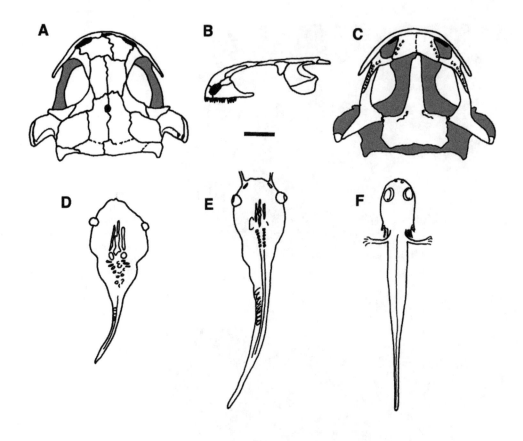

Figure 9.15. (above) (A–C) Dorsal, lateral, and palatal views of the skull of the branchiosaur Apateon. Scale bar = 10 mm. (D, E) Drawings of larval specimens of Amphibamus as preserved. (F) Reconstruction of a branchiosaur "tadpole." From Boy (1974) and Milner (1982).

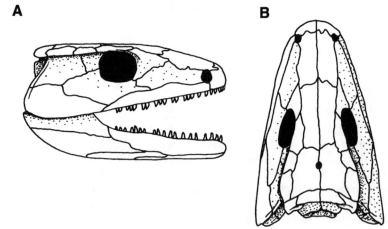

Figure 9.16. (right) The skull of Eoherpeton in (A) right lateral and (B) dorsal views. Skull about 150 mm long. From Smithson (1985b).

been considered to be a primitive character related to the kinetic skull roofs of osteolepiforms, although the story may not be so simple. Very often, there is a knob, or small button, or in some groups, a bladelike process growing out from the back of the tabular bone, called the tabular horn. This is particularly prominent in the embolomeres.

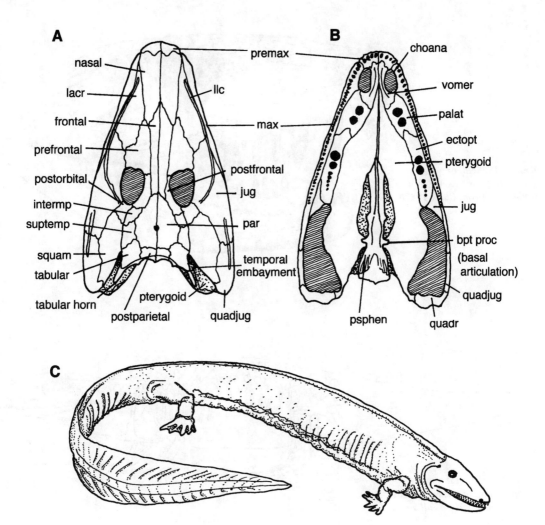

Figure 9.17. Drawing of skull and skeleton of Pholiderpeton.
(A) Skull in dorsal view. About 300 mm long.
(B) Skull in ventral view.
(C) Whole body reconstruction. Animal up to 4 m long.

The embolomeres were long-bodied, crocodilelike piscivores, whose remains are known from sites in the United Kingdom and the eastern United States (Panchen 1972b). The most primitive embolomere is *Proterogyrinus*, known from several sites in the United States and also Scotland (Holmes 1980, 1984; Smithson 1986). Its vertebrae, like those of *Eoherpeton*, were not the fully embolomerous type found in more typical forms such as *Pholiderpeton* (Clack 1987a) (Fig. 9.17). There seems to be a morphocline in vertebral construction among embolomeres, from animals in which neither the pleurocentrum nor intercentrum were complete rings, to those in which both were (Panchen 1966, 1967). The significance of this is discussed in Chapter 10. *Anthracosaurus* was probably the most fearsome predator of its day, but it is only known from its skull (Clack 1987b; Panchen 1977) (Fig. 9.18). In embolomeres the form of the tabular horn is distinctive, and quite different from the prong on the tabular found in *Acanthostega*. Only one genus of embolomere, *Archeria*, is known from the Early Permian of the United States (Holmes 1989a) (Figs. 9.20, 9.21), and a few fragmentary specimens show that anthracosaurs persisted as

The Late Carboniferous • 253

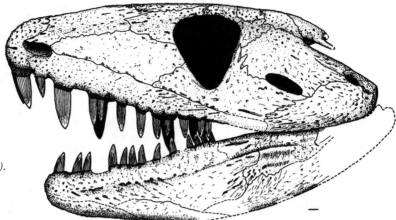

Figure 9.18. Skull of Anthracosaurus russelli, *about 400 m long (from Clack 1987b).*

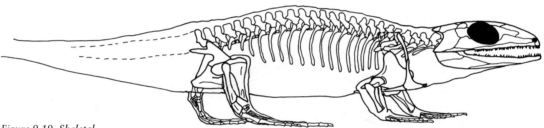

Figure 9.19. Skeletal reconstruction of Gephyrostegus *(from Carroll 1970). Skull about 50 mm long.*

Figure 9.20. Skull and neck vertebrae of Archeria crassidisca *(AMNH 7117). Scale is a one-cent coin. Photograph from UMZC archives.*

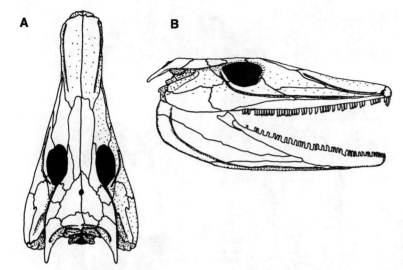

Figure 9.21. Skull of Archeria in (A) dorsal and (B) lateral views. Skull about 200 mm long (from Holmes 1989a).

rare, specialized members of the fauna as late as the Early Triassic of what is now Russia (Novikov et al. 2000).

Only two or three other genera of anthracosaur are known from substantial remains, and of these, *Gephyrostegus* is perhaps the most significant (Fig. 9.19). In the early 1970s, it was seen as being a possible ancestor to "reptiles" (Carroll 1970). Its body form is very like that of some of the early amniotes from Joggins, and anthracosaurs in general share many features with early amniotes that have been considered to show a phylogenetic relationship between them (Panchen 1972a, 1972b). Amniotes also, for example, show tabular–parietal contact in the skull roof, although they have lost the intertemporal. Anthracosaurs and amniotes have gastrocentrous vertebrae in which the pleurocentrum is the dominant element (see Chapter 10), and most of them have an ilium with two dorsal blades. Anthracosaurs are still considered to be more closely related to amniotes than they are to temnospondyls, although the relationship is now considered to be more distant than was once thought (see below). There is another group of early tetrapods to which anthracosaurs are considered to be quite closely related, and these animals were once thought actually to be reptiles. These are the seymouriamorphs, known from the Early Permian of the United States and more recently discovered in Europe.

Seymouriamorphs

Seymouria baylorensis, the best-known seymouriamorph, is known from many specimens found from the red-beds of the Late Permian deposits in Texas and other southern states of the United States. Its name derives from two counties in Texas, Seymour and Baylor, where many of the first specimens were discovered. For several decades during the 1920s to 1960s, a debate took place over whether *Seymouria* was a reptile or an amphibian (Watson 1919; White 1939). The reason for the confusion was that although its skull resembled that of an anthracosaur in some ways, its postcranial skeleton was very like that of some early reptiles (Figs. 9.22, 9.23). The skull retained an intertemporal, a primitive feature, and the skull table showed tabular–parietal contact. It is no longer clear whether the amniotes

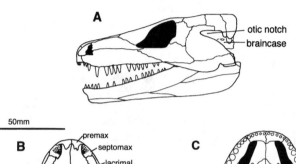
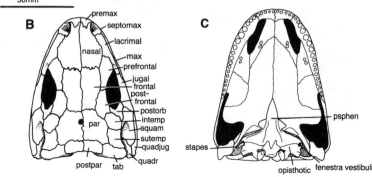
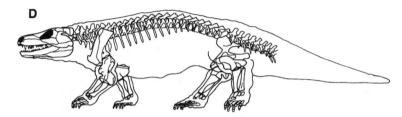

Figure 9.22. Drawings of Seymouria. (A) Skull in lateral view. (B) Skull in dorsal view. (C) Skull in ventral view. (D) Whole skeleton. A, C, and D from White (1939); B from Berman et al. (1992). Scale bar for skulls only.

inherited this contact from an anthracosaur ancestor or they acquired it separately, because amniotes have lost the intertemporal.

The seymouriamorphs also share with anthracosaurs a palatal structure, which, like those of very early tetrapods, is closed—that is to say, it lacks the palatal vacuities of temnospondyls. However, the closed palate is clearly a primitive feature, not one indicative of close relationship.

The postcranial skeleton is quite heavily built with robust limbs. The most amniotelike feature of the skeleton lies in the structure of the vertebral centra and neural arches. The centra are composed mainly of a solid pleurocentrum, attached firmly to the overlying neural arch by a suture. The intercentrum is small in seymouriamorphs and lost in most amniotes. The neural arches of seymouriamorphs are broader than they are long, and the interarticulations are set on horizontally placed, broad zygapophyses (see Chapter 10). The neural spines are domed rather than the more commonly found elongate rectangle. This domed shape is found in some groups of early amniotes, but not all, so its significance is a puzzle.

The feature that for some people decided the issue was the ear region. Seymouriamorphs have deeply embayed otic notches, and the stapes is interpreted (the evidence is not good) as a slender rod (Fig. 9.22). Although the otic region of the braincase is unique and specialized, the configuration of the notch and stapes suggested an ear like that of a temnospondyl rather than of an early amniote. Thus seymouriamorphs were interpreted to be amphibians. However, this is not the whole story. On the one hand, phylogeny suggests that the ear structure of seymouriamorphs is probably an independent development, not derived from that of a temnospondyllike

Figure 9.23. Model of Seymouria skeleton (UMZC specimen). Photograph by S.M.F.

arrangement (see also Laurin 1996, 2000). On the other hand, new evidence suggests that larval seymouriamorphs retained external gills like most modern amphibians and like temnospondyls.

Several species from eastern Europe were described from well-preserved remains showing traces of the body wall and, more significantly, of external gills like those of larval salamanders (Klembara 1995, 1997, 2000). These animals, called discosauriscids after the best-known genus, *Discosauriscus,* were eventually recognized as related to the seymouriamorphs of North America (Fig. 9.24). Good growth series of fossils from discosauriscids now show that the adult forms are more or less indistinguishable from *Seymouria* itself. This shows conclusively that seymouriamorphs had an amphibian-type life history with aquatic young stages that lost their external gills after metamorphosis. The lifestyle similarity between seymouriamorphs and temnospondyls cannot suggest a relationship, however, because reproductive strategies are very strongly influenced by ecology, and in any case, larvae with external gills are found in many fishes, so the attribute is probably a primitive one. Seymouriamorphs and anthracosaurs are usually considered to belong to the amniote stem lineage (see below), but seymouriamorphs were definitely not amniotes themselves.

Lepospondyl Groups: Microsaurs, Lysorophids, Aïstopods, and Nectrideans

In older classifications, the groups of larger tetrapods such as temnospondyls and anthracosaurs were placed together in a group called the *labyrinthodonts* (Romer 1947, 1966). Current thinking has split the labyrinthodonts into disparate, more or less unrelated groups. Collateral to the

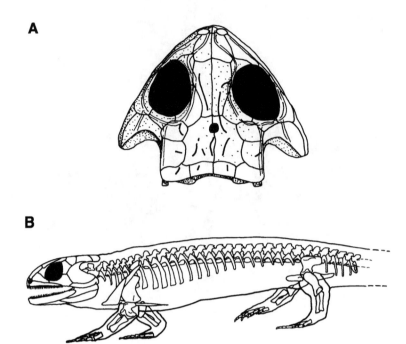

Figure 9.24. Discosauriscus.
(A) Skull in dorsal view, about 27 mm long.
(B) Skeletal reconstruction.
From Klembara (1997, 2000).

labyrinthodonts was a second assemblage, this time of small animals, called the *lepospondyls* (Carroll 1987; Carroll et al. 1998; Romer 1966). Opinions about the relationships of the lepospondyls to one another and to other tetrapods differ widely, and one recent suggestion based on a large cladistic analysis of morphological characters is that not only are they closely related to one another but that they are also closely related both to modern amphibians and to amniotes (see below) (Laurin and Reisz 1997). This controversial idea has yet to be tested by time. It is hard to tease apart morphological similarities deriving from relationship and those deriving from lifestyle and ecological similarities among small animals. Common lepospondyl features that appear to be shared specializations include the absence of any kind of temporal notch, the reduction in the number of skull bones (although not exactly comparable conditions in each), and the consolidation of the vertebrae. In all, there is a large, cylindrical centrum that is usually the sole central element. A few genera are known in which a possible intercentral element is present. The neural arches are either sutured or fused to their respective centra, making each vertebra a coherent unit.

Microsaurs

The earliest member of this group is known from the upper part of the Viséan, whereas the last microsaurs were Permian. Microsaurs comprise two rather different suborders: the predominantly terrestrial tuditanomorphs and the mainly aquatic microbrachomorphs (Figs. 9.25, 9.26). The two groups have different skull table bone patterns. Each family is quite diverse and shows a variety of body forms, some lizardlike, some like salamanders, and several with long bodies and short limbs that may have been burrowers (Carroll 2000a; Carroll and Gaskill 1978; Carroll et al. 1998).

Among the first microsaurs to be recognized were several genera from the tree stumps at Joggins, where their remains were jumbled together with

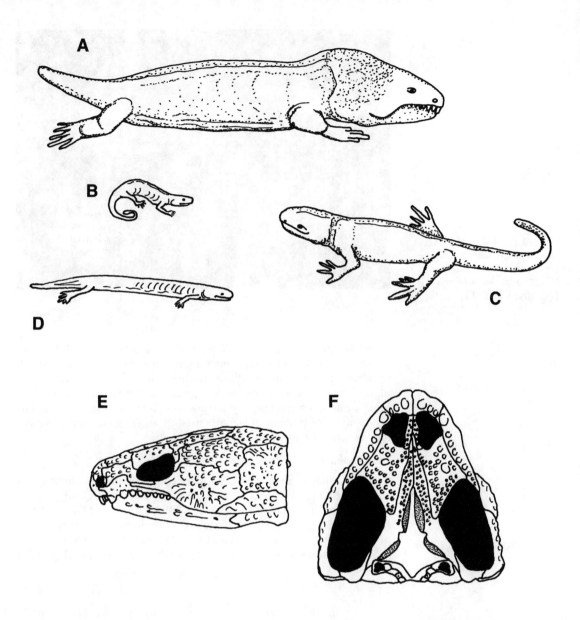

Figure 9.25. Selection of microsaurs.
(A) Pantylus. *Skull length 80 mm.*
(B) Saxonerpeton. *Skull length 10 mm.*
(C) Tuditanus. *Skull length 30 mm.*
(D) Microbrachis. *Skull length 10 mm.*
(E) *Lateral and*
(F) *ventral views of the skull of* Pantylus. *From Carroll and Gaskill (1978).*

those of genuine amniotes. This created confusion. The term *microsaur* means "little reptile," and microsaurs are very like early amniotes in their anatomy, which was one of the sources of confusion; another source of confusion occurred because fossils of the two types were difficult to disentangle physically. The similarity between microsaurs and early amniotes has led to suggestions over the years that they are closely related, with the recent cladistic analysis being the latest of these. However, there are several characters in which microsaurs differ substantially from reptiles, among them the pattern of skull roofing bones. Microsaurs have a skull table pattern involving a single temporal bone where early amniotes have several, and of the two families, the amniote pattern more closely resembles that of the aquatic microbrachomorphs. Microsaurs have a specialized atlas–axis vertebra that fits into a unique form of occipital condyle not

The Late Carboniferous • 259

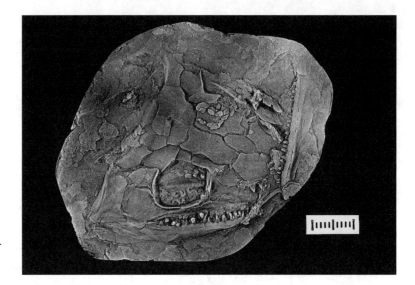

Figure 9.26. Skull of the microsaur Crinodon limnophyes. *Photograph of a replica in UMZC. Scale bar = 10 mm. Photograph by S.M.F.*

found in reptiles. It bears more similarity to the occipital joint of temnospondyls. Headless skeletons of microsaurs and early amniotes are difficult to tell apart, one of the few distinguishing features being that where amniotes have five toes on the forelimb, microsaurs never have more than four. In this they again resemble temnospondyls rather than amniotes, one of the pieces of evidence sometimes used to suggest that microsaurs and temnospondyls are closely related (see below).

Lysorophids

These animals seem to be highly derived microsaurs, with long bodies containing up to 100 presacral vertebrae, very small limbs, and a unique pattern of skull roofing bones (Carroll 2000a; Carroll and Gaskill 1978; Carroll et al. 1998). Fossilized "cocoons" of lysorophids have been found in the Early Permian red-beds of the United States, showing that at least one group of tetrapods dealt with periodic aridity by estivating (Fig. 9.27) (Olson 1971).

Aïstopods

The first known aïstopods date from the Early Carboniferous of Wardie (Chapter 7) and East Kirkton in Scotland (Chapter 8), but by the Late Carboniferous, two families had become quite common and widespread (Carroll 2000a; Carroll et al. 1998). The two families had rather different skull morphologies. In one, the eyes were very far forward and small, and the sides of the cheek were covered by a mosaic of small plates (Fig. 9.28). These would have been flexible enough to accommodate the huge jaw-closing muscles housed in the enormous adductor chamber that the skull incorporated. In the other family, the skull proportions were more conventional, but the cheekbones formed a slender framework attached to a solidly ossified braincase. Both groups had snakelike proportions with up to 230 presacral vertebrae and no vestiges of limbs. Each had at least some ribs that were shaped like a letter K, and it has been suggested that these were used for rib walking, as in modern snakes. It is not clear whether aïstopods were all aquatic or whether some might have been terrestrial burrowers.

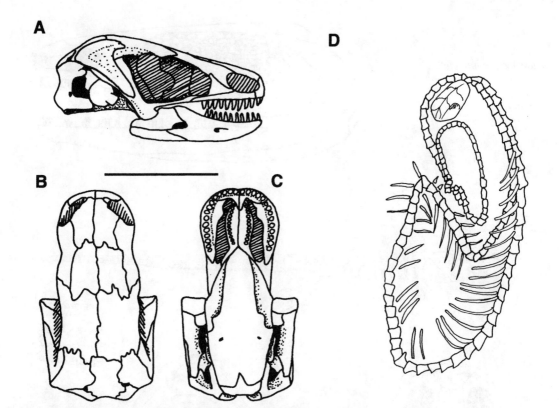

Figure 9.27. (A) Lysorophid skull (Brachydectes elongatus) in lateral, (B) dorsal, and (C) ventral views (from Carroll et al. 1998). Scale bar = 10 mm. (D) Drawing of a specimen preserved curled up in an estivation burrow (from Olson 1971). (E) Photograph of lysorophid specimen (UCLA VP2802). Photograph by A.R.M.

Nectrideans

The earliest nectrideans are not found until the early part of the Westphalian, but by their demise in the Early Permian, they had evolved into some of the strangest tetrapods of all (Carroll 2000a; Carroll et al. 1998; Milner 1980a). The implication from the diversity that appears among the Westphalian forms is that their origins must have been much earlier. Among their distinguishing features are the form of the neural and hemal arches of the tail. These are hatchet-shaped from front to back, and

The Late Carboniferous • 261

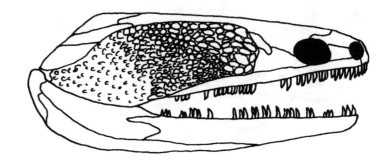

Figure 9.28. (right) Skull reconstruction of the aïstopod Oestocephalus. *Skull length 52 mm. From Milner (1994).*

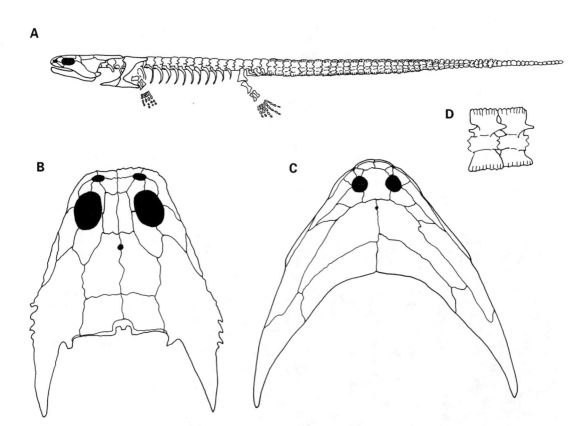

Figure 9.29. Drawings of nectrideans. (A) Keraterpeton. *Skull length 40 mm. (B)* Diceratosaurus. *(C)* Diplocaulus. *Skull width 60 mm. Maximum skull width 200 mm. (D) Nectridean tail vertebrae (cf. Fig. 9.25). A–C from Carroll et al. (1998).*

the margins are marked by crinkled ornament, known as pie crust decoration as a result of its similarity to the way the edges of pastry cases are sealed (Figs. 9.29, 9.30). The effect of expanding the neural and hemal arches is to produce a tail flattened but strengthened to produce lateral bending used for propulsion in swimming. Other features of the skeleton are fairly conservative; the trunk is short and the limbs relatively long and slender.

Early nectrideans lost the intertemporal bone but acquired an anthracosaurlike tabular parietal contact, although they are not usually considered to be anthracosaur relatives. Later nectrideans developed palatal vacuities like temnospondyls, but again, this is considered to be a convergent acquisition. Nectrideans elaborated the temporal region of the skull

Figure 9.30. Tail of the nectridean Ctenerpeton (Carnegie Museum specimen CM 44777, from a replica in UMZC). Scale bar = 10 mm. Photograph by S.M.F.

into horns or prongs, which in the totally aquatic Early Permian forms was taken to extremes (Fig. 9.29). The corners of the skull were drawn back to an exaggerated degree, so that the skull formed a boomerang shape. One suggestion for the function of this shape was that it provided an airfoil section to the skull, providing lift for the animal as it attempted to lunge upward from the bottom of the stream or river where it lived to capture prey passing overhead (Cruickshank and Skews 1980). More recently, trace fossils of nectridean resting places have shown that the corners of the skull were attached by sheets of skin to the body, which left the legs free. The function of the skin flap is not known (Walter and Wernerberg 1988).

For at least one species of nectridean, there is a well-documented growth series showing how the horns change shape and orientation as the animal reaches certain sizes. This may indicate a period of metamorphosis or a change of lifestyle. From the subsequent growth pattern, it is possible to suggest that these nectrideans kept on growing, rather than reaching a determinate size (Rinehart and Lucas, in press). For other lepospondyls, it seems clear that growth was determinate—in other words, the adults reached a final size and stopped growing. For these animals, only small representatives have ever been found.

The Earliest Known Amniotes and the Beginnings of Modern Amniote Groups

It is during the Westphalian that the first remains of animals that can confidently be called reptiles are found—that is, basal members of the amniote radiation. (Note that the term *reptile* has recently been used in a cladistic sense to mean amniotes other than the synapsid/mammal lineage. In popular usage, it has meant amniotes that are neither mammals nor birds.) These animals share a number of osteological characters with each other and with modern amniotes that have been used as the basis for their classification, but whether they shared the physiological characters of the

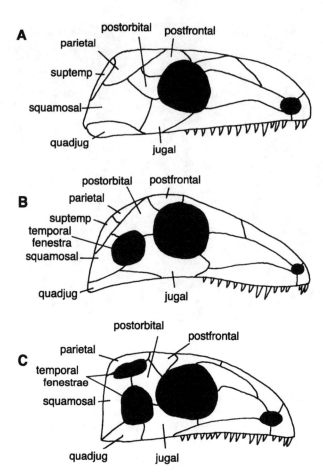

Figure 9.31. Amniote skull types. (A) Primitive type with no temporal fenestrae (anapsid) as in Captorhinus. (B) Single temporal fenestra (synapsid) as in Archaeothyris (mammalian lineage). (C) Two temporal fenestrae (diapsid) as in Petrolacosaurus.

egg, which the term *amniote* implies, can never be established from fossils. The word is derived from one of the three membranes that surround the egg of an amniote, whether the egg is shelled, like those of crocodiles, turtles, or birds, or not, as in most mammals. It is unlikely that it will ever be known whether many of these early amniotes laid shelled eggs or not, whether some of them retained reproductive life histories like seymouriamorphs, or whether some of them employed mammallike egg retention. The best assessments of phylogeny of some of these entirely extinct groups must rely on skeletal characters, and integrating these studies with those from living forms has produced controversial results over the last many decades.

Modern amniotes can be divided into three main supergroups: the synapsids, to which the mammals belong; the diapsids, to which lizards, snakes, crocodiles, and birds belong; and the parareptiles, to which turtles belong (Fig. 9.31). The earliest representatives of both synapsids and diapsids can be traced to the Westphalian, although the closest relatives of turtles are controversial. Some have seen them as derived from a group called captorhinids (Gaffney 1980), but others see them as belonging to a separate lineage with some extinct groups called procolophonids and pareiasaurs (Fraser 1991; Lee 1995, 1997; Reisz and Laurin 1991). More recent studies have suggested that they could be modified diapsids (Rieppel and DeBraga 1996).

Hylonomus is generally considered to be one of the first animals to be a true amniote. It is one of the creatures found in the Joggins tree stumps, making it Westphalian A in age (Fig. 9.32). It was more advanced in its terrestrial features than *Casineria*, for example in having an even more gracile humerus, and unlike *Casineria*, its skull is well known. It was a small, lightly built animal, rather lizardlike in form, and it is one of several similar genera that appear at about this time or only a little later. At the back of its skull, the bones meet to form a single sheetlike cheek region. Many of its contemporaries had this form of cheek region, and they have been called anapsids because there were no holes in the side of the head (the "aps" in *anapsids* actually means the arches bounding the holes) supporting these holes. Figure 9.33 shows the early amniote *Captorhinus* to illustrate this kind of skull structure. This condition is the same as that found in earlier tetrapods and is primitive, so it cannot be used to support any phylogenetic relationship between the animals possessing it. However, only a little later than *Hylonomus*, two forms appear that are superficially similar to it in skull shape and body form. One of them, *Archaeothyris*, has a single hole piercing the side of the cheek (Reisz 1972, 1986). This is the earliest member of the synapsids whose Triassic descendants produced the mammals, and the other is *Petrolacosaurus*, which had two holes (Reisz 1977) (Fig. 9.31). This is the earliest known diapsid and its descendants include not only modern lizards, snakes, crocodiles, and birds, but also the dinosaurs, the pterosaurs, and a wealth of early diapsid amniotes that dominated the tetrapod faunas of the land and sea throughout the Mesozoic.

Some morphological characters unite all the Paleozoic amniote forms, although some of these characters may be paralleled in other nonamniote groups, such as the microsaurs. Some of these are exemplified by the Permian amniote *Captorhinus*, which, although anapsid, is thought to be quite closely related to diapsids (Laurin and Reisz 1995). The most conspicuous skeletal feature of many amniotes is a flange of the pterygoid that is turned downward into the back of the throat and frequently bears a suite of teeth. The upper surface of this pterygoid flange is thought to have provided an attachment point for the pterygoideus muscle that closes the jaw (Fig. 9.33) (see also Chapter 10). Early amniotes were exclusively carnivorous or insectivorous feeders on land, and a powerful bite required a reconfiguration of the mouth and jaws to facilitate this action. At the back of the skull, the skull table was reduced until first the postparietals and then the supratemporals and tabulars became small bones exposed mainly on the occipital surface of the skull. A midline ossification of the braincase called the supraoccipital developed. The cheek outline was smooth, with no hint of an otic notch, and the stapes, rather than being a slender upwardly directed rod, was a stout downwardly or horizontally directed strut running between the braincase and the quadrate (Fig. 9.33). It appears that early amniotes did not have any form of specialized ear for terrestrial hearing, unlike the temnospondyls or seymouriamorphs.

In contrast to the paired exoccipital condyles of temnospondyls, amniotes had a convex condyle formed by the basioccipital (see Fig. 10.7), which attached to a specialized atlas–axis vertebra, similar to but different from that of microsaurs. In the postcranial skeleton, the most conspicuous amniote invention was found in the ankle, where three of the small tarsal bones fused into a single structure called the astragalus. This is presumably correlated with increasing facility for terrestrial locomotion. Associated with these adaptations, the limbs are gracile and slender, and the muscles' attachment points are close to the joints. Gone is the flattened, L-shaped humerus of early tetrapods, replaced with something that looks much more

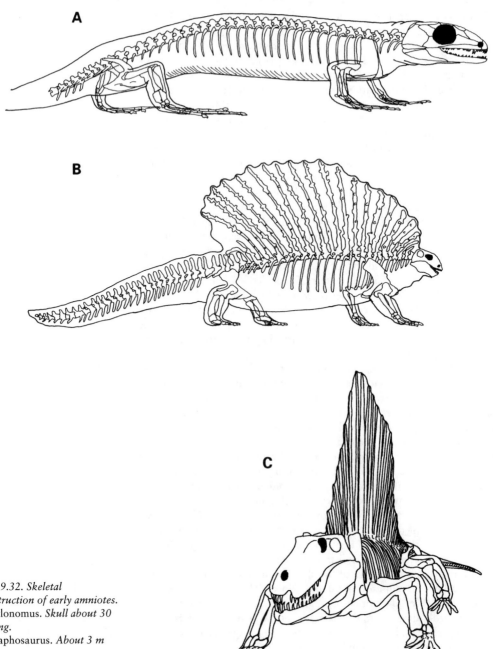

Figure 9.32. Skeletal reconstruction of early amniotes.
(A) Hylonomus. Skull about 30 mm long.
(B) Edaphosaurus. About 3 m long.
(C) Dimetrodon. About 3 m long.

like the humerus of a modern amniote. The vertebrae have cylindrical pleurocentra, firmly attached or sometimes fused to the neural arches; the intercentra are small; and the ribs are long and slender, and encompass most of the abdomen.

It was not until the Late Carboniferous that skeletal adaptations for full herbivory appeared in tetrapods, although plant material could have formed part of the diet of earlier forms (Hotton et al. 1997). One of the first

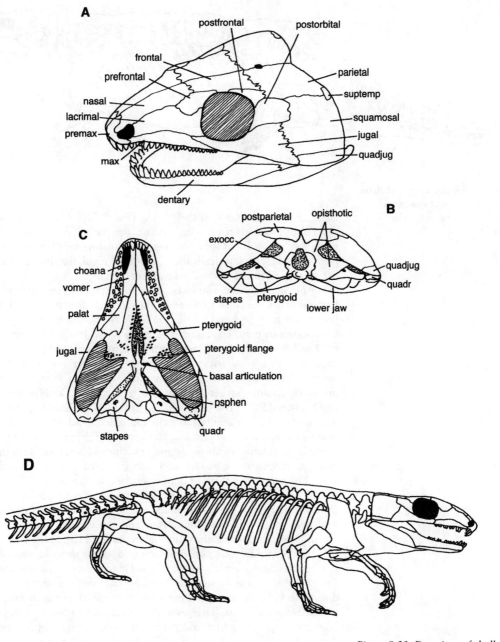

Figure 9.33. Drawings of skull and skeleton of Captorhinus.
(A) Skull in oblique view.
(B) Skull in occipital view.
(C) Skull in ventral view.
(D) Whole skeleton. Skull length 80 mm. A–C from Heaton (1979).

groups to produce herbivorous forms was the synapsid lineage, and *Edaphosaurus* developed a battery of grinding teeth on the lower jaws and palate. It has been suggested that herbivory could not be evolved easily among tetrapods until parental care in the form of feeding allowed the gut flora necessary to digest plant food to be passed directly from adult to young (Modesto 1992). Another suggestion is that each of the several lineages of amniote to have evolved herbivory seems to have arisen from insectivorous ancestors, and the required symbiotic gut flora for cellulose digestion could have been picked up during this phase (Sues and Reisz 1998). *Edaphosaurus* and some of its Early Permian relatives such as *Dimetrodon* carried

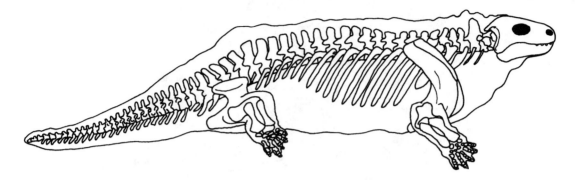

Figure 9.34. Drawing of skeletal reconstruction of Diadectes, *about 3 m long (from Case 1907).*

elongated neural spines on their backs (Fig. 9.32B, C). It is thought that these supported a fin of skin, the function of which is most plausibly explained as a radiator for collecting or dissipating heat from the sun. This may be the first skeletal indication of thermal regulation in the lineage that eventually led to the internally heated mammals.

Among Paleozoic amniotes are a number of other entirely extinct forms that are not closely related to any modern survivors, as well as some forms that may be basal to all amniotes; these include the large Permo-Carboniferous form *Diadectes* (Fig. 9.34) (Case 1907). The relationships of *Diadectes* to other amniotes remain controversial, and a recent analysis suggests a link with synapsids (Berman 2000; Berman et al. 1992). In contrast, other studies have suggested that this animal is after all the most primitive amniote, rather than the smaller creatures like *Hylonomus* (Lee and Spencer 1997). The large, clumsy-looking *Diadectes* appears to have been one of the first animals to have succeeded in adopting herbivory. Its barrel-shaped body and grinding teeth speak of the ability to process and digest large volumes of fibrous, poorly nutritious plant food. Like the other major amniote groups, its earliest member evolved in the Late Carboniferous, although unlike them, it did not outlast the Permian.

By the Early Permian, many new forms of amniote had evolved, including the small procolophonids and large cowlike pareiasaurs, small lizardlike millerettids in Gondwana, and the earliest aquatic amniote, *Mesosaurus,* mentioned above in the section on biogeography (Fig. 9.6). All these forms are thought to belong to a group of amniotes called the parareptiles and seem to be mainly a Gondwanan radiation (Modesto 1999). Small anapsids such as *Captorhinus* were common worldwide during the Early Permian, and the diapsids by this time had diversified into a wide range of forms. They even produced strange forms with elongate bony struts sprouting from the trunk that supported flaps of skin on which they could glide (Evans and Haubold 1987). By the end of the Permian, many of the amniote innovations had evolved and the foundations laid for the subsequent history of the group.

At the end of the Permian, however, some disaster struck the Earth, and 95% of its fauna was wiped out. The end-Permian extinction event may not have been a single event or one caused by a single phenomenon, but it nevertheless produced the most catastrophic series of extinctions in Earth's history, apparently within half a million years. The most recent ideas of the causes of this extinction are summarized in Erwin et al. (in press) and include evidence for anoxic conditions in both shallow and deep waters, massive eruptions of flood basalts in Siberia, and global warming at the end of the Permian. Some evidence also suggests a possible asteroid impact that might have contributed to the extinction event. As far as the tetrapods are concerned, most of the temnospondyls, most of the anthracosaurs, all of the lepospondyls, and many of the stranger forms of amniote disappeared during this event. The tetrapods reradiated after the event from the few families that survived. In the case of the amniotes, they evolved with new vigor, and by the end of the Triassic, they had produced many of the lineages whose legacy is familiar today. But that is part of another story.

Interrelationships of Paleozoic Tetrapods and Origins of the Modern Tetrapod Groups

Phylogeny of Paleozoic Tetrapods

Ideas about the interrelationships of early tetrapods have changed significantly over the last 20 years or so, largely as a result of the advent of cladistic analysis and more rigorous assessments of primitive versus derived characters. Up to about the early 1980s, most people accepted that almost all early tetrapods (or fossil amphibians, as they would have been more commonly called) fell into one of two large groups, called labyrinthodonts and lepospondyls. The scheme was formalized and made popular by A. S. Romer (see above) and became widely known not only among paleontologists, but also among scientists remotely interested in any aspect of vertebrate evolution. It is worth discussing this classification because it still persists and has influenced ideas about many issues in the subject.

The term *labyrinthodont* derives from the structure of the teeth found in many early tetrapods, in which the enamel of the tooth crown has a complex folded appearance in cross section. It was first applied to the group that is now known as temnospondyls, because temnospondyls were among the first fossil amphibians to be found and recognized. Later, more creatures were discovered with this tooth structure, and they were also placed in the group. They included the animals now called anthracosaurs, colosteids, baphetids, and the problematic genera *Crassigyrinus* and *Ichthyostega*. In addition to sharing similarities in tooth structure, these animals also shared the following characters: multiple parts to their vertebral centra; skull roofs with ornamented bone; a fairly standard bone complement showing no openings other than those for the eyes, nostrils, and parietal organ, usually with lateral lines carried in open grooves; and often heavily ossified braincases. Most of them also had closed palates covered with denticles, although it was recognized that temnospondyls were distinguished by their large palatal vacuities. Most of these creatures were relatively large—desktop-sized, so to speak (Romer 1947, 1966).

The lepospondyls were distinguished first by their type of vertebral centra, which were usually formed of solid centra, spool-shaped, and lacking any of the separate elements seen in the labyrinthodonts. None of them had labyrinthodont teeth, and most of them were relatively small and laptop-sized by comparison. The group included microsaurs (when these were eventually separated from early amniotes), aïstopods, nectrideans, adelogyrinids, and lysorophids. Skull construction was rather different in each of the groups because each appeared to lack one or more of the standard complement of bones found in labyrinthodonts (Romer 1966).

With the advent of cladistic analysis, it became obvious that at least the labyrinthodonts were united only by characters that were primitive for tetrapods, most of which were found also in the stem lineage of fish such as *Eusthenopteron*. This was as true of the tooth structure as it was of the vertebral and palatal construction, although the skull roof pattern among the labyrinthodonts showed consistent differences from the fish that created difficulties of homology between the two groups (as mentioned in Chapter 6). The skull roof pattern is now seen to distinguish tetrapods but does not serve to define a subgroup within tetrapods. The distinction between labyrinthodonts and lepospondyls might almost be thought of as one of size, except that, for example, some microsaurs were then discovered to have intercentra in the vertebrae (Carroll and Gaskill 1978). As ideas about early tetrapod classification have changed, so it has become more obvious that lumping all the large forms together under the same collective title has camouflaged important differences between the individual groups. The danger lies in assuming that members were basically all the same in aspects of their anatomy and biology, which now seems not to be the case at all. A case in point will be discussed in the next chapter.

One of the first people to try to place the classification of early tetrapods on a cladistic footing was Smithson (1982), who looked at the construction of the occiput. He noted that temnospondyls and microsaurs shared an apparently derived feature in which the exoccipital bones became very important and were the main bones to tie the back of the braincase to the skull roof (Fig. 9.35). In most other tetrapods, the exoccipitals were small, and the opisthotics, part of the otic capsules, provided this anchorage. This is explained in more detail in the next chapter, but the observation led him to separate the microsaurs from the other lepospondyls and place them in a clade with the temnospondyls and the colosteids. Thus, labyrinthodonts and lepospondyls became paraphyletic groups. He distinguished two lineages: first, the temnospondyls, colosteids, and microsaurs shared their occipital construction with modern amphibians and were part of a batrachomorph lineage; and second, anthracosaurs were placed with seymouriamorphs into a reptiliomorph lineage along with amniotes (Fig. 9.36). In this lineage, the occipital construction was essentially primitive, but the members shared features of the vertebral column (Smithson 1985b). These conclusions were corroborated by a follow-up study by Panchen and Smithson (1988), although in this article, even the two coauthors could not agree on the details of interrelationships within the reptiliomorph lineage. One intriguing suggestion by these authors was that *Ichthyostega* was more closely related to the batrachomorph lineage than to the reptiliomorphs. This conclusion has not been borne out by later analyses, but the phylogenetic position of *Ichthyostega* remains unclear.

In the years that followed, further cladistic analyses were carried out after the discovery of new genera, especially from the Late Devonian. Lebedev and Coates (1995) attempted the next large-scale analysis, and they corroborated part of Smithson's scheme by showing a dichotomy between the reptiliomorph and batrachomorph lineages (Fig. 9.37). They suggested that the split between these groups occurred in the Late Devonian and that *Tulerpeton* was the earliest known member of the reptiliomorph clade. That leads to the inference that modern amphibians and amniotes belong to separate lineages going back to the Late Devonian. Lebedev and Coates also showed that genera such as *Acanthostega* and *Ichthyostega* were stem tetrapods and had no especially close relationship either to each other or to the Carboniferous forms, other than the general one of being tetrapods. Like Smithson, they found that *Greererpeton* and the microsaurs grouped with temnospondyls, a conclusion reinforced in Coates's later work (1996) on the postcranial skeleton of *Acanthostega* (Fig. 9.37).

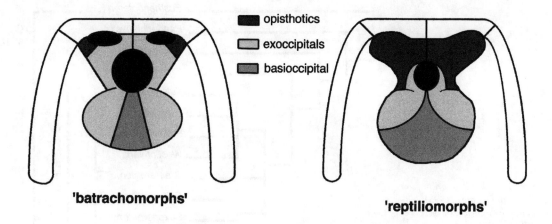

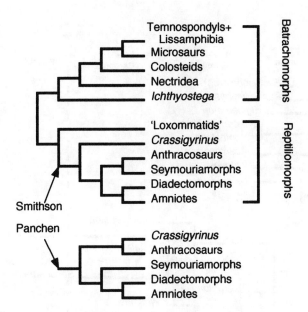

Figure 9.35. (above) Occipital construction: batrachomorphs and reptiliomorphs, showing alternative contributions of the otic capsule (opisthotic) and occipital arch (exoccipital) components of the occiput according to Smithson (1982).

Figure 9.36. (left) Cladograms from Panchen and Smithson (1988), showing the two authors' alternative groupings of the reptiliomorphs.

A very large-scale cladistic analysis was performed by Carroll in 1995 in which a wide range of taxa and a large number of characters were included (Fig. 9.38). One of the main findings of this study was the re-establishment of the clade known as lepospondyls because they appeared to share many derived characters that united them into a monophyletic clade. The study placed the lepospondyls as the sister group to amniotes, with seymouriamorphs and anthracosaurs as successively more primitive taxa on that branch of the cladogram. As in Smithson's analysis, colosteids grouped with temnospondyls.

More recently, in an attempt to work out the relationships of genera such as *Crassigyrinus* and *Eucritta,* Clack (1998a, 1998b, 2001a) performed analyses of most of the better-known early tetrapod groups, although most of these analyses excluded any lepospondyl or early amniote

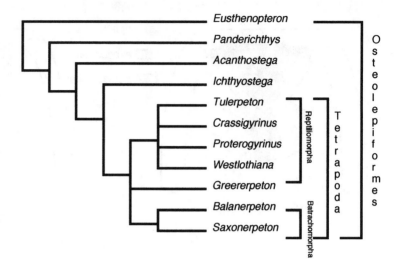

Figure 9.37. (A) Cladogram from Lebedev and Coates (1995) showing the division into reptiliomorphs and batrachomorphs.

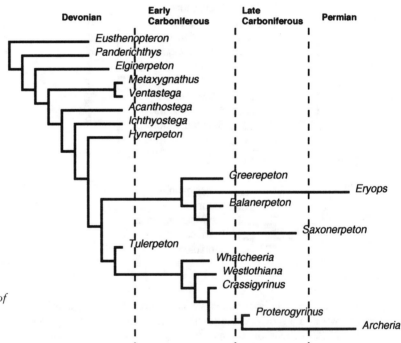

(B) Cladogram from Coates (1996), a further development of the analysis, showing the chronological relations of the animals.

taxa. This is because it was considered that the lepospondyls and amniotes would probably not affect the more basal parts of the cladogram that were the main focus of interest. In these studies, *Acanthostega* consistently appeared as the most primitive tetrapod, with *Ichthyostega* and the colosteid *Greererpeton* as successive plesions. *Crassigyrinus* usually appeared in the anthracosaur lineage, as did *Whatcheeria*, although these two taxa were among those whose placement was found to be most unstable. The baphe-

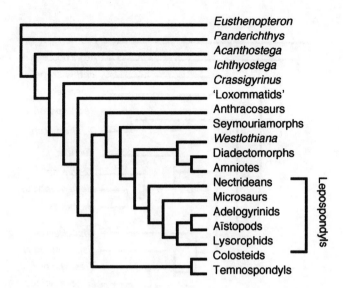

Figure 9.38. Cladogram from Carroll (1995), one of the first large-scale cladistic analyses to include lepospondyls.

tids were also unstable, sometimes falling into the clade with temnospondyls and sometimes with anthracosaurs. This accords with the conflicting set of characters seen in *Eucritta,* which does appear more or less consistently as a basal baphetid (Fig. 9.39).

A study to investigate the position of *Casineria* by Paton et al. (1999) also suggested a similar stem group sequence, and like Carroll's analysis, it placed microsaurs alongside amniotes and seymouriamorphs. This study placed *Westlothiana* alongside the amniotes, but new studies may well suggest a different position, possibly among the microsaurs.

In recent years, a larger cladistic analysis was performed by Laurin and Reisz (1997) based on Carroll's (1995) scheme but including modern forms alongside fossil ones. It produced controversial results, which will be discussed in the next section.

In Chapter 10, the phylogeny I use is that which unites temnospondyls with frogs and modern amphibians, and anthracosaurs with amniotes. The implications of alternative phylogenies will be considered where appropriate, but in many instances, the conclusions are unaffected by which phylogeny is chosen. In describing the evolution of tetrapods from fish and the subsequent acquisition of terrestriality, certain events must necessarily have taken place. A well-founded phylogeny can help to show when these changes might have occurred and in what order or how many times, but the fact of their occurrence can be inferred even without a phylogeny. In fact, the phylogeny of early tetrapods is still very uncertain and controversial, so that the ideal of showing when and how often certain changes took place is still a long way off.

Origin of Modern Amphibians

Modern amphibians, or Lissamphibia, are represented by three surviving groups, the anurans (frogs, toads), the urodeles (newts, salamanders), and the caecilians (apodans). Until the advent of phylogenies based on molecular sequences, it was unclear whether these three formed a monophyletic group, but now it seems most likely that they are all each other's closest relatives. (There is still no agreement, despite sequence data, about which group is sister group to which, but that is a separate problem

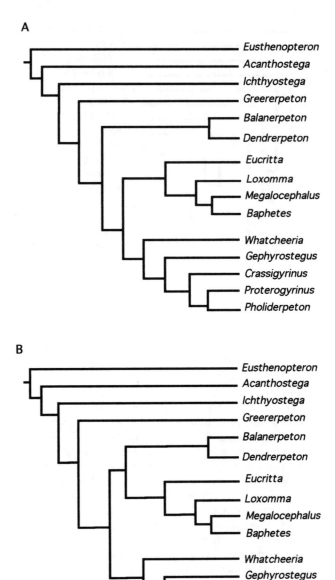

Figure 9.39. Cladogram from Clack (2001a) with Eucritta *and the baphetids shown grouping with the (A) anthracosaurs or with the (B) temnospondyls.*

and will not be further dealt with here.) The earliest fossil record of any animal that can be associated with any of the lissamphibian groups is dated as Early Triassic and bears little resemblance to any Paleozoic form (Rage and Rocek 1989). Because the amniote lineage is known to have appeared by at least the Late Carboniferous and is monophyletic to the exclusion of lissamphibians, it can be inferred that the lineage leading to lissamphibians must also have been distinct by that time. This leaves a long period of evolutionary history of the lissamphibians for which there is no fossil record. This is surprising, given the good fossil record for amniotes throughout that same period and given the fact that one would expect amphibians

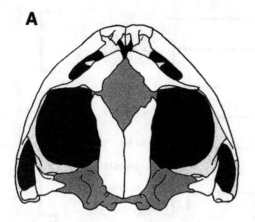 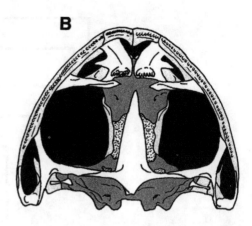

Figure 9.40. (A) Dorsal and (B) palatal views of a frog skull for comparison with that of a branchiosaur such as Apateon and the dissorophoid Amphibamus. *In A, the bones of the palate are light shaded; in B, the bones of the skull roof are light shaded. In both, braincase elements are dark shaded. Spaces shown black.*

to have lived and died in waterside or swamp conditions, thus providing a relatively good chance of preservation (Milner 1993b). Part of the problem may reside in the small size of the animals involved: they could have been overlooked in the past. Certainly more clues in the form of tiny broken and disarticulated elements belonging to lissamphibianlike animals are now beginning to come to light. However, because of this dearth of fossils, the origin of lissamphibians remains controversial and is now by far the least well documented and understood of any of the major vertebrate evolutionary transitions. Because the origin of the lissamphibians must have occurred during the Late Paleozoic, it is appropriate to consider some of the issues here.

D. M. S. Watson was one of the first people to point out similarities between frogs and small dissorophoid temnospondyls from the Late Carboniferous, specifying the genus now known as *Amphibamus* (Fig. 9.13) (Watson 1940). Other studies followed in the 1960s, and these too concluded that temnospondyls were the most plausible group to have given rise to frogs (at least) (Parsons and Williams 1963). In modern terminology, temnospondyls share a number of characters with frogs especially, although some of them are also seen in salamanders and caecilians. Among the most striking are the large palatal vacuities, particularly well developed in frogs and dissorophoid temnsopondyls, and the fact that the occipital condyle in all of them is formed by facets formed only from the exoccipitals. Other similarities have been noted in recent years, such as the shortness of the ribs in temnospondyls generally, the similarities of the ear region in temnospondyls and dissorophoids, the fact that some dissorophoids (such as *Amphibamus* and its close relative *Platyrhinops*; Fig. 9.13) have the shortest vertebral columns of any tetrapods except frogs, and the fact that among dissorophoids are some families of animals, the branchiosaurs, that show tadpolelike larvae and evidence of pedomorphosis in their development. Some larval branchiosaur skulls are remarkably like those of frogs and salamanders, with many of the same bones reduced or lost and with very similar skull proportions (Figs. 9.14, 9.15, 9.40). Similarities in lifestyles and physiology between some dissorophoids and lissamphibians do not seem improbable, such as their mode of buccal pumping, their metamorphic life histories, and the prevalence of heterochrony in their ontogeny. In recent years, this hypothesis has been put on a more formal cladistic

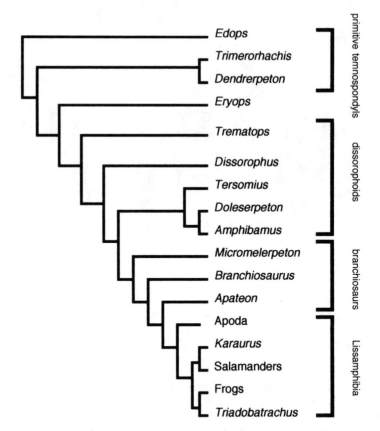

Figure 9.41. Trueb and Cloutier's (1991) cladogram of temnospondyls and lissamphibians. *Karaurus* and *Triadobatrachus* are the earliest fossil salamander and frog, respectively. Apoda is an earlier name for caecilians.

footing by studies such as those of Milner (1988, 1993b) and Trueb and Cloutier (1991), who have shown a nested series of forms in which the attributes of lissamphibians are progressively acquired (Fig. 9.41).

Other opinions coexisted with this temnospondyl theory for a monophyletic lissamphibian origin, and these focused on a microsaur ancestry for either caecilians, urodeles, or both. Carroll and Holmes (1980) pointed out a number of specific resemblances between one particular microsaur genus and one particular caecilian genus, which are compelling until the hypothesis is considered more widely, whereupon it appears that these similarities do not seem to apply to microsaurs or caecilians in general. The two chosen genera could equally likely resemble each other because of convergent evolution. However, Carroll's view has received some support recently from the discovery of a Late Triassic/Early Jurassic caecilian, *Eocaecilia*, that retains limbs as well as a few other primitive characters (Jenkins and Walsh 1993). It was interpreted as showing some microsaurlike features of its skull (Carroll 2000b).

Even if urodeles or caecilians were eventually shown to be related closely to microsaurs, Lissamphibia could still be a monophyletic group in one sense if, as Smithson (1982) suggested, microsaurs and temnospondyls are closely related.

The cladistic analyses that suggested temnospondyl affinities for lissamphibians were all less than comprehensive in at least one important way. All omitted any lepospondyl groups. Subsequently, a series of studies has included not only lepospondyls but also representatives of most early

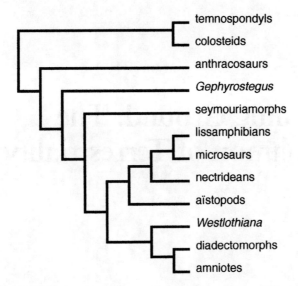

Figure 9.42. Laurin and Reisz's (1997) cladogram, which places the lissamphibians and amniotes both close to lepospondyls, with the temnospondyls and anthracosaurs as primitive tetrapods.

tetrapod groups and some extant representatives of both lissamphibians and amniotes. Carroll's treatment in 1995 has been mentioned above, with his reestablishment of the lepospondyls as monophyletic. This was used as the basis of another study that added the living groups into the equation. This was by Laurin and Reisz (1997), and their results were in many ways surprising (Fig. 9.42). They found that lepospondyls were not strictly monophyletic but in fact gave rise not only to lissamphibians but also to amniotes. Temnospondyls, they suggested, had nothing to do with lissamphibians, and furthermore, anthracosaurs had nothing to do with amniotes.

These results are intriguing, but they remain contentious and are not universally accepted despite the apparent rigor of the cladistic analysis that produced them. One of the problems is that the cladograms imply that not only are a large number of characters convergent between lissamphibians and temnospondyls, such as those connected with breathing and hearing, but they also require many reversals of fundamental characters. Most of the members of this lepospondyl–lissamphibian clade are small animals that show characters associated with miniaturization, and with forms so widely separated in time, even cladistics is hard put to distinguish resemblances based on relationships from those resulting from convergent evolution (see also Chapter 3). Studies are now in progress to test Laurin and Reisz's hypothesis (Fig. 9.42), which, if corroborated, will force a fundamental rethinking of the biological aspects of the origin of amphibians, as well as that of amniotes.

Ten
Gaining Ground: The Evolution of Terrestriality

Steps toward Terrestriality

This extended survey of the anatomy and lifestyles of tetrapods throughout the Paleozoic has explored the evidence and speculation bearing on the advent of tetrapods onto land. This final chapter now goes on to consider the evolution of several key aspects of their biology and how they became truly adapted to terrestriality. The solutions that these early tetrapods arrived at laid the foundations for terrestrial living in a huge group of vertebrates that have ultimately become a highly conspicuous part of the fauna of the planet. How these changes were achieved over that time has influenced the anatomy, morphology, and evolutionary pathways of all subsequent tetrapods and is still reflected in our own anatomy.

Many of the changes took place very gradually and can be followed from the early fossil lobe-finned fishes, through the earliest tetrapods, into the explosion of terrestrial forms in the late Paleozoic. Some of the aspects that can be followed in most detail include those related to the interconnected mechanisms of feeding, hearing, breathing, and locomotion, and this chapter draws together threads from new discoveries in paleontology and in the biology of living animals. The story of skeletal evolution is continued in this chapter from *Acanthostega* and its Devonian contemporaries and shows how parts of the skeleton were gradually adapted by

Carboniferous forms for terrestrial living, how changes to one part affected those to another, and how different groups made these changes in different ways and at different times. The origin of tetrapods is demonstrably separate from the origin of terrestriality in that tetrapods as a group appear in the fossil record in the Late Devonian, whereas the ensuing adaptations for terrestriality are first seen in the latter part of the Early Carboniferous, appearing bit by bit throughout the rest of the Late Paleozoic.

Cranial Modifications for Terrestriality

Changes to the Skull Roof

What is known of the skulls of Devonian tetrapods shows them to be essentially conservative. There are a number of consistent features that mark them as different from their fish relatives, such as the reduced number of snout bones, the pattern of the cheek and skull table, and subtle characters of the lower jaw, but in other ways, they show little variation. The Devonian forms all seem to be rather flat-headed, with spade-shaped snouts and eyes placed at about the midpoint in the skull, looking more or less dorsally. Only *Ichthyostega* shows much in the way of elaboration of the inside of the skull roof in that it has produced elaborate flanges that attach to the braincase at the back and sides. In *Acanthostega,* there are no such flanges, and the internal surface of the skull roof is essentially smooth. The only parts of its skull roof to show specially thickened bone are two ridges running up the frontals and prefrontals above the eye. From the little that is known, *Ventastega* is much the same.

Among Carboniferous forms, even those from the earliest parts, some striking modifications can be seen to the skull architecture. Several early forms such as *Whatcheeria* and the Tournaisian animal have skulls that are more deep than broad. *Whatcheeria* has specially thickened ridges inside the skull anterior to the eyes, and *Crassigyrinus* (Fig. 7.8) has taken both these modifications to extremes, with a particularly deep skull, elongated posteriorly, and with greatly thickened ridges running up the snout, between the eyes, and inside the prefrontal. Its eyes are relatively far forward, and furthermore, the orbits are rhomboidal. Even in more conventionally shaped skulls such as those of baphetids, the eye sockets acquired forwardly extending embayments (Fig. 9.7).

Other examples of extreme modifications can be seen among some of the lepospondyls, in which the skull roof pattern of early tetrapods is scarcely recognizable. Many of the skull roof bones are lost, in some forms to be replaced by numerous small plates and in others by struts and braces. Eyes in some lepospondyls moved extremely far forward (Figs. 7.4, 7.5, 9.27, 9.28). The development of narrower-snouted and deeper skulls may have facilitated the development of binocular vision, especially in some of the smaller forms such as microsaurs and amniotes.

Another region of the skull to undergo change was the external nostril. Rather a small opening in *Acanthostega* (Fig. 5.20) and *Ichthyostega* (Fig. 5.10), little different from that of *Panderichthys* (Fig. 3.17), it soon became modified, apparently quite rapidly, into the larger opening found consistently among Carboniferous forms. The maxilla and premaxilla both contributed, but the lacrimal and sometimes the nasal were also involved in forming the margin of the external nostril. In many forms, a bone known as the septomaxilla was located within the narial opening. In some modern tetrapods, the septomaxilla is a bone that bears muscles that open and close the soft tissue of the nostril.

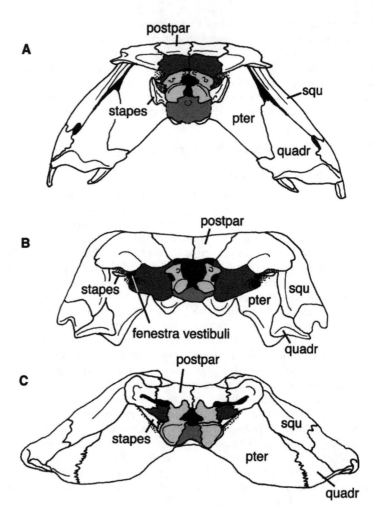

Figure 10.1. Occipital views of Pholiderpeton *(an anthracosaur)*, Seymouria *(a seymouriamorph)*, and Eryops *(a temnospondyl)*. *Opisthotic, dark shading; basioccipital, medium shading; exoccipital, light shading.*

One of the most basic forms of this bone might be that found in baphetids, where a dermal bone occludes a large part of the narial opening (e.g., *Megalocephalus,* Fig. 9.7). It is ornamented in the same manner as other dermal bones of the skull roof. In these forms, little distinguishes the bone from the anterior tectal, found in fishes such as *Panderichthys,* and so called in *Acanthostega* (Fig. 5.16). There have been arguments about the homologies of this bone between fish and tetrapods parallel to those concerning the skull table bones, and these arguments have not been satisfactorily resolved. Although there seems no compelling reason to doubt the homology between the baphetid bone and that of *Acanthostega,* in other tetrapods, the situation is more confused.

One such confusing animal is *Crassigyrinus*. The floor of its nasal opening includes a thickened, cushion-shaped bone, whose surface, judging by the bone type, appears not to have been exposed on the outside of the animal. Identification as a septomaxilla seems the most likely possibility, but whether this bone is a modified anterior tectal or something different is at this stage difficult to say (Fig. 7.8). Septomaxillae are found

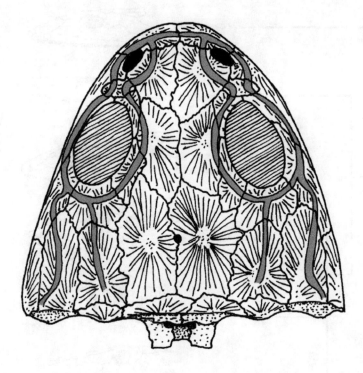

Figure 10.2. Skull roof of Dvinosaurus, *a temnospondyl with no otic notch. From Bystrow (1938).*

in many early tetrapods, usually internal to the nostril. In some cases, such as some early amniotes, they are simple curled or conical structures (Heaton 1979), and this seems to be paralleled in the temnospondyl family known as trematopids, terrestrial members of the dissorophoids. The feature is associated with a narial region that seems adapted for dry conditions (Dilkes 1993).

Another region where there is some diversity to be seen among early tetrapods is the way the skull roof and the palate are joined together at the back along the line formed where the back of the squamosal and quadratojugal meet the pterygoid. In *Acanthostega,* for example, there is no suture between these two plates of bone, which simply butt up against one another down the length of the cheek. In other forms, although there may have been a firmer union, it appears that it may have been a loose and possibly moveable one—for example, as has been suggested for baphetids as part of a skull kinetic mechanism (Beaumont 1977). In some embolomeres such as *Pholiderpeton,* there was hardly even a meeting between the two (Fig. 10.1) (Clack 1987a). In other forms, such as colosteids, some temnospondyls such as *Eryops,* and animals such as *Seymouria,* there was a firm seam running down the back of the cheek joining skull and palate in a solid union (Fig. 10.1). In microsaurs and some early amniotes, there was no ossified junction at all between these bones (see Fig. 9.33 for *Captorhinus* and Fig. 10.19 for a microsaur), but a large gap was left that may or may not have been filled with cartilage. Some of these differences may be connected with size: the junction seems to be absent in small forms. However, not all large forms show the sealed construction. Where the junction is sealed, it must mean that the skull was well on the way to being autostylic.

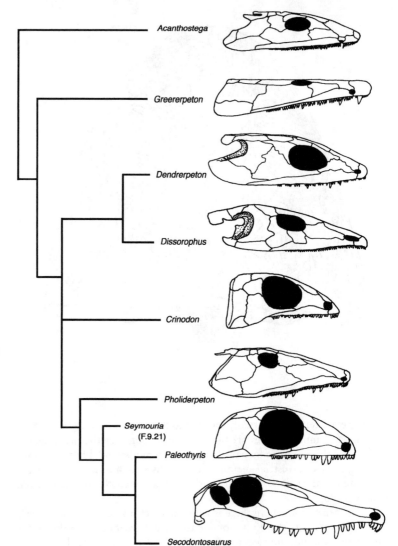

Figure 10.3. Distribution of the temporal notch among a selection of early tetrapods; the Devonian tetrapod Acanthostega; the stem tetrapod Greererpeton; the primitive temnospondyl Dendrerpeton, the dissorophid temnospondyl Dissorophus, the anthracosaur Pholiderpeton, the microsaur Crinodon, the early amniote Captorhinus related to diapsids, and the synapsid amniote Secodontosaurus related to mammals. Not to scale. (Secodontosaurus from Reisz et al. 1992.)

The presence of a spiracular notch at the back of the skull in early tetrapods has been mentioned before, and this constitutes another region of the skull where variability is seen early on in tetrapod history. In most forms, it is easy to say whether a notch was present or absent, although there are one or two exceptions. The exceptions are creatures such as the anthracosaurs *Eoherpeton* (Fig. 9.16) and *Gephyrostegus* (Fig. 9.19), where the slope of the back of the cheek is shallow. However, notches do seem to be present in most anthracosaurs and their postulated relatives. Notches are consistently present in most of the earliest tetrapods but absent in *Greererpeton* (Fig. 7.7), most lepospondyls (Figs. 9.25–9.28), all early amniotes (e.g., *Captorhinus*, Fig. 9.33) and a few temnospondyls, such as *Dvinosaurus* (Fig. 10.2). Putting the presence or absence of a notch onto a cladogram tells an interesting story (Fig. 10.3).

Only two Devonian tetrapods are known from enough skull material to show the temporal region, and both of these—*Acanthostega* (Fig. 5.20) and *Ichthyostega* (Fig. 5.10)—had notches at the back. It is true that the notch structure is rather different in detail in each of these two, but in general terms, they occur in equivalent positions.

Among the next earliest tetrapods, the Tournaisian animal (Fig. 7.11) and its relative *Whatcheeria*, each have deep notches, as do *Crassigyrinus* (Fig. 7.8), *Eucritta* (Figs. 8.14, 8.15), *Balanerpeton* (Fig. 8.10), *Silvanerpeton*, and other early temnospondyls and anthracosaurs. The enigmatic adelogyrinids also have large, rounded notches occupying the whole back margin of the skull (Fig. 7.5). Interpretation of the function of the notch has changed in recent years, and part of the story is told in a later section. However, its primitive fishlike function may have been as a persistent spiracle, as suggested in Chapter 6.

Phylogeny suggests that possession of a notch is primitive for tetrapods, whatever its function may have been. However, animals lacking notches occur at several points in tetrapod phylogeny (Fig. 10.3). In a number of cladistic analyses, *Greererpeton* comes out as the next most primitive tetrapod after *Ichthyostega*, suggesting that colosteids lost their notches early. Notchlessness may be one character that unites lepospondyls (with the exception of adelogyrinids), and it seems also to unite early amniotes. Some phylogenies suggest that it unites lepospondyls and amniotes (Laurin and Reisz 1997). However, there may be functional correlates that suggest that convergence rather than shared history accounts for the lack of a notch in some groups. Furthermore, it is present in the most primitive lepospondyls, the adelogyrinids.

The notch is lost in perennially aquatic temnospondyls such as some of the Late Permian forms such as *Dvinosaurus* from Russia (Fig. 10.2). These are known to have been neotenous forms with well-developed gill bars (Bystrow 1938). Supposing the notch to have been originally spiracular rather than part of an ear, *Dvinosaurus* and colosteids could each have lost the notch because they no longer relied on air breathing and did not use the spiracle for exhalation. Gill rakers are known from colosteids. These are mineralized projections from the gill bars into the mouth. They are used in some fishes as a kind of filtering device for preventing food particles entering the gill chamber and damaging gill filaments, for catching minute food particles for ingestion, or for helping to seal the buccal chamber during suction feeding. Their presence in fossil tetrapods suggests that these animals used their gills for feeding in water in some way, and perhaps for breathing as well. On the other hand, small forms such as microsaurs and amniotes may have begun the evolution of costal ventilation, using body wall muscles for inhalation as well as exhalation (see below and Chapter 6) and gradually abandoning the buccal pumping of early tetrapods. In their case, the spiracular notch may have become redundant for a different reason from that in colosteids and some temnospondyls. Implication of changes to breathing mechanisms may be drawn.

In summary, changes to the skull shape and architecture, changes to the shape and position of the orbit, and changes to the construction of the nostril follow soon after, but they do not quite accompany the origin of limbs with digits in appearance. It is as if after an initial period of tetrapod conservatism, some constraint had been lifted that allowed the evolution of a much more radical suite of skull modifications. That constraint may have been connected to reduction of reliance on gill breathing, which may in turn have been affected by the final size to which the animals grew. Smaller animals may have been able to limit their reliance on gill breathing in favor of cutaneous gas exchange or buccal pumping. In its turn, the changes affected the evolution of the neck and consequently the position of the heart (see below).

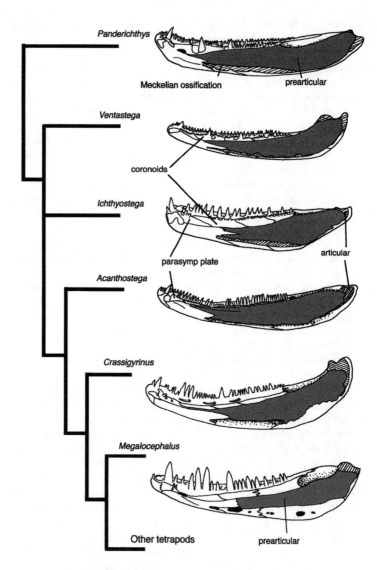

Figure 10.4. Early tetrapod lower jaws; the stem tetrapodomorph Panderichthys, *the Devonian tetrapods* Ventastega, Ichthyostega, *and* Acanthostega, *and the primitive tetrapods* Crassigyrinus *and* Megalocephalus. *The prearticular is shaded; Meckelian ossifications, including the articular, are hatched. Not to scale. From Ahlberg and Clack (1998).*

Jaws, Dentition, and Feeding

Lower jaws constitute the most commonly preserved parts of the anatomy of Devonian and early Carboniferous tetrapods, and a fairly orderly series of changes to the lower jaws can be demonstrated among them. However, this was followed by a radiation of lower jaw shapes, which suggests a rapid diversification among mid–Late Carboniferous forms (Ahlberg and Clack 1998). Lower jaw changes show a parallel story in this respect to those already noted to skull shape and architecture; many of them had to do with changes to feeding mechanisms. Figure 10.4 places the lower jaws of several Devonian and Carboniferous tetrapods onto a cladogram of their relationships.

Most fishes today feed by means of a suction mechanism of some kind. There are exceptions, such as the garpike, which has evolved long, narrow jaws for slicing quickly through water, but the majority of fishes have elaborate mechanisms to draw in as much food as possible with the water they suck in as they open their mouths. This often involves widening the gape as far as possible, brought about by loosening the connections between their skull bones to form a flexible framework for the jaw muscles. It is thought that even early fishes used this method, although their skull adaptations toward it were not as extreme as in later ray-finned fishes. The intracranial hinge (Figs. 2.1, 3.4) found in early lobe-fins is thought to be an example of an adaptation to make suction feeding more efficient, although it tends to disappear in longer-snouted forms (Thomson 1967, 1969).

Both *Panderichthys* and *Elpistostege* had lost the hinge mechanism and appear to have been rather unspecialized gulpers; they presumably still fed in water. It seems that the lineage leading to tetrapods had found other methods for feeding, which might have been related to a possible adaptation to bottom-dwelling life or life in shallower waters (Ahlberg and Milner 1994).

Changes to feeding mechanisms will naturally be reflected in changes to jaw structure and operation, so that the lower jaw, often the only clue there is to the existence of Devonian tetrapods, offers evidence of this shift. Accordingly there are many similarities in the jaw structure and dentition of the earliest tetrapods to those of *Panderichthys* and its close osteolepiform relatives. Indeed, many specimens of lower jaws now known to have belonged to tetrapods were once thought to be those of osteolepiform fishes. The similarities are sufficient to suppose that the earliest tetrapods were, like these fishes, aquatically feeding animals. All of them would have been carnivorous, feeding on other fishes and invertebrates, or scavenging anything edible that became available. The early tetrapods were probably gulpers like *Panderichthys,* and with their flattened, broad heads, they may have behaved like the giant Japanese salamanders of today.

In all primitive bony vertebrates, including early tetrapods, the lower jaw consisted of internal and external faces forming a tubular structure (Fig. 2.1). The number and arrangement of bones was the same in early tetrapods and osteolepiforms, with teeth on the dentary and coronoids forming parallel rows. However, changes to the teeth and jaw structure in the evolution of tetrapods can be traced in such features as the relative size and number of teeth in each of these rows (Fig. 10.4). The bones and teeth of the palate change in concert with those of the lower jaw.

In animals such as *Eusthenopteron* and *Panderichthys,* the outer marginal row on the dentary was made up of numerous small teeth, often a double row with even smaller teeth in an external row (Figs. 2.1, 10.4). The coronoid teeth, however, consisted of very large fang pairs and their replacement pits, with perhaps a row of small toothlets or denticles running alongside, external to the fangs. In the early tetrapods, the emphasis shifted away from the coronoid teeth, so that the large tusk pairs were replaced first by rows of small teeth in the earliest tetrapods, and then by a shagreen field of denticles in later ones such as *Crassigyrinus* and the embolomeres. Most baphetids had lost even the shagreen (Fig. 10.4). Ultimately, the coronoids lost their teeth, as in microsaurs and amniotes, and the number of coronoid bones was reduced, disappearing completely in some later tetrapod groups.

Among early tetrapods, a gradual strengthening of the tubular structure of the lower jaw started with loss of the spongy endochondral Meckelian bone that filled the tube in lobe-fins, present in *Panderichthys*, reduced in *Ventastega,* and absent in *Crassigyrinus* and *Megalocephalus* (Fig. 10.4). In its place, internal and external faces became sutured together along the back and lower edges. This can clearly be seen in the sequence from *Acanthostega*, through *Crassigyrinus* and the baphetids, and continued through into the anthracosaurs, temnospondyls, and amniotes.

There was also a change in the angle at which the jaw articulated, and the rather backwardly oriented joint in osteolepiforms gradually became more forward-facing. This would have affected the muscles that opened the jaw and the angles at which they were most effective. The implications of these changes for jaw movements and function have not yet been explored.

The cladogram in Figure 10.4 shows these changes quite clearly. They were very general to begin with, but among the tetrapods of the mid- to Late Carboniferous, many more specializations can be seen that must relate to increasing adaptation to terrestrial feeding and their radiation into different niches in the terrestrial habitat. They correspond to the wider range of overall skull shapes seen among Carboniferous forms.

In temnospondyls, for example, the lower jaw remained generally long and narrow, although it became gradually more consolidated, with some of the coronoid series being lost. In this case, changes to the lower jaw structure may have been constrained by its use in buccal pumping, where a light, narrow jaw may have been an advantage in this amphibian lineage (Szarski 1962). Nonetheless, temnospondyls were the first group of tetrapods to produce a radiation of long, narrow-snouted forms, resembling the modern gharial in shape (Fig. 9.11). This shape is exploited by animals that feed in water by rapid snaps. A narrow snout encounters less resistance to the water than a broader one. Like the modern garpike, the long-snouted temnospondyls must have been specialized aquatic snappers of fish.

The tusk pairs of the palate also ultimately disappeared, but more gradually. For example, *Acanthostega* retained massive vomerine teeth like those of osteolepiforms, but the other palatal teeth were small (Fig. 5.20). Many early tetrapod groups such as temnospondyls (Figs. 8.10, 9.10) and early anthracosaurs (Fig. 9.17) still had palatal tusk pairs, although again, they were usually relatively small. A gradual reduction can be traced in the palatal teeth, seen especially clearly in the amniote lineage. No early amniote has palatal tusks or teeth on the vomers or palatines, and the ectopterygoid bone itself disappears entirely (Fig. 9.33). The sheet of denticles found on the pterygoid of almost all early tetrapods also gradually disappears. In early amniotes, a few strips of denticulated tissue remain, and later, most of these disappear. All that remain are a few teeth in rows on the pterygoid, but importantly, some of these are ranged across the downwardly turned pterygoid flange (Fig. 9.33). This is one of the few osteological characters that most early amniotes share and is one of the few skeletal clues to distinguish an amniote from a nonamniote. Its function will be discussed below.

In contrast to the coronoid and palatal teeth, the marginal dentition in tetrapods became more important. In the osteolepiforms, the small, numerous dentary and maxillary teeth probably served mainly to anchor prey impaled on the massive palatal and coronid fangs. In tetrapods, from even the earliest forms, the dentary and maxillary teeth had become larger, although fewer, and the palatal fangs grew smaller and more even-sized, with only the odd exception, such as *Crassigyrinus,* reevolving a massive and specialized palatal dentition. The inference is that although these early tetrapods were largely aquatic animals still feeding mainly in water, they were using slightly different methods from their fish forebears, or they were eating different prey types. *Crassigyrinus* may represent another instance of this kind of adaptation, with its narrow and elongated skull adapted to slashing through water.

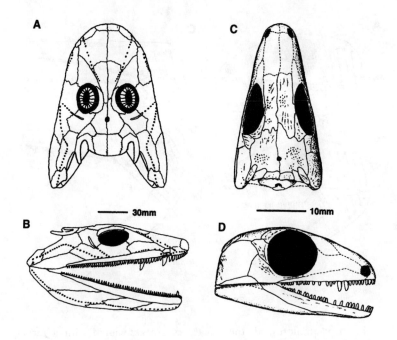

Figure 10.5. Early tetrapod skull shapes. (A) Dorsal and (B) lateral views of the skull of Acanthostega. (C) Dorsal and (D) lateral views of the skull of the early amniote Paleothyris. Note the difference in size between these two skulls.

Whereas in most of the very early tetrapods and most temnospondyls the skull was broad and flat, in the reptiliomorph lineage of tetrapods, including the anthracosaurs, a different type of skull architecture is seen. Here, the skulls tend to be high and narrow, with deep cheeks (Fig. 10.5). The contrast may be related to the interconnected needs of breathing and feeding on land.

On land, two modes of jaw closure can be recognized, which are summarized in Figure 10.6. In animals with long skulls that have the jaw joint right at the back, the lower jaw is swung closed by muscles whose greatest force acts as the jaw begins its travel. By the time it closes, the jaw is moving with considerable speed and inertia so that the teeth impale the prey rapidly. The wider the mouth can be opened, the better. This is called the kinetic inertial system. In the other mode, called the static pressure system, the jaw muscles are arranged so that they exert their greatest force when the jaw is nearly closed and after it has shut. In this case, the prey can be held tightly and squeezed by the mouth and teeth. To achieve this, the point at which the muscles insert has to be moved further forward on the jaw, so that the muscles act at right angles to the articulation when the jaw is closed. Shorter tooth rows and a crest for muscle insertion behind the teeth are characteristic of this mode.

Some anthracosaurs show some modifications toward this static pressure jaw system, including a crest on the surangular bone, but it is best seen in the amniote and microsaur lineages. These animals may represent the first indications of specialized terrestrial feeding. Some microsaurs evolved crushing teeth, which look as though they might have dealt with the shells of the first land snails, for example (Fig. 9.25), or thick-cuticled arthropods. Both groups produced small forms with very short tooth rows and reduced the number of coronoids to one. In the amniotes, the coronoid was produced backward into a crest onto which jaw closing muscles inserted. The down-turned pterygoid flange that bore teeth also provided a point of origin for the powerful pterygoideus muscle, important in the static pressure method of jaw closure (Fig. 9.33).

Gaining Ground • 287

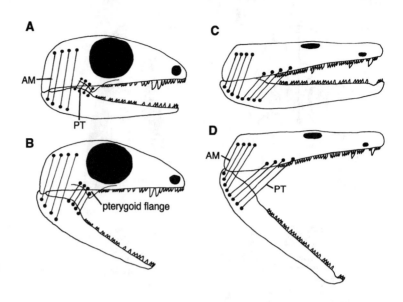

Figure 10.6. Jaw closing mechanisms in early tetrapods. (A, B) Static pressure system in an early amniote. (C, D) Kinetic inertial system in an animal such as Greererpeton. AM = adductor mandibuli muscle; PT = pterygoideus muscle. Note the angle that the pterygoideus makes with the lower jaw when it is open in each case.

Figure 10.7. (opposite page) (A) Occipital view of diapsid skull (e.g., Paleothyris) to show shape of skull and orientation of stapes. Cheek and braincase are linked to each other via the stapes. Gray shading posttemporal fossae, of uncertain extent.
(B) Occipital view of the basal synapsid Ophiacodon. The paroccipital process (opisthotic) meets the cheek, but the stapes is still strutlike.
(C) Occipital view of a lizard ("advanced" diapsid) skull. The paroccipital process meets the cheek and the stapes is a slender rod. Gray shading, posttemporal fossae and other skull fenestrae.
(D, E) Occipital and ventral views of the skull of the "advanced" synapsid Cynognathus. The braincase, palate, and skull roof are closely integrated by sutures, and the role of the stapes as a bracing strut is reduced. Gray areas are the posttemporal fossae. Ophiacodon, Cynognathus and lizard from Romer (1966).

In early amniotes, all these features are associated with a rather boxlike skull. None of these animals had an embayment at the back of the cheek like temnospondyls or many other early tetrapods, and instead of being a somewhat triangular shape in occipital view, the amniote skull was often rectangular (Fig. 10.7). The stapes ran from the braincase, approximately at the center of the rectangle, to the lower corners of the skull, contacting the quadrate. In the early amniote skull, the basal articulation remained much like those of early tetrapods and probably acted as a shock absorber. The stapes was thus called upon to help tie the braincase and skull roof together, as well as to counteract some of the inwardly directed forces acting at the jaw joint produced by the static pressure feeding system. In early amniotes, therefore, the stapes was a structural member of the skull. In other words, they were still partially hyostylic. In diapsid groups, the stapes' role as a bracing member seems to have been gradually replaced by the development of the strutlike paroccipital process that gained a firmer and firmer attachment to the bones of the cheek at its distal end (Fig. 10.7C). In mammals, by contrast, the skull roof and braincase became more firmly tied to each other by elaboration of the occipital plate, and the palatal bones also gained firm attachments to the braincase in place of the old basal articulation (Fig. 10.7B, D, E). Both groups became effectively autostylic but did so in contrasting ways. Only when this process was well under way could the stapes be finally freed from its structural role to be freely used as a hearing ossicle (Clack 1997b).

Returning to the lower jaw, the mammals, evolving in the Triassic, continued some of the changes that were begun by their Carboniferous ancestors and took them to further extremes. For example, they reduced the number of bones further, to just a single one on each side: the dentary. It articulated to the squamosal bone, and part of it formed an equivalent to the coronoid process seen in earlier amniotes. At the same time, consolidation of the skull architecture allowed the stapes to become reduced in size, and the bones of the former jaw articulation, the quadrate and articular, were incorporated with it into the middle ear (Allin and Hopson 1992). This helped to form the mammals' unique hearing system, once more connected to changes in skull structure (Fig. 10.8).

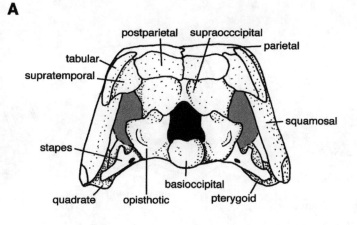
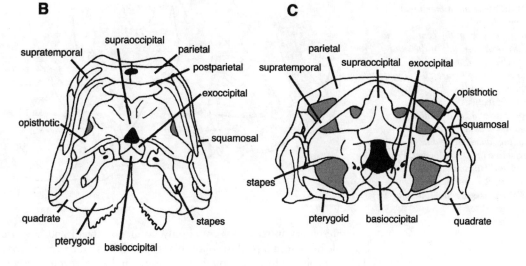
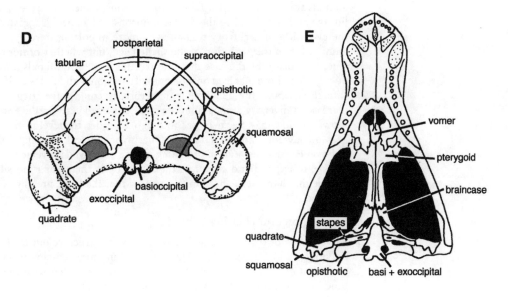

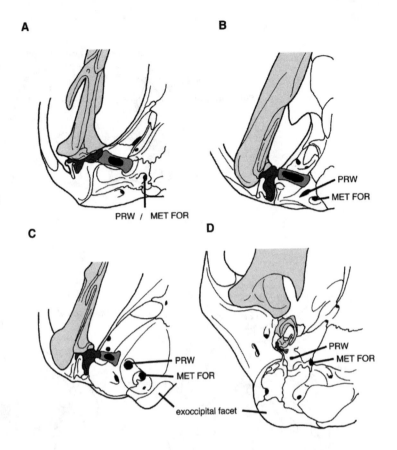

Figure 10.8. Mammal ear evolution.
(A) Thrinaxodon, *a cyndont synapsid*.
(B) Pachygenelus, *a trithelodontid synapsid*.
(C) Morganucodon, *a primitive mammal*.
(D) An opossum, *a marsupial mammal*.
The figures show the ventral view of the posterior half of the skull with the lower jaw in place. Compare with Figure 10.7. Light shading, lower jaw bones, and their derivatives (in D, the tympanic bone and malleus); medium shading, stapes; dark shading, quadrate and its derivative (in D, the incus). From Allin and Hopson (1992). MET FOR = metotic foramen; PRW = pressure relief window.

Herbivory in tetrapods could only occur once teeth and jaws were capable of being adapted to collect plant food. The static pressure system had to evolve before this could be achieved, and herbivorously adapted animals are first found in the Late Carboniferous, some among the synapsid lineage and some among the diadectomorphs. It has been suggested that the evolution of herbivory was an even more important development in amniotes than the evolution of the shelled egg, but whether or not that is true, it seems to be the case that herbivorous clades of animals are more diverse than those that feed only on animal food (Sues and Reisz 1998), and something about the acquisition of herbivory could have triggered the increase in diversity among amniotes that occurred early in the Permian. However that may be, it is possible that in order to develop the static pressure jaw mechanism that allowed herbivory, buccal pumping first had to be replaced by aspiration breathing, which employed movements of the ribcage to expand and compress the lungs. Only then was it possible to change the shape of the head in the manner that the static pressure system required.

Ear Region and Hearing

Because of the interrelated changes to skull structure outlined in the previous section, a discussion of jaws and feeding connects closely with the evolution of hearing airborne sound. Both are connected with the early function of the stapedial bone, used in most modern tetrapods as a hearing ossicle.

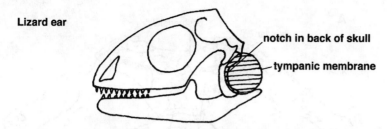
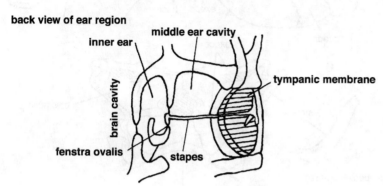

Figure 10.9. Diagram of a lizard ear to show the elements of a tympanic ear.

To capture sound energy traveling through air, an animal must evolve the equivalent of an acoustic transformer. Because the air is more transparent to sound than the animal's body, most of the sound waves would normally bounce straight off the surface, so that an amplifier is needed to make the best use of what is left. A flexible membrane (the tympanum) collects the sound waves that impinge on it, and a freely vibrating lever system (the stapes) magnifies and transmits them across an air space (the middle ear cavity), where they are concentrated into the fenestra vestibuli (or ovalis), the opening into the animal's braincase. These are the main components of a tympanic ear (Fig. 10.9), which the ear of a lizard exemplifies well. Once inside the animal's head, the sound wave energy is picked up by receptors in the fluid of the inner ear and transmitted to the brain, where it is processed into usable information.

Early Tetrapod Ears and the Fossil Record

The history of ideas about the evolution of a terrestrial hearing mechanism in tetrapods illustrates a number of important principles in paleontology. First, it emphasizes how changed perceptions of phylogeny can radically alter views of the evolution of an organ system and thus the functional interpretation of skeletal parts—and even how the relationships of the skeletal parts to one another are seen. It illustrates how a system such as the ear, which appears similar in different groups, may, on deeper examination, differ in ways that have been at first overlooked. Finally, it provides a good example of how changed search images may suddenly reveal things that, although they had been there all along, were not recognized for what they were without additional information. For this reason, it is worth spending a little time on the history of the middle ear in paleontological debate.

Because the middle ear is one of the only two sensory systems to leave a positive fossil record in tetrapods (the other is the lateral line, absent in amniotes), it has been of interest to paleontologists more or less since the

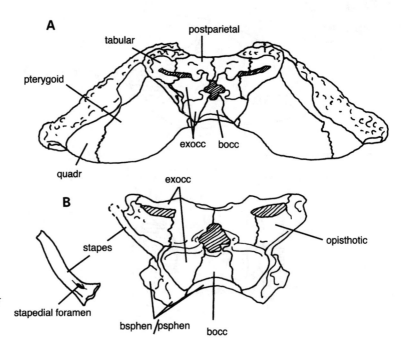

Figure 10.10. (A) Occipital view of Eryops skull.
(B) Close-up of occiput and ear region to show the relationship of the stapes to the braincase; inset, the stapes.

science began (for a review, see Clack 1993). One of the earliest observations concerning the ear region was that many early tetrapods had an embayment at the back of the skull in a similar position to that of the old osteolepiform spiracular cleft. At first the embayment was interpreted as an otic notch, which would have supported a tympanic membrane, and by implication the rest of the components of a tympanic ear. What are now known as temnospondyls were originally called labyrinthodonts, and they showed this ear structure well. In some, a rodlike stapes had a footplate that fitted into the animal's braincase at the fenestra vestibuli, and it passed dorsally toward the center of the otic notch through the air space of the middle ear cavity, toward a tympanic membrane that was presumed to have stretched across the notch. *Eryops* provides an illustration of this (Fig. 10.10), and now *Balanerpeton* is known to be similar. The whole arrangement resembles the ear region of many frogs and lizards. As discoveries of early tetrapods continued, other animals were added to the labyrinthodonts (see Chapter 9), including *Ichthyostega* and anthracosaurs such as embolomeres (Fig. 9.17) and *Seymouria* (Fig. 9.22). An ear region apparently similar to that of temnospondyls was present in *Seymouria*, although the structure of its braincase in the otic region is quite unlike that of any temnospondyls, and its stapes is rather poorly ossified and horizontally positioned (Figs. 9.22, 10.1). Anthracosaurs, which also had otic notches, were assumed to have had stapes like those of temnospondyls in a typically labyrinthodont ear. For many years, no anthracosaur stapes had been found, but they were reconstructed along these lines. The term *labyrinthodont ear* attests to the perceived phylogenetic relationships between the groups in question. Other early tetrapods with labyrinthodontlike characters also had notches at the back of the skull, such as baphetids, *Ichthyostega* (Fig. 5.10), and *Crassigyrinus* (Fig. 7.8), and these too were placed among the labyrinthodonts. Their stapes were also unknown at this stage.

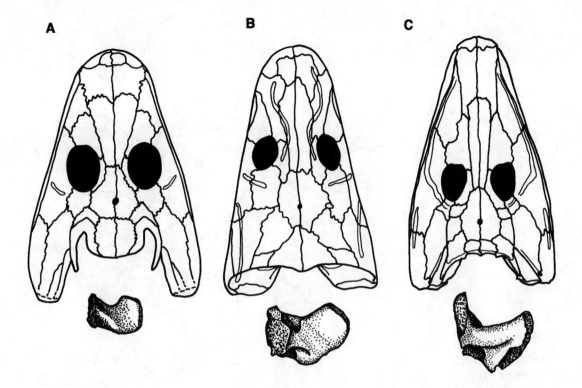

Figure 10.11. Skulls and stapes of early tetrapods, with skull above and right stapes below.
(A) Acanthostega.
(B) Greererpeton.
(C) Pholiderpeton (not to scale).
From Clack (1992).

Because of its wide distribution in a range of primitive tetrapods, the labyrinthodont ear was perceived as a primitive feature common to them all, and the inference was that the tympanic ears found in modern animals would have been derived from this original early tetrapod invention. Other assumptions were linked with this one. Terrestrial hearing would have evolved right at the outset of tetrapod evolution because the animals were seen as having been terrestrial from the outset. The labyrinthodont ear would have evolved early in tetrapod history, as soon as the stapes' role in operating gills and the opercular system had been jettisoned. Once freed from this role, the stapes would also have become redundant as a lever tying the palate to the braincase. The tetrapod skull was considered to have become autostylic during the earliest days of their evolution.

Many recent discoveries about early tetrapods have refuted each one of these assumptions. Early tetrapods such as *Acanthostega* have been shown to be largely aquatic—indeed, on the basis of current evidence, limbs and digits arose while the animals still spent most of their time in water. Most importantly, from the early 1980s onward, it has become evident from finds of the stapes of the colosteid *Greererpeton* (Smithson 1982), the anthracosaurs *Pholiderpeton* and *Paleoherpeton* (Clack 1983), and from *Acanthostega* (Clack 1989, 1992) that very early tetrapods did not have stapes that were lightly built or apparently freely vibrating, but were rather stout with flattened or wing-shaped ends (Fig. 10.11). With this knowledge, it was discovered that the stapes of the anthracosaur *Paleoherpeton*, closely related to *Pholiderpeton*, had been present on the specimen since its first discovery in the 19th century, but it had been unrecognized as such using the search image of a rodlike bone (Fig. 10.12) (Clack 1983).

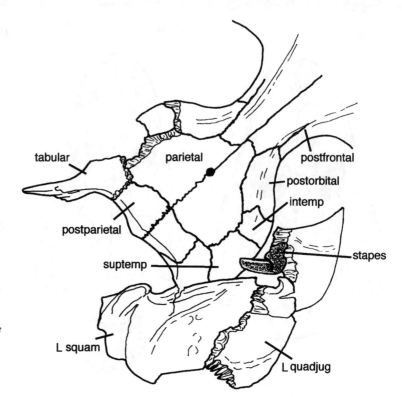

Figure 10.12. Specimen drawing of Palaeoherpeton *showing where the stapes was preserved. The bone is very similar to the stapes of* Pholiderpeton *(see Figure 10.14). Stapes is about 15 mm long.*

Other reassessments suggested that the skulls of the earliest tetrapods were not after all fully autostylic but that their stapes seem still to have been acting as a brace between braincase and palate (Carroll 1980; Smithson 1982); the basal articulation was probably mobile, and even connections between the skull roof and palate were loose or even kinetic. *Acanthostega* is now thought to have had functional internal gills, and the stapes may still have been involved with breathing (Coates and Clack 1991). Finally, and importantly, the group known as labyrinthodonts is not now seen as valid phylogenetically (see Chapter 9).

Along with the new paleontological discoveries went a revised appreciation of the fine-detailed anatomy of the ear regions in modern tetrapods. It became clear through the work of Wever (1978) that in modern amniotes, although superficially similar, the tympanic ears of lizards, crocodiles, and turtles showed differences not only in construction of their stapes and tympanic membranes (see also Presley 1984), but also in the inner ear and brain wiring. Birds show similarities to crocodiles (see also Carr 1992), whereas mammals are quite different from any other amniote group. These findings began to suggest the separate evolution of the tympanic ear in modern animals.

In 1979, at about the same time that the stapes of *Greererpeton* and *Pholiderpeton* were first discovered, Lombard and Bolt (1979) made a fresh study of the evolution of the tympanic ear in the light not only of Wever's work, but also Lombard's own work on frog ears. It had become clear that the inner ear of frogs and amniotes showed fundamental differences that made it difficult to see how one could have evolved from the other (see the section on the inner ear in Chapter 6). In addition to differences in the course of the perilymphatic duct (see Chapter 6), one of the

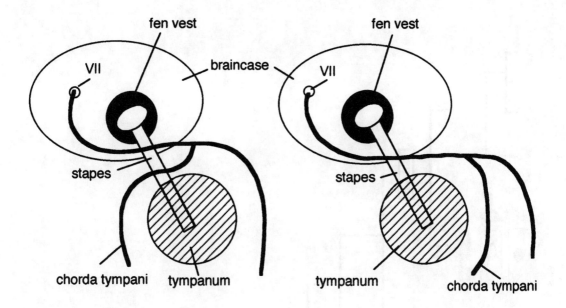

Figure 10.13. Diagram showing the contrasting course of the chorda tympani nerve in amphibians and amniotes.

nerves from the brain to the jaw follows a comparably incompatible route through the ear in frogs and amniotes (Fig. 10.13). Lombard and Bolt's study suggested that it was at least as parsimonious to assume a separate origin for the tympanic ear in several tetrapod groups as to assume a single origin, as in the old view.

Since then have come the discoveries of the aquatic nature of early tetrapods; of the flattened, robust form of the primitive tetrapod stapes; of the likelihood of the role of the stapes in skull architecture; and of the breaking up of the labyrinthodonts into a tetrapod stem lineage that gave rise to amniote and amphibian lineages. Following from this, a revised view of the evolution of the ear can be put forward (Fig. 10.14).

Apart from the fact that in tetrapods, the stapes fits into the fenestra vestibuli in the side wall of the braincase, the otic regions of early tetrapods and osteolepiform fishes are rather similar. The inner workings of the brain, including the receptors for sound, were also probably much the same, so it seems unlikely that tetrapods would have had the neural mechanisms for keen differentiation of sound stimuli (however, there is evidence that even modern fishes with no obvious adaptations to hearing in water do appear to perceive sound stimuli, and the same may therefore have been true of early tetrapodomorph fishes). By virtue of the close proximity of the stapedial footplate to the saccular region of the braincase, there was doubtless some auditory function, perhaps to capture low-frequency waterborne or groundborne vibrations, but by and large, the best model for the hearing capacity of early tetrapods such as *Acanthostega* is probably provided by lungfishes and primitive ray-finned fishes rather than any modern tetrapod.

The robust, flattened stapes of many of the earliest tetrapods, such as

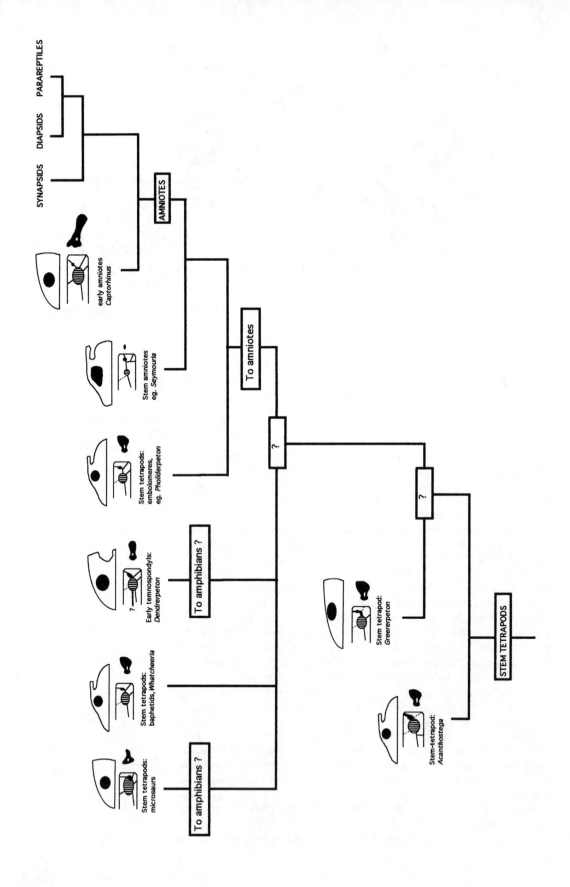

Acanthostega, Greererpeton, baphetids, and embolomeres, appear to have been connected to the palatal bones in an echo of the way the osteolepiform hyomandibula was. The braincases of these animals were not well ossified, but it seems clear that they did not have many of the inner ear specializations of modern animals with tympanic ears. It has been suggested that instead of an otic notch, the embayment positioned in the region of the old spiracular cleft, in fact still retained this original function and housed a persistent spiracular opening to the outside in forms such as *Acanthostega* (Carroll, in Panchen 1985; Clack 1989) (see Chapter 6). Exhaled air could have been expelled this way as in the modern but primitive ray-finned fish *Polypterus* (see Chapter 2). Because of this, the notch is better called something other than otic. The term *spiracular notch* is a possibility, but because its function cannot be ascertained and may in any case have been different in different groups, a neutral term such as *temporal notch* is probably preferable (Godfrey et al. 1987).

The first occurrence of what may have been a true tympanic ear is seen in the temnospondyl lineage, with *Balanerpeton* from the Viséan of East Kirkton (see Chapter 8). It had a broad embayment in the back of the cheek, but also the stapes is known to have been a relatively small bone. It is only known from the end nearest the braincase, but this suggests a relatively light bone directed toward the center of the embayment (Milner and Sequeira 1994). The arrangement bears close comparison with the ear region of frogs, and in Chapter 9, it was suggested that modern amphibians are the descendants of temnospondyls. If this is correct, the ears of frogs are a direct inheritance from this group of Carboniferous ancestors (see also Bolt and Lombard 1985). Evolution of a terrestrial hearing mechanism in temnospondyls may be consequent upon their development of the interpterygoid vacuities. If temnospondyls specialized as buccal pumpers, the vacuities could have allowed the roof of the palate a greater flexibility and facility for expansion. Given this, the basal articulation might have become redundant and immobile; thus, the skull became increasingly autostylic, eventually freeing the stapes from its bracing role to be used instead as a specialized hearing ossicle (Clack 1992).

Alternatively, if lissamphibians are related to lepospondyls rather than to temnospondyls, the middle ear in frogs represents yet a further iteration of the development of a tympanic ear because no such modification is seen in any lepospondyl (Laurin and Reisz 1997). In this context, the otic region of *Eucritta* deserves mention (Figs. 8.14, 8.15). In the large baphetids *Megalocephalus* and *Spathicephalus,* the stapes is known to be of the stubby "winged" form seen in *Acanthostega, Greererpeton,* and *Whatcheeria,* a morphology not likely to be associated with a terrestrially adapted ear and tympanic membrane. However, the otic notch of *Eucritta* is of much the same shape and relative size as that of *Balanerpeton,* whose stapes is known to be like that of other small terrestrial temnospondyls that are assumed to have had tympanic ears (Fig. 8.10). If *Eucritta* really is a baphetid, two explanations are possible. The first is that the earliest baphetids had tympanic ears like those envisaged in temnospondyls, and that in secondarily aquatic forms, this structure was lost. Alternatively, the notch of *Eucritta* was not otic but still spiracular. This carries the possible implication that the notches in early temnospondyls, very similar in appearance to those of *Eucritta,* were not otic either. Unfortunately, the best way to test the strength of either of these hypotheses is to examine the stapes of *Eucritta.* It has not yet been found.

Some recent information on temnospondyls may have a bearing on the question. In some juveniles of the Triassic temnospondyl *Parotosuchus*

Figure 10.14. (opposite page) Cladogram of early tetrapods and their ear characters. The fenestra vestibuli, shown hatched, starts as a hole bounded by otic capsule and occipital arch bones, and the dermal parasphenoid. The lateral otic fissure is open but the basi- and exoccipitals are coossified. The stapes, shown black, is a flat, broad plate in early tetrapods and is converted to a strut in early amniotes. The metotic foramen is the remnant of the lateral otic fissure found in later tetrapods and diminishes further through their evolution.

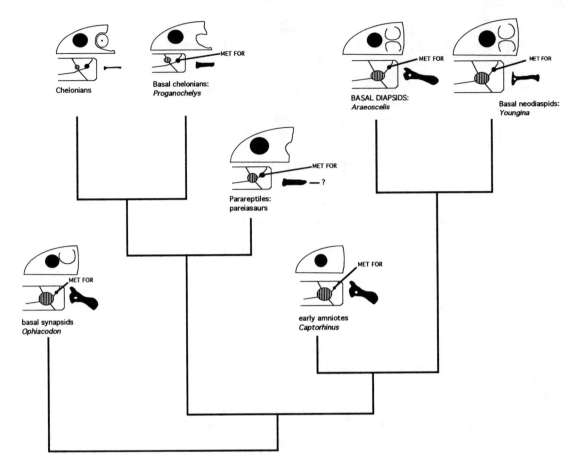

Figure 10.15. Cladogram of amniotes and their ear characters. Same conventions as Figure 10.14. The stapes reduces in diameter through the evolution of turtles and diapsids independently. MET FOR = metotic foramen. For mammalian ear evolution, see Figure 10.8.

aliciae, Warren and Schroeder (1995) have suggested that there was a persistent spiracle because of the association between the stapes and an oblique ridge running just in front of it beneath the cheek. This creates a wall that they suggested formed the boundary of a spiracular cleft, rather than a middle ear cavity. Second, some late Triassic temnospondyls with otic notches have been discovered to have gill rakers preserved in their pharynxes (Van Hoepen 1915). The implications of this are that the animals may have been feeding in water by use of gills. Whether or not they were breathing with the gills is unknown, but perhaps the existence of a tympanic ear in such cases is unlikely.

As described in the section on jaws, in all the early amniotes there was no embayment at the back of the skull, and the stapes was acting as a structural brace between the jaw joint and the brain. The story of the changes to the stapes, braincase, and skull architecture in amniotes is a complex one, but the fossil record of these features can be followed through evolutionary time in several different lineages; some of these are summarized in Figures 10.15 and 10.16 (Clack 1997b). Turtles, mammals, and diapsid reptiles can be shown to have evolved many crucial components of the tympanic ear separately from beginnings that all resemble those of early, effectively earless forms.

A critical stage in each lineage was the freeing of the stapes from its early bracing function. This cannot be achieved without consolidating the

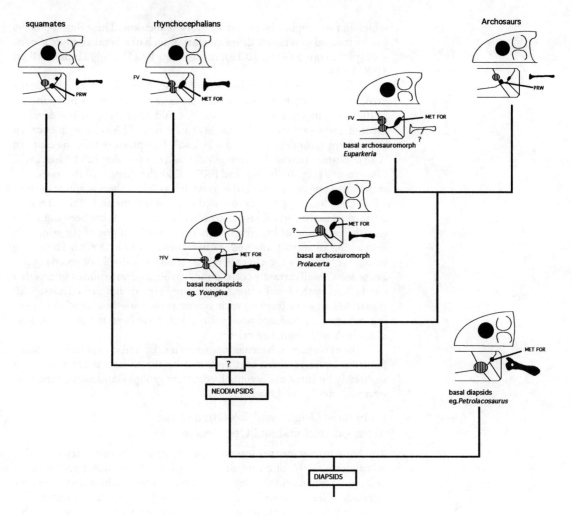

Figure 10.16. Cladogram of diapsids and their ear characters. Same conventions as Figures 10.14 and 10.15.
PRW = pressure relief window. A reduced fenestra vestibuli and the PRW have appeared in archosaurs and lepidosaurs independently.

skull by some other means, and these have been outlined in some amniote groups in the section on feeding.

Paralleling the changes to the stapes, there is a consistent pattern in the evolution of the fenestra vestibuli to be seen in the fossil record (Figs. 10.14–10.16). In the earliest tetrapods, the footplate of the stapes is a relatively large component occupying a considerable part of the braincase wall. It contacts several of the braincase bones, including those of the otic capsule, the occipital arch, and the sphenethmoid region. In fact, it lies more or less on the line of the embryonic metotic fissure that separates these discrete embryonic elements (see Chapter 2 and below). This condition is found throughout early tetrapods, whether they are part of the stem lineage, reptiliomorphs, amphibian lineage, or lepospondyls, and there can be little doubt that it was primitive for tetrapods. Followed through the fossil record, which is especially good in amniotes, the relative size—but more importantly the position—of the fenestra vestibuli changes. It gradually shrinks relative to the braincase wall, and eventually its relationship to the occipital arch and sphenethmoid bones is broken and it becomes a hole that is completely defined by otic capsule bones. A small fenestra vestibuli indicates a small stapedial footplate, which implies a more slender stapes,

Gaining Ground • 299

which in turn implies increased auditory refinement. These developments can be tracked as separate developments in each of several amniote groups —diapsid amniotes (Fig. 10.16), turtles (Fig. 10.15), and mammals (Figs. 10.8, 10.15).

A further stage in the refinement of higher-frequency reception was the evolution of a mechanism by which pressure waves, set up by sound vibrations impinging on inner ear fluids, could be relieved by the development of a separate window in the braincase wall. This is sometimes called the fenestra rotunda, or "round window." This pressure relief mechanism is an important part of the tympanic ear, and its evolution can be seen in the fossil record (Fig. 10.16, marked PRW). Like the changes to the role of the stapes, this one can be tracked separately in each of these amniote groups. All of them were apparently initiated only toward the end of the Permian, or possibly in the early Triassic, in each of the groups. It has been suggested that the evolution of buzzing insects may have been one of the stimuli in developing an aerially adapted ear in amniotes (Clack 1997b). The timing of the evolution of the characteristics of an ear capable of receiving airborne sound well certainly corresponds to the huge radiation of insects that can be seen in the fossil record of the late Permian and early Triassic. Alternatively, in some lineages such as mammals, with the evolution of nesting behavior, an additional stimulus might have been the ability to hear offspring's higher-pitched cries.

The evolution of hearing in air was a slow and complicated process; far from being achieved and perfected by the earliest tetrapods, it was separately invented many times by different groups—and a long time after tetrapods first gained the land.

Embryonic Origins and Evolution of the Tetrapod Occipital and Otic Regions

Some insight into the formation of the tetrapod occipital region and its relationship to the otic capsule can be gained from consideration of the embryonic origins of the several braincase regions and how they have been affected differently during the fish–tetrapod transition. As outlined in Chapter 2, the braincase forms from a number of discrete components that contribute to the whole (Fig. 2.2). After their initial formation, the components may fuse to each other completely except for the necessary foramina for nerves and blood vessels, or gaps and fissures may remain even into the adult structure. The extent and pattern of fusion varies among different groups and taxa, but some key landmarks are usually identifiable, such as the ventral cranial fissure between the front and rear halves of the braincase in many early forms. In many lobe-fins, this fissure extends dorsally to the roof of the skull to contribute to the cranial hinge characteristic of certain groups (e.g., *Eusthenopteron*, Figs. 2.1, 2.12, 6.9). In the earliest tetrapods, the ventral cranial fissure can still be identified (e.g., in *Acanthostega*, represented by a suture in Fig. 2.12), although it is eliminated in most later ones. The lateral otic fissure is another such landmark, visible during the embryonic development of all modern vertebrates and evident in many fossils, including fish such as *Eusthenopteron* and early tetrapods such as *Acanthostega*. Figure 10.17 shows the braincases of *Eusthenopteron*, *Acanthostega*, and *Greererpeton* for comparison.

The fissures appear to correspond to the junctions between tissues of different embryonic origins (Clack 2001b). In fossil forms, although the embryonic development of the animals cannot be directly observed, some of the boundaries are clear enough for one to infer that the same embryonic

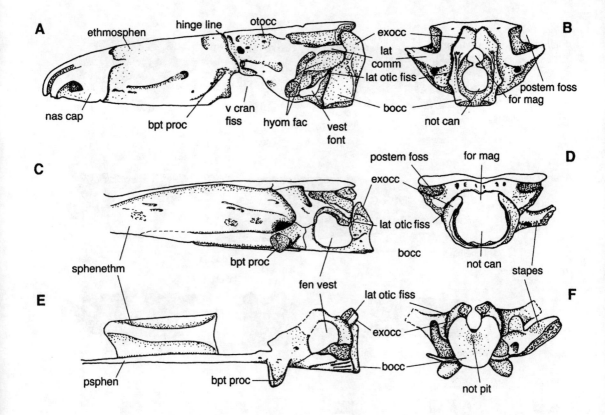

Figure 10.17. Braincases of Eusthenopteron, Acanthostega, *and* Greererpeton *showing comparable structures.*

tissue contributed to the corresponding parts as in modern forms. Thus, for example, it appears that the otic capsule was formed by a type of mesoderm associated with purely cranial development (cephalic mesoderm), whereas the underlying parachordal cartilages and occipital arch material were formed from mesoderm that is more closely associated with the rest of the body (somitic mesoderm). The anterior parts of the braincase, including the trabecular cartilages, the sphenethmoid, and the basisphenoid, are formed from neural crest, as of course is the underlying dermal bone, the parasphenoid. The hyomandibula or stapes is also largely formed from neural crest, being a branchial arch element, and it appears that some of the otic capsule surrounding the footplate of the stapes might also be of neural crest origin, although this is debated (see Chapter 6). Figure 10.18 shows the inferred domains of these components in three animals, *Eusthenopteron, Acanthostega,* and *Greererpeton.*

In Chapter 6, the point was made that in the sequence *Eusthenopteron–Panderichthys–Acanthostega,* changes to the otic region of the braincase appear to have occurred rapidly between *Panderichthys* and *Acanthostega.* Although *Panderichthys* is much like *Eusthenopteron* in the relationships of the otic capsule and hyomandibula, *Acanthostega* is essentially tetrapodlike. At the same time, it is apparent that *Acanthostega* had an occipital construction in most ways just like *Eusthenopteron.* This region remained conservative, changing little despite the radical restructuring going on in parts of the otic capsule. If the embryonic origins of these two regions are taken into account, some inferences can be drawn. Those

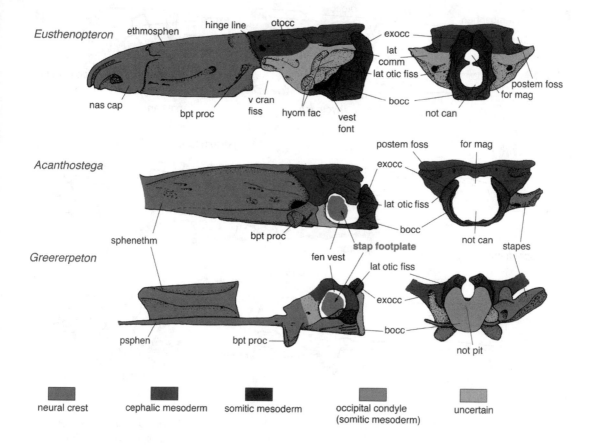

Figure 10.18. Embryonic components of the braincases of Eusthenopteron, Acanthostega, *and* Greererpeton *(from Clack 2001).*

parts of the braincase that underwent the most radical change are those associated with neural crest. Along with changes to the otic region affecting the hyomandibula/stapes are others to the front and mid parts of the braincase. The cranial hinge is eliminated, the basisphenoid joins up with the basioccipital, and large bulbous basipteryoid processes are produced, all during the transition represented by *Panderichthys–Acanthostega*. These parts too are all of neural crest origin, an observation that carries the implication of increased activity of neural crest tissue in the early stages of tetrapod evolution.

If another early tetrapod is added to the sequence—and only *Greererpeton* is known in sufficient detail to be useful—further developments of this change can be seen (Figs. 10.17, 10.18). The parasphenoid has elongated to underlie both parts of the braincase, and the anterior portion of the braincase itself is longer with respect to the posterior, an effect also explored in Chapter 6. But there has been one further development. In *Greererpeton*, the occipital condyle has developed and become a solid articulating surface.

In modern animals (and caution must be exercised here, because only birds have been examined in this context), the condylar part of the occipital complex is actually added at a later stage during the embryonic development of the braincase than the rest of the basioccipital. It forms from an extra segment of the somitic mesoderm that is incorporated into the braincase. Could the contrast between *Acanthostega* and *Greererpeton* reflect this addition of the extra segment, with *Acanthostega* lacking it, and later tetrapods from *Greererpeton* onward having gained it? Are the contrasts in occipital formation seen in the two ancient lineages reflected in the embryonic development of animals today? Such questions can only really be approached by looking at the embryonic development of modern animals—for instance, by investigating how amniotes might differ from amphibians.

One of the critically affected regions at the fish–tetrapod transition was that where the head of the hyomandibula and its articulations to the braincase became transformed into the stapes and fenestra vestibuli (see Chapter 6 and Figs. 2.12, 6.9). The fenestra vestibuli is a gap in the bones lying just at the boundary between otic capsule, occipital arch, and neural crest domains. The subsequent evolution of this region and how the gaps between these tissues are sealed are the key to the evolution of the tetrapod auditory apparatus. Because it involves modifications to the occipital region, the ear region was strongly affected by changes to the neck joint (Clack 2001b). For example, the size and extent of the exoccipitals and their relationship to the opisthotic bones affect the formation of the round window in diapsids. Other relationships between the occipital region and the ear are outlined in the next section.

Neck Joint

The evolution of a neck joint may well have gone hand in hand with increasing reliance on air breathing in early tetrapods and their forerunners. For example, the operculogular region in the tetrapodlike fish *Panderichthys* was relatively reduced, and that in modern lungfishes has disappeared. The ability to raise the head and widen the mouth cavity to gulp air is increased by this means. *Acanthostega*, with an occipital region like that of *Eusthenopteron* (see Figs. 2.12, 10.17), had not developed the flexible neck joint seen in later tetrapods, but it had lost the operculogulars and was probably an air gulper, a bit like a lungfish, which did not spend much time out of water.

It is possible to track changes to the neck region among early tetrapods, with interesting results. First, there is a consolidation of the braincase so that the front ethmosphenoid becomes more firmly attached to the rear occipital region. The notochord in fish such as *Eusthenopteron* penetrates to about the midpoint of the braincase; in the earliest tetrapods, in which the posterior part of the braincase is much shortened, the notochord necessarily runs for a shorter distance inside it. Among Carboniferous forms, the space for the notochord becomes a shallow cone in shape. Eventually, only a dimple is left to mark the position of the front of the embryonic notochord. The resulting concave surface is used as the articulation surface, which is matched by a convex surface on the most anterior centrum (Fig. 10.19).

Above the basioccipital lie the exoccipitals, a pair of bones that flank the hole through which the spinal cord enters the brain. In some ways, they resemble paired halves of a neural arch. In *Eusthenopteron, Acanthostega,* and many early tetrapods, they are quite small and are loosely attached to the otic region of the braincase. In some early tetrapods, they articulate with the arch of the atlas vertebra or with the proatlas. Following the fate of these occipital arch elements shows that in the anthracosaurs and in the amniote lineage, they stay as relatively small, loosely attached bones. The

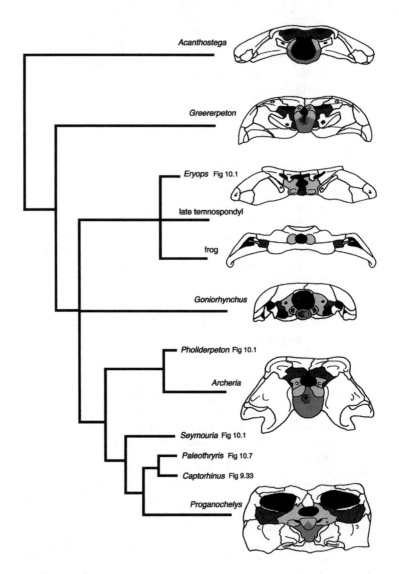

Figure 10.19. Occipital construction in tetrapods. In Acanthostega, *there is no occipital condyle, and basi- and exoccipitals are coossified; in the early tetrapod* Greererpeton, *the concave condyle is formed from the coossified basi- and exoccipitals. In temnospondyls, the occipital condyle is formed mainly or entirely from the exoccipitals, which contact the skull roof and exclude the opisthotics from the occiput. In frogs, the basioccipital is not distinguishable, and the paired condyles are formed from the exoccipitals. In microsaurs, the condyle is formed from basi- and exoccipitals in roughly equal proportions. In anthracosaurs and seymouriamorphs, the concave condyle is formed mainly from the basioccipital with small exoccipitals that do not contact the skull roof. In the early amniotes* Paleothyris *and* Captorhinus, *the convex condyle is basioccipital, as in diapsids, but in turtles it is formed from basi- and exoccipitals. Opisthotic, dark shading; basioccipital, medium shading; exoccipital, light shading. (*Proganochelys *from Gaffney 1990.)*

exoccipitals do not make contact with any part of the skull roof, so that contact between the braincase and skull roof is made via the otic capsules (Figs. 9.35, 10.1, 10.19). Through the gaps that remain separating the exoccipitals from the otic capsules, nerves such as the vagus and blood vessels such as the jugular vein pass out to the body. In early members of the amniotes—and in some modern ones such as turtles—part of each exoccipital joins with the basioccipital to form the condyle with which the atlas vertebrae articulates. The condyle in amniotes is always convex, although even here, sometimes there is a small dimple representing the old course of the notochord. In mammals, the basioccipital's role diminishes, and the condyle becomes a double one formed only by the exoccipitals (Fig. 10.19). In all amniotes, it is the otic capsule, especially the opisthotic paroccipital process, that comes to form the main part of the rear of the skull bearing the muscles that run to the vertebral column.

In the amphibian lineage of tetrapods, something different happens. The exoccipitals come to be important from an early stage. They increase in size among temnospondyls so that they grow over much of the rest of the occipital surface, masking the otic capsule from behind and growing upward to join the skull roofing bones (Figs. 10.1, 10.19). In these animals, material from the occipital arch provides much of the anchorage between the braincase and the skull roof, whereas the otic capsules remain relatively poorly ossified. The basioccipital becomes gradually reduced in temnospondyls until it leaves a gap between the two exoccipital facets, which are left to form the condyle (Watson 1926). The atlas vertebra has a matching paired articular surface. Here the back of the skull was strengthened by quite different means from that in amniotes (Smithson 1982) (see Chapter 9). A similar occipital formation is found among modern amphibians, which also have a double condyle like temnospondyls. Among microsaurs, the form of the occipital condyle is characteristic and highly specialized as a concave, horizontally configured figure-eight shape (Fig. 10.19) (Carroll 2000a) described in more detail below.

One inference from this is that the neck joint started from a structure that was relatively unstable for a terrestrial animal and became strengthened gradually as the animals became more and more adapted to life on land. It seems to have been achieved by a different method in each of the two groups of animals that have left living descendants. The way the back of the skull became strengthened to receive neck muscles also had an effect on the role the stapes played and thus on the evolution of hearing. For example, as outlined in the section on jaws and feeding above, in diapsids, the stapes began as a strengthening device for the back of the skull, but with the development of a paroccipital process, the stapes could be freed from its former role to become a freely vibrating ossicle. In mammals, the occiput was strengthened by increasingly firm connections between the opisthotics and palate, as well as between the opisthotics and skull roof. The fossil record shows an inverse relationship between the extent of these connections and the size of the stapes (Allin and Hopson 1992). In temnospondyls, the occiput was strengthened by a different method: the expansion of the exoccipitals and their intimate connection to the skull roof. In addition, the palate and skull roof became firmly tied to each other at the basal articulation and around the margins. Breathing by buccal pumping was enhanced by the palatal vacuities. The formation of an occipital condyle gives another example of a terrestrial adaptation that was achieved convergently several times among tetrapods, with extant and extinct groups showing a range of patterns.

Postcranial Modifications for Terrestriality

Axial Skeleton: Centra, Atlas–Axis, Vertebrae, and Ribs

It might be supposed that one of the parts of the skeleton to have been most affected at the fish–tetrapod transition would be the vertebral column. The vertebral column supports the weight of the body when it is suspended between the pillars of the four limbs, so that in a terrestrial animal, adaptations to take this weight might be expected. In fact, the vertebral column seems to have changed very little in the early stages of tetrapod evolution, which speaks of the relatively gradual way in which terrestriality was achieved.

Centra

In fishes such as *Osteolepis* and *Eusthenopteron*, the vertebral construction is that called rhachitomous—that is, the main element is a horse-

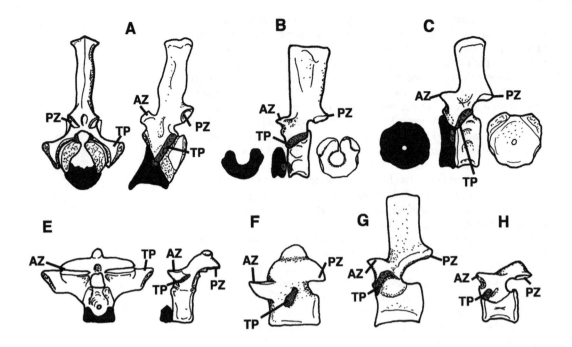

shoe-shaped structure ventrally, usually single but in some circumstances with a split between two halves at the midline. This is the intercentrum. Dorsal and posterior to that lie smaller paired elements called pleurocentra. These elements lie beside the notochord and sheathe it to a greater or lesser extent, although the notochord would still have been present and relatively unconstricted (Figs. 2.3, 2.4). These elements appear to have ossified from a cartilage precursor that formed in the sheath around the notochord (Carroll et al. 1999). Often, although their external surfaces would be covered in smooth periosteal bone, the internal surfaces remained "unfinished" where the surfaces were covered only in cartilage. In the tetrapods, the intercentrum bore a laterally placed facet onto which one of the rib heads articulated.

This kind of construction is found more or less universally in the earliest tetrapods (Fig. 10.20). *Acanthostega* shows a textbook form (Fig. 6.10), although in *Ichthyostega,* the intercentra are relatively tall and the pleurocentra rather smaller than is often found in this type. *Whatcheeria* and the newly discovered Tournaisian animal also have more or less classic rhachitomous centra, and the style is found throughout the majority of Paleozoic temnospondyls. There seems to be no doubt that this kind of centrum is primitive for tetrapods, and at least in the earliest forms, it was associated with an aquatic lifestyle.

Developments to the centra during the transition seem to lag behind those to the neural arches and ribs. Although *Acanthostega* has barely developed zygapophyses and has short, straight ribs (Figs. 5.19, 6.10), *Ichthyostega* and most other early tetrapods have more or less robust ribs and conspicuous zygapophyses. These are presumably associated with changes to locomotory patterns, although that may not be the whole story.

There are other forms of vertebral construction found among tetrapods from the late Early Carboniferous onward, and there seems to be a phylogenetic as well as a functional component to their distribution. Figure 10.20 summarizes the different types of vertebral construction seen in early tetrapods. The reptiliomorph lineage, exemplified by anthracosaurs, seymouriamorphs, and amniotes, have what are called gastrocentrous vertebrae. In this form, the dominant element is the pleurocentrum, which is again a horseshoe-shaped structure, and the more anterior intercentrum is a small ventral wedge. The neural arch sat securely on the pleurocentrum supported by broad facets, or even sutured to it. In early anthracosaurs, the pleurocentrum was incompletely ossified dorsally, but in more specialized forms, it became a complete disk or cylinder. In some members of the lineage, the intercentrum became less and less conspicuous and eventually disappeared. Like the rhachitomous form, the gastrocentrous type ossified from a cartilaginous precursor, and the distribution of periosteal to unfinished surface is comparable (Carroll et al. 1999).

Another type of vertebral construction is found among the lepospondyls. Here, centra are almost universally single cylinders, often spool-shaped, and always sutured or even fused to the neural arch, and this is called "holospondylous" (Fig. 10.20). In a very few forms, an additional small, wedge-shaped ventral element, presumed to be an intercentrum, is present, but these are exceptions. It is also notable that in later amniotes, the form of vertebral construction approaches that of the holospondylous groups (Carroll et al. 1999).

It is natural to try to attribute some functional interpretation to these varieties of vertebral types. Among the earliest was that by Parrington (1967b), who built models of the various types and subjected them to a range of tests. He concluded that the rhachitomous type allowed the vertebral column to twist at the same time as bending laterally. Panchen (1966, 1967) also looked at the matter and suggested that the rhachitomous form was associated with use of the axial musculature for lateral movements and swimming. He postulated that in swimming or laterally flexing animals, the axial musculature, founded on the ribs, required good support from the column. With the emphasis on the intercentrum, the rhachitomous column would provide this better than the gastrocentrous, in which the intercentrum was small. Conversely, in the gastrocentrous column, emphasis would have been on the foundation for the neural arch, that is to say on the pleurocentrum, and this made more sense in an animal that raised its body off the ground and relied on limb (appendicular) muscles for its locomotion. Thus, the rhachitomous type might be expected in more aquatic forms, such as *Acanthostega* and many temnospondyls, whereas the gastrocentrous type might be expected in more terrestrial forms, such as amniotes and their kin.

The idea fits in with a number of observations in early tetrapods and is tested by some exceptions found in both temnospondyls and anthracosaurs. For example, in the secondarily aquatic embolomeres, the emphasis has almost been shifted back again, so that the intercentra become larger once more and they form complete disks like the pleurocentra (Fig. 10.20). This is known as embolomerous. The vertebral column becomes composed of a series of disklike central elements. By contrast, in some of the larger and more permanently aquatic temnospondyls, the pleurocentrum is reduced in size until it disappears in the later forms, and the intercentrum becomes a complete, often solid disk. These centra are described as stereospondylous. In the colosteids, there appears to have been another instance of the evolution of almost complete ring centra, where the pleurocentra are enlarged in an otherwise more or less rhachitomous column (Godfrey 1989a). *Crassigyrinus* seems to have had two more or less equal parts to its

Figure 10.20. (opposite page) Types of vertebrae in early tetrapods. (A) Eryops, a rhachitomous temnospondyl, posterior view at left to show paired pleurocentra, lateral view at right. (B) Proterogyrinus, a gastrocentrous anthracosaur, intercentrum at left, pleurocentrum at right. (C) Archeria, an embolomerous anthracosaur, intercentrum at left, pleurocentrum at right. (E) Seymouria, a gastrocentrous seymouriamorph, anterior view at left to show small intercentrum, lateral view at right. (F) Captorhinus, a gastrocentrous (or holospondylous) early amniote with domed neural spines. (G) Archaeothyris, a gastrocentrous (or holospondylous) early amniote with bladelike neural spines. (H) Pantylus, a gastrocentrous (or holospondylous) microsaur. All are lateral views shown with anterior to the left. Intercentrum is dark shaded, rib articulations light shaded. Not to scale. Note the sloping zygapophyses in the animals in A–C and the almost horizontal ones in the animals E, F, and H. AZ = anterior zygapophysis; PZ = posterior zygapophysis; TP = transverse process.

centra, although both seem to have remained cartilaginous dorsally (Panchen 1985). The evolution of disklike centra seems to have occurred commonly among swimming forms and is also found in the much later marine reptiles such as ichthyosaurs. By contrast, in some of the more terrestrial temnospondyls, such as a peculiar animal called *Peltobatrachus* (Panchen 1959) and some of the dissorophoids including *Doleserpeton* (Bolt 1969), there has been the evolution of something approaching the gastrocentrous form of centrum. The little *Casineria* from the mid-Viséan of Scotland is the earliest known example of gastrocentrous vertebrae with cylindrical pleurocentra and small intercentra, and the creature appears to have been fully terrestrial (Paton et al. 1999) (Fig. 7.3). Fully gastrocentrous vertebrae are also found in the amniotelike creature *Westlothiana* from East Kirkton (Fig. 8.17).

Of course, there are some animals that do not fit into this neat scheme, one of the best known being *Eryops* (Fig. 9.12). This large temnospondyl appears to have been capable of land locomotion and of carrying its body weight entirely by its limbs, but its vertebrae are rhachitomous (Moulton 1974). The picture is probably more complex than that painted above because several important factors, such as the angle that the zygapophyseal facets make to the column, have been left out of the equation (Holmes 1989b).

Because some growth series are known for a few early tetrapods, especially temnospondyls, it is possible to say a little about the way their vertebral centra became ossified. Centra in the earliest tetrapods, whether pleurocentra or intercentra, began as paired elements that often became coossified in the midline. The centra of younger individuals enclose a larger space for the notochord, and the ossification of the centra gradually encroaches into the space to occlude the notochord to a greater or lesser degree. From the distribution of unfinished versus periosteal surface, it seems reasonably certain that the centra formed first from cartilage that was gradually replaced by bone during the animal's growth, as in typical endoskeletal development. The pattern can be described as ossification from the inside to the outside (Carroll et al. 1999).

Furthermore, it is also clear that ossification of the vertebrae began at the front of the column and increased backward as the animal grew. That order is in accord with the way the vertebral column is initiated in the embryo, with embryonic segmentation and somite formation starting at the front and working backward. There is evidence for this kind of development and vertebral form in most early tetrapods that at one time would have been called labyrinthodonts. At the same time, it is interesting to note that the reverse order applies to the fin rays and spines of the tail in *Eusthenopteron* in that they start to ossify from the rear forward. Use of the tail in swimming from an early stage may be the operating factor here, emphasizing the changes that can be brought about by the needs of function in some cases (Cote et al., in press).

In the lepospondyls, the situation is quite different. No lepospondyl is known in which the centra are not obviously cylindrical, even in the smallest and presumed youngest individuals ever found, and even in those forms where what appears to be an intercentrum is present. This is not simply a matter of size or maturity. Some of the smallest temnospondyls and anthracosaurs are smaller than many lepospondyls, and some lepospondyls are larger than some small temnospondyl species, but still this consistent difference pertains. It does seem, however, that although small, young lepospondyls had cylindrical centra, the centra were not well ossified. In the flattened fossil vertebrae, a clear X shape is often seen, representing the walls of a spool-shaped element in which the cone-shaped recesses for the notochord are outlined. The outer bony skin of the centrum was present, but the inside was still soft. This indicates a contrasting pattern of ossification from either temnospondyls or anthracosaurs, that proceeds from the outside to the inside. It is not clear whether they were preformed in cartilage or ossified directly around the notochord, but the latter is the case in modern salamanders. In frogs and amniotes, although the vertebrae are in some ways similar in final appearance to those of lepospondyls, the centra form first in cartilage. However, the details of ossification and formation of vertebral centra are complex and different among many of the modern groups, so that homologies between the centra of fossil tetrapods and modern ones are almost impossible to be sure about. Although cylindrical centra in lepospondyls are something they all show in common, it remains uncertain whether this represents a single shared derived character. Its appearance among several other groups might give cause to suspect that it is a parallelism and that the constraints of making vertebrae in a small animal might be involved in all of them (Carroll et al. 1999).

It is true that modern amniotes end up with centra that are very similar in appearance to those of some lepospondyls, although none shows anything very similar to the centra of early tetrapods. This is one of the arguments that has been used in the debate about the relationships between amniotes and some lepospondyl groups. There are, nonetheless, sufficient similarities between the centra of early amniotes and animals such as seymouriamorphs to suggest that even though their developmental patterns might have diverged, the amniotes nevertheless had ancestors whose vertebrae were typically gastrocentrous.

Something that has not been generally considered when trying to assess the function of early tetrapod vertebrae is the possible connection between vertebral construction, body wall musculature, and the movements of breathing and locomotion. In early tetrapods, the axial musculature, supported by the ribs, acts largely in locomotion. The same is true of modern salamanders. While it is so engaged, the part that both the muscles and the ribs can play in the operation of aspiration breathing may be constrained. Thus, it may be that rhachitomous animals that used body wall muscles for locomotion had to remain buccal pumpers. With the development of gastrocentrous vertebrae, the freeing of axial musculature from their role in locomotion allowed the ribs and the intercostal muscles to be used in the movements of costal ventilation.

Atlas–Axis

There is one region of the vertebral column that becomes differentiated from the rest early on in tetrapod history, and that is the most anterior part. The two most anterior vertebrae—the atlas and axis—become modified to provide support and flexibility for the neck joint, and their development echoes that of the occipital region of the skull (Fig. 10.21). In *Acanthostega*, even though there is little to distinguish most of the vertebrae from one another along on the column, at the front, the atlas and axis are already recognizable (Coates 1996). The atlas arch is very small, whereas the axis arch has a neural spine that is longer than average front to back, although it is shorter than the other trunk neural spines. There seem to be no atlas pleurocentra, but the intercentrum is slightly longer than average, and furthermore, where other intercentra are usually formed from two separate halves, the atlas intercentrum is a robustly built single element.

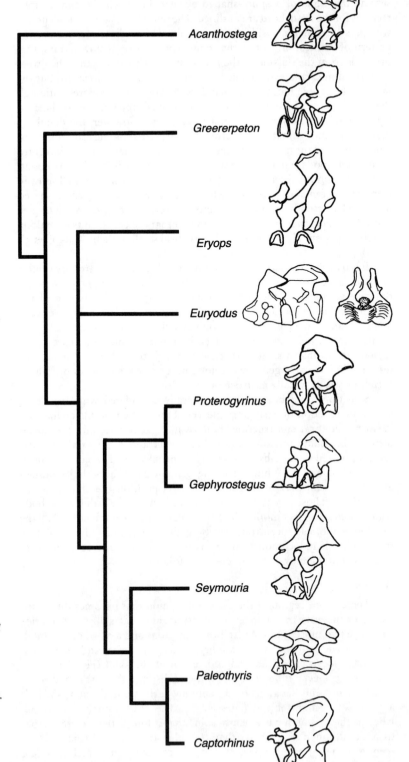

Figure 10.21. Atlas–axis construction in early tetrapods; the early tetrapods Acanthostega *and* Greererpeton, *the temnospondyl* Eryops, *the microsaur* Euryodus, *the anthracosaurs* Proterogyrinus *and* Gephyrostegus, *the seymouriamorph* Seymouria, *and the early amniotes* Paleothyris *and* Captorhinus. *Anterior is to the left in all. In* Acanthostega, *the third vertebra has been included to show its longer spine. The atlas–axis complex of the microsaur* Euryodus *is shown in anterior view at right. Not to scale.*

Between *Acanthostega* and the tetrapods of the mid-Carboniferous, there is very little information on the development of the atlas–axis complex. That of *Ichthyostega* is still unknown. The next early tetrapod that gives any clues is *Greererpeton* from the Namurian (Godfrey 1989a). In this animal, the atlas arch was relatively a little smaller than in *Acanthostega*, and its position was shifted so that the axis arch was clasped between each half (Fig. 10.21). There was a small proatlas; although this might also have been present in *Acanthostega*, it has not been found. Proatlases are such small elements that they are often lost in disrupted specimens, but it is thought that they articulated with the exoccipitals and the atlas arch. The axis arch in *Greererpeton*, as in most other subsequent tetrapods, had both its halves firmly fused together, reflecting its role in supporting occipital musculature. The atlas and axis centra looked little different from other trunk centra. There was no expanded bony surface to correspond to the large occipital condyle present in the skull of *Greererpeton*, so that it must be supposed that the joint was partly cartilaginous. There are very few atlas–axis vertebrae fully described from early temnospondyls, one of the exceptions being, not surprisingly, *Eryops*. The reason for this may simply be lack of preservation, or it may be that the region was poorly ossified or only loosely articulated in the intact animal. In any case, there seems to be a contrast here with the amniote lineage.

In the lineage leading to amniotes, along with loss of the atlantal neural spine went the fusion of the atlantal pleurocentrum to the axial intercentrum to form a compound central element supporting the atlas arch, whereas the axial neural arch became fused to its pleurocentrum (Fig. 10.21) (Sumida and Lombard 1991; Sumida et al. 1992). Some early amniotes still retained a proatlas articulating with the exoccipitals, for example in *Paleothyris, Captorhinus*, and the synapsid *Ophiacodon*. Axial neural arches in the amniote stem lineage in general tend to be taller in height and longer from front to back than other trunk arches. As in many early tetrapods, the facet on the atlas arch did not match precisely the shape of the occipital condyle of the braincase with which it articulated, so that there must have been cartilage present to complete the articulating surface.

One group, the microsaurs, is more or less defined by the unique structure of the atlas–axis region and its occipital joint (Fig. 10.21). In this group, the occipital facet on the skull is a narrow horizontal figure eight in shape and is divided into three approximately equal parts. The outer two parts are contributed by the exoccipitals, the center part by the basioccipital. This central part is recessed relative to the outer parts. Corresponding to this the facet provided by the atlas–axis complex is also a narrow oval, but its central part is produced into a projecting knob called the odontoid that fits into the basioccipital recess on the skull. The facet is provided by the atlantal centrum, and in these animals, there is a close match between the shape of the occipital condyle and the atlantal facet. The atlantal centrum is fused or sutured to the atlas arch, whereas the axis arch and its centrum are also sutured or fused together (Carroll et al. 1998; Carroll 2000a).

With each modification of the atlas–axis vertebrae came increased flexibility of the neck joint, accompanied by increasing independence of the head from the shoulder girdle. Development of the joint may well be read in parallel to loss of reliance on the musculature and blood system used in fishlike buccal pumping.

Neural Spines and Zygapophyses

In most early tetrapods, other regions of the vertebral column remain very little differentiated from one another, quite unlike the obvious demarcation into cervical, thoracic, lumbar, and sacral seen in mammals (e.g., Fig. 2.3). The way in which the neural arches articulate to each other and the development of an attachment between the pelvic girdle and the vertebral column are two of the areas of most importance in the evolution of terrestriality in tetrapods. The sacral region will be treated in the section on the pelvic girdle, but next, developments to the neural arches will be considered.

Fishes such as *Eusthenopteron* not only do not have much differentiation into regions along the column, as many later tetrapods do, but their neural spines do not articulate with one another (Fig. 2.4). However, they do support the axial muscles and tend to be long and at a steeper angle to the column than in tetrapods. The steep angle also reflects the cone-shaped segments into which the axial muscle blocks are arranged to facilitate control of swimming movements. If early tetrapods are examined with this in mind, it seems that the tail in Devonian forms remained essentially in the fishlike condition. Gradually, the fishlike attributes of the column are found to begin further and further back along the tail in the evolution of the group, until they are eliminated altogether. This can be seen if *Eusthenopteron* is compared with *Acanthostega* and an anthracosaur such as *Proterogyrinus* (Fig. 10.22). In *Acanthostega*, an oblique fishlike angle to the neural spines is initiated at a level just past the sacral vertebra, whereas in *Proterogyinus*, it does not begin until caudal vertebra 15. The point at which the caudal supraneural radials and lepidotrichia begin also seems to move backward in the sequence *Eusthenopteron–Acanthostega–Ichthyostega*. This has been used to suggest that tetrapod-type features of the vertebral column were initiated toward the anterior end of the column and gradually came to extend further and further back along the column (Coates 1996), which would fit with the anterior–posterior embryonic development seen in most tetrapod vertebrae. As a test for this hypothesis, it would be of great interest to know more about the embolomere in which caudal supraneural radials have been found.

Zygapophyses, the joints between the neural arches, consist of paired oval facets facing upward at the front of the arch and downward at the back. As with the slope of the neural spine, so the number of caudal vertebrae bearing zygapophyses also increases from about five or six in *Acanthostega*, through about 12 in *Greererpeton* to 22 or more in *Proterogyrinus* and other later tetrapods (Coates 1996). As one of the most primitive tetrapods, *Acanthostega* has barely differentiated zygapophyses, and in that respect, it is the most fishlike tetrapod. Those of *Ichthyostega* are more pronounced, but more detailed work is required on this animal to determine their exact nature. In *Crassigyrinus*, the anterior set of zygapophyses are most conspicuous; the posterior ones are no more developed than in *Acanthostega*.

In animals with well-developed sets of zygapophyses in both front and rear, the angle at which the facets lie becomes important and can constrain the range of movement in the vertebral column (Holmes 1989b). Most early tetrapods, such as the temnospondyl *Eryops,* have the facets set at an angle to the vertical plane, sloping in toward the midline, although in some animals—such as *Seymouria,* some early amniotes, and some microsaurs—the angle is almost rigidly horizontal (Fig. 10.20). Presumably this facilitates lateral bending at the same time as restricting rotation. This kind of horizontal arrangement is usually found in conjunction with solid spool-shaped centra and low, dome-shaped neural spines. The functional implications of this are not well understood, but it seems to have arisen several times in early terrestrial tetrapods.

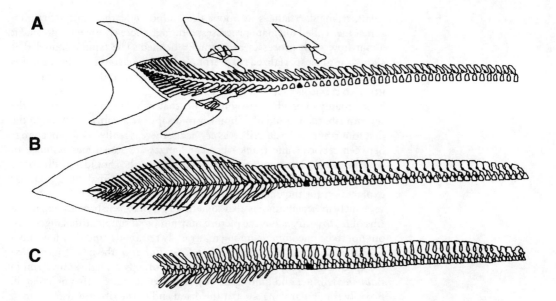

Figure 10.22. Regional variation in the vertebral column in fish and tetrapods.
(A) Eusthenopteron.
(B) Acanthostega.
(C) Proterogyrinus. *The sacral vertebra is indicated in black.*
From Coates (1996).

Ribs

Substantial ribs attaching to the vertebral column via the neural arch and centrum are characteristic of tetrapods and are known even in the earliest forms. However, there is more contrast in rib morphology between *Acanthostega* and *Ichthyostega* than there is between any other two Paleozoic tetrapods, so it is difficult to draw general conclusions about what function the ribs may have had in these very early forms. *Acanthostega*, with its short and for the most part undifferentiated ribs, seems more like what might be expected of a primitive tetrapod, in contrast to the broad overlapping ribs of *Ichthyostega*, which seem like a specialized anomaly. There is a great contrast here between the lack of differentiation of the vertebrae in these early tetrapods and the extremes sometimes seen along the series of ribs, even in animals such as *Ichthyostega*.

Rib patterning among early tetrapods shows what appears to be an overall consistency, with one or two notable exceptions that prove (i.e., probe) any "rule" that might be imagined. Even among apparently closely related forms, the ribs often differ remarkably. Figure 10.23 shows a selection of rib morphologies among early tetrapods. Among temnospondyls, the majority have short, more or less straight ribs that do not show much curvature around the body (e.g., *Platyrhinops*, Fig. 9.13; *Eoscopus* [Daly 1994]; Fig. 10.23), and in fact probably only extended over the dorsal part of the body wall. Most do not show much differentiation along the column, except that the cervical and immediately presacral ribs are somewhat shorter than the rest. Some appear to have no ribs on the axis or atlas vertebrae, but apart from that, no cervical ribs can readily be distinguished. However, by contrast, a few, such as *Dendrerpeton*, have slightly broadened ribs associated with at least some cervical vertebrae (the sequence is not fully known) (Fig. 2.9). The dissorophids show more variation in rib morphology than most temnospondyl families, perhaps associated with their production of more terrestrial forms such as *Dissorophus* itself (Fig. 10.23). In that animal, although there are no cervical ribs showing expansions, ribs associated with vertebrae 7 through 13 show small, triangular flanges at about the midpoint of the shaft. These are sometimes called uncinate processes, although there is some disagreement about how such processes are actually defined. The temnospondyl that shows the most specialized rib morphology is *Eryops*. Its ribs show a series of flanges and expansions, starting with short ribs on the atlas, axis, and first two cervicals that bear spiky processes (Fig. 9.12). Thereafter, the ribs bear broadly expanded triangular ends that increase in size to about the seventh rib, then gradually diminish to disappear at rib 13. Although the first four ribs may be described as cervicals, there is really no abrupt change between cervicals and trunk ribs. The flanges are sometimes assumed to have been associated with the muscles attaching the shoulder girdle to the body in what is taken for a more or less terrestrial animal, with a similar explanation for the uncinate processes of *Dissorophus*.

Such an explanation is also sometimes given for the broadly expanded ribs of *Ichthyostega*, but the picture may not be so simple. Rib flanges turn out not to be uncommon among many early tetrapods, and they have now been discovered in *Whatcheeria* (Fig. 7.6), *Eucritta*, the new Tournaisian tetrapod (Fig. 7.11), and (from the evidence of only a single known rib) in *Baphetes* (Milner and Lindsay 1998), although they are different in detail. Broadly triangular flanges set at the distal end of the ribs are also found in the seymouriamorph *Kotlassia* (Fig. 10.23) (Bystrow 1944). As far as one can tell, these animals were more aquatic than terrestrial, although there is some doubt about the mode of life of *Kotlassia*. Other seymouriamorphs such as *Seymouria* itself and *Discosauriscus* show variations. The latter has slightly expanded anterior thoracic ribs, whereas in *Seymouria*, it is the cervicals that appear to be expanded (Fig. 10.23). No studies have yet been made of the possible function of such flanges, which in some cases are bizarre in form. They may have been concerned with musculature involved in axially driven locomotion and the added strain of performing such movements on land rather than in water.

Yet another variation is found in some animals such as *Greererpeton* (Figs. 6.1, 7.7, 10.23). Although like the majority of temnospondyls, it has no distinctive cervical ribs, most of its trunk ribs bore small processes about midway along the shaft, somewhat similar to the uncinate processes of *Dissorophus*. However, *Greererpeton* was an aquatic animal, as was *Acanthostega*, where very modest processes appear on about vertebrae 10 to 13 following a slight expanded series more anteriorly. Again, the function of these processes is unknown, although in such predominantly aquatic animals, the needs of axial locomotion might be involved.

In some early tetrapods, there is an abrupt change in rib morphology defining a cervical region. These are animals that have been traditionally placed in the reptiliomorph lineages such as anthracosaurs (see Coates 1996). *Archeria* and *Gephyrostegus* are among these (Fig. 10.23), although the seymouriamorphs *Seymouria*, *Kotlassia*, and *Discosauriscus* do not show an abrupt transition like this. In many of these forms, three or four ribs following the atlas and axis vertebrae bear ribs that are set at a narrow angle to the vertebral column, sometimes almost parallel to it. They are often broader than the trunk ribs, expanding gradually down the shaft to the distal end, although not showing distinct processes such as those in *Eryops*. This kind of differentiation is also seen in a range of early amniotes such as *Coelostegus* and *Anthracodromeus* (Carroll and Baird 1972) (Fig. 10.23).

Microsaurs show an intermediate condition, in which in most forms, the first three vertebrae only are distinguishable as cervical. They may or may not carry ribs, but thereafter, the ribs tend to be simple curved bars

Figure 10.23. (opposite page) Rib patterns in the cervical and thoracic region in early tetrapods; the early tetrapods Acanthostega *and* Greererpeton, *the temnospondyls* Dendrerpeton, Eryops, Dissorophus, Platyrhinops, *and* Eoscopus, *the microsaurs* Tuditanus *and* Microbrachis, *the anthracosaurs* Gephyrostegus *and* Archeria, *the seymouriamorphs* Seymouria, Discosauriscus, *and* Kotlassia, *and the early amniotes* Anthracodromeus, Captorhinus, *and* Coelostegus. *Anterior is to the left in all. In* Dissorophus *and* Eoscopus, *the atlantal and probably the axial ribs are missing. Not to scale.* Eoscopus *from Daly (1994).*

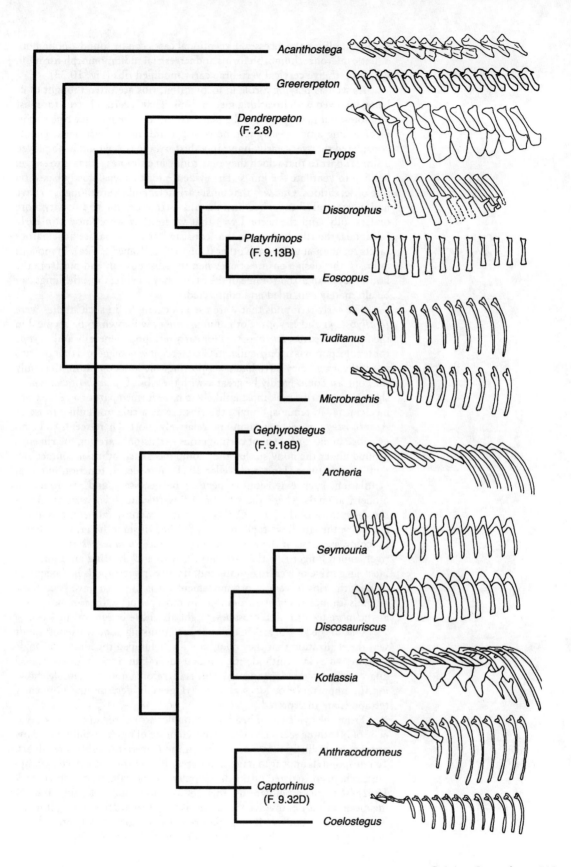

with no additional processes. Exceptional cases can be found among both aquatic microbrachomorph forms and terrestrial tuditanomorph forms, in which up to five cervical vertebrae carry modified ribs (Fig. 10.23).

As well as playing a role in locomotion, ribs are often thought of as being involved with breathing mechanisms. That is certainly true for most amniotes, but perhaps this is the place to consider the possible role of ribs in breathing among early tetrapods in general. In the past, most people considered that very early in tetrapod history, the ribs would have played a similar part to that which they play today in amniotes—that is to say, in helping to ventilate the lungs. The evidence for this was mostly based on two observations. One was that most early tetrapods were thought to have had substantial ribcages in which the ribs surrounded the body cavity more or less fully, and the second was that where they were known in early tetrapods, the ribs had heads that consisted of two separate articulatory facets set at an angle to each other (Gans 1970; Romer 1972a). In modern animals, this design ensures that when the ribs move in and out from the body, they change the volume of the cavity they enclose; thus, the lungs can be alternately expanded and contracted.

The early tetrapods that were used as examples in such studies were *Ichthyostega* and *Eryops*. Both animals are now known to be atypical in the form of the ribs, *Ichthyostega* for early tetrapods generally and *Eryops* for temnospondyls in particular. In Chapter 5, it was suggested that the ribs in *Ichthyostega* may have had a supporting role in locomotion. Certainly the contrast could hardly be greater with the ribs of *Acanthostega*, whose short, straight ribs look most unlikely to have formed part of a ventilatory mechanism. In temnospondyls, the ribs are as a rule not unlike those of *Acanthostega* in being straight and relatively short. There seems to be no evidence in most of them of cartilaginous extensions carrying the rib ends around under the body so that they could meet each other or contact any midline structures (Janis and Keller 2001). In order to function in costal ventilation, such extensions appear to be necessary and are found in modern animals where the ribs have this role. In *Ichthyostega, Acanthostega*, most of the Early Carboniferous tetrapods, and temnospondyls generally, the ribs have tapering ends finished in smooth periosteal bone. By contrast, some of the anthracosaurs such as *Pholiderpeton* have relatively much longer ribs that are curved, with ends bearing an unfinished facet suggestive of a cartilaginous end. If any early tetrapods had begun to employ the ribs in ventilatory movements, it is likely to have been these animals, thought to be stem amniotes by most people. Other animals, such as the more terrestrial microsaurs, might also have begun this process.

One of the key factors in the change of thinking about early tetrapods was the realization that they could not all be lumped together as a single unit known as labyrinthodonts, but that several lineages should be teased apart and treated independently. This is therefore another example showing the importance of ideas about phylogeny in determining how early tetrapods are interpreted.

From the evidence of head and shoulder structure, and from knowledge of breathing mechanisms and musculature of modern animals, it now appears more likely that most early tetrapods were probably air gulpers. Temnospondyls appear to have elaborated this system into a buccal pumping mechanism, assisted by the development of the palatal vacuities (Clack 1992). Modern amphibians use this method too. Not only do their breathing mechanisms not involve the use of ribs to ventilate the lungs (although they do involve body wall musculature in exhalation; see Chapter 6), but in most frogs, the ribs have been lost altogether. In the previous section on vertebrae, the connection between the structure of the centrum and the role of the neural arch and ribs in locomotion was pointed out, and the ideas mesh very well with these new interpretations of breathing mechanisms in which rib musculature was not fully involved with aspiration breathing. Rhachitomous vertebrae, short ribs, and buccal pumping seem intimately related.

In amniotes, by contrast with temnospondyls, ribs are usually quite long and curved. Ribs may have become in some way involved in breathing among amniotes at an early stage in their evolution, bearing opposing sets of muscles to open and close the ribcage as they do in modern mammals. However, there is evidence that these early animals did not rely on them entirely to ventilate lungs. In modern amniote groups, lizards, crocodiles, turtles, and mammals all use different methods of lung ventilation that may not use ribs at all. In turtles, for example, the ribs have become incorporated into the shell internally, so it is hard to see how the turtles' ancestors could have done this if the lungs had been ventilated exclusively by rib movements. However, amniotes do have certain features in common that suggest at least the beginnings of costal aspiration may have been present in all of them. These features include intercostal musculature for moving the ribcage. It may be that early tetrapods, even including early amniotes, were using all possible means by which to aid gas exchange—air gulping, buccal pumping, and skin breathing as well as an early version of rib movement.

A possible role for the expanded ends of the ribs in the earliest tetrapods may be one associated with the development of the transverse abdominal muscle, the extra muscle layer that tetrapods developed to aid exhalation during aspiration breathing. This appears to have arisen in the common ancestor of amphibians and amniotes and may therefore have been present in some of the earliest tetrapods. It is the innermost of the body wall muscle layers in salamanders and presumably was so in early tetrapods too. Its appearance may herald the initiation of axial muscle involvement with ventilation. The flanges may have given this muscle an increased length of edge onto which to attach. At present, all that can be said is that these expanded trunk rib flanges are found in a number of the earliest tetrapods, as well as in some more derived ones, and that although they are more often found in aquatic forms than terrestrial ones, even some forms thought to be at least semiterrestrial, such as *Eryops,* have enlarged trunk flanges on their ribs.

Rib and vertebral patterning in birds and mammals is known to be under the control of *Hox* genes. In these animals, the cervical series is characterized by the lack of ribs and associated structural correlates. Cervical vertebrae occur where *Hoxc6* operates in the absence of *Hoxc8* (Burke et al. 1995), and this also specifies where the limbs will be formed. However, in early tetrapods, often all the most anterior vertebrae carry ribs of some form. As detailed above, in some amniotes and amniotelike tetrapods, they are clearly distinct from other ribs in orientation or shape, but in other groups, they are often little different from trunk ribs. The pattern may indicate that the *Hox* gene control was less rigid in early tetrapods and the interactions between the two *Hoxc* genes less clearly defined.

Evolution of the Cervical Region and the Brachial Plexus

Movements of the limbs are controlled to a large degree by a complex of nerves that come together to form the brachial and sacral plexuses in the spinal column, in the cervical and sacral regions. In modern amphibians, spinal nerves near the anterior end of the column are involved; in frogs,

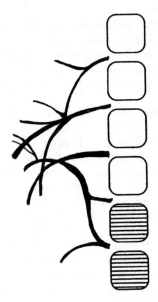

Figure 10.24. The brachial plexus of a lizard. Cervical vertebrae, plain; thoracic, hatched. Nerves to the pectoral region coalesce and intertwine to form a ganglion around cervical 3 and 4. Amphibians have no complex brachial plexus.

spinal nerves 2 and 3 (corresponding to cervical vertebrae 2 and 3) and in salamanders nerves 3 and 4 (with a few fibers also from 2 and 5) contribute to a simple brachial plexus. There is little or no enlargement of the spinal cord itself at this point in modern frogs (Huxley 1882) or salamanders (personal observation). Caecilians, which lack limbs, also lack the plexuses.

In amniotes, the situation is much more complex and variable. In lizards, for example, there are usually five (or more) nerves that contribute to the brachial plexus (Fig. 10.24). The most posterior of these nerves exits the vertebral column between the last cervical and first thoracic vertebrae, which means that the nerves pass out between cervical vertebrae 4 through 8. In crocodiles, the five nerves of the brachial plexus are associated with spinal nerves 7 through 11 (Giffin 1995).

The size and complexity of the brachial and sacral plexuses reflect the degree of coordination, range of movements, and refinements of limb control among amniotes. The variability seen in amniotes is also paralleled by the degree of differentiation seen along the length of the column. For example, cervical vertebrae in amniotes are usually distinct from those of the thoracic region, marked out by features such as the position of rib attachments (Giffin 1995, and above). The situation contrasts with that in salamanders, where usually only the atlas vertebra, lacking a rib, is clearly distinct from any of the others. In its lack of distinct regions, the vertebral column of, say, a salamander such as *Megalobatrachus* is similar to that of the early tetrapod *Acanthostega*. It is likely that, as in salamanders, there was little elaboration of the brachial plexus in early tetrapods such as *Acanthostega*, and that the nerves serving the forelimbs were associated with more anterior vertebrae, as in salamanders and frogs.

The position of the brachial plexus is of some interest because it reflects changes to the position of the shoulder girdle and forelimbs. Among modern animals, amniotes contrast with amphibians in having much longer cervical regions as well as having more differentiation of this region. A long neck is characteristic of amniotes (Janis and Keller 2001). In early tetrapods, the shoulder girdle remains tucked close up behind the head, so that the interclavicle often underlies the back part of the skull, between the lower jaw joints, and the clavicle and cleithrum likewise almost touch the back of the head. In *Acanthostega,* the shoulder girdle retained one of its fishlike functions in providing a seal for the opercular flap (Fig. 5.22). Even in animals where there is no clear evidence of internal gills, this close relationship between head and shoulders is still maintained. The reason may have to do with the retention of buccal pumping, whose musculature was essentially similar to that used in gill ventilation and was to a large extent inserted onto the shoulder girdle. A further implication is carried for the position of the heart. In fishes, the heart lies more or less amid the gill basket, between the operculars. Judging by the position of the shoulders, it still rested in much the same place in early tetrapods.

The inferences may be carried further to suggest that only once buccal pumping had been abandoned could the shoulder girdle move further back, the circulatory system lose its primitive gill-related structure, the neck lengthen, and the brachial plexus increase in complexity. Thus, perhaps only in amniotes could fine control of the forelimbs be achieved. In this context, it is notable that frogs, which have maintained buccal pumping, have elaborated the function of the hindlimbs but reduced the importance of the forelimbs. In salamanders, both sets of limbs are of relatively minor importance in locomotion and in some species have been reduced or lost altogether.

Appendicular Skeleton: Pectoral and Pelvic Girdles, Limb Bones, and Joints

Pectoral Girdle

Post-Devonian tetrapods show a consistency of construction in the shoulder girdle. After major changes during the fish–tetrapod transition, such as enlargement of the interclavicle and increase of the scapulocoracoid, most of the subsequent developments consist of a further general reduction of the dermal girdle. From the pattern seen in most Devonian tetrapods, in which the cleithrum and scapulocoracoid are coossified, in the later forms this connection is lost. The cleithrum and clavicle retain their interconnection, but the scapulocoracoid becomes a separate, and much larger, entity. It no longer has a bony connection to the dermal elements, as it had in fishes. There are no facets on the internal surface of the cleithrum to receive it, as there were in fishes, and it must have been held in place by soft tissue connections. Eventually, the scapulocoracoid seems to have taken over the area formerly occupied by the cleithrum and even ends up with a similar shape in some forms (cf., e.g., the scapulocoracoid of *Dendrerpeton* with the cleithrum of *Eusthenopteron*; Figs. 2.9, 6.4).

As was mentioned in Chapter 6, the dermal elements in tetrapods became reduced in size and the connections between cleithrum and clavicle became less intimate. The cleithrum is the bone that undergoes the most radical reduction, becoming first a narrow splint, as found in most post-Devonian tetrapods (Fig. 6.4), and eventually disappearing altogether in many amniote groups. This could well be associated with the loss or reduction of branchial musculature and its decoupling from breathing. Clavicles in most early tetrapods remain remarkably conservative in form, only undergoing significant reduction among the more terrestrially adapted amniotes and microsaurs. In some such cases, the clavicle is reduced to a narrow splint more or less at right angles to the body and often associated with an interclavicle that has taken on a T-shaped appearance. In such animals, the way was open for the development of movements of the shoulder girdle, so

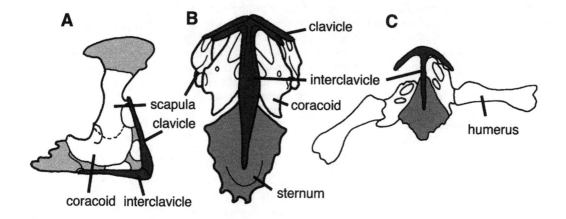

Figure 10.25. (above) Pectoral girdle of a lizard.
(A) Lateral view.
(B) Ventral view.
(C) Relative movement between coracoid and sternum during the stride. This movement helps lengthen the stride and is only possible because of the flexible attachments between the elements and the relative reduction of the dermal bones. Dermal bones, dark shading; sternum, medium shading; cartilaginous elements, light shading. From Carroll (1987).

Figure 10.26. (opposite page) Pelvic girdles in early tetrapods (lateral views); the early tetrapods Acanthostega and Greererpeton, the temnospondyls Dendrerpeton and Eryops, the microsaurs Goniorhynchus and Ricnodon, the anthracosaurs Eoherpeton and Archeria, the seymouriamorph Seymouria, the early amniote Coelostegus, and the diapsids, a lizard and a crocodile. Anterior is to the right in all. Not to scale.

that not only could the girdle move relative to the body during each step, but this movement could be amplified by movements of the bones relative to one another (Fig. 10.25). This kind of movement, called girdle rotation, is an important component of the locomotory cycle in animals such as lizards and was presumably first developed among the early amniotes whose girdles were essentially similar in shape to those of lizards.

Pelvic Girdle

A more or less orderly sequence can be seen in the evolution of the tetrapod pelvic girdle and sacrum, once the enlargement of the girdle bones typical of tetrapods had taken place. No pretetrapod lobe-fin shows any hint of the broad medially sutured pair of plates seen in even the earliest tetrapods; although admittedly the pelvis of *Panderichthys* is unknown, given its small pelvic fins, the girdle seems most likely to have been *Eusthenopteron*-like. The earliest known tetrapodlike pelvic girdle elements belong to *Elginerpeton,* and these are similar in size and probably morphology (they are only known in part) to those of *Ichthyostega* (Ahlberg 1998). *Acanthostega* has a pelvic girdle that appears to be relatively much smaller than that of *Ichthyostega,* although as suggested in Chapter 5, the body proportions of *Ichthyostega* are uncertain. If the proportions were as envisaged by Coates and Clack (1995), the pelvic girdle appears overengineered for the size of hindlimb as compared with later tetrapods (Fig. 5.12).

These early tetrapod pelvic girdles were typically single ossifications as they are found in the fossil record, which differs from the trinity of bones found in later forms: ilium, pubis, and ischium presumably were represented, but the limits of their respective regions cannot be distinguished in the earliest forms (Fig. 10.26). Some later tetrapods such as the anthracosaur *Eoherpeton* also had a unitary pelvis like the Devonian forms, but by and large, in most Carboniferous forms, the three separate ossifications are clear. The means by which this separation came about remains speculative. One possibility is a process of pedomorphosis, whereby in early tetrapods, the three contributory bones fuse early in life so that in the adult, only the solid structure is left. As tetrapod evolution proceeded, pedomorphosis delayed the fusion of the three into later and later stages, so that in some the fusion was never complete. To delay the fusion of separate bones in this way may have allowed for longer periods of growth in the larger forms.

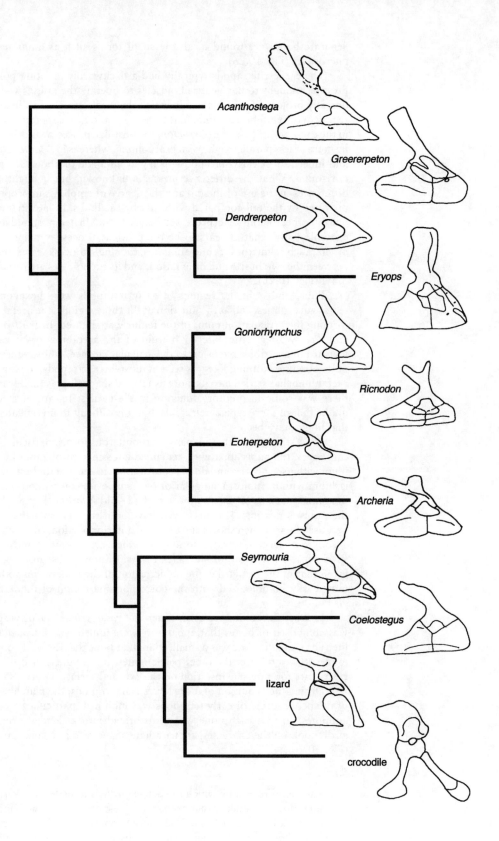

Separate bones are found in all the small forms such as amniotes and microsaurs (Fig. 10.26).

The earliest tetrapods typically had a backwardly pointing postiliac process, thought to have carried muscles to operate the tail, as well as a dorsally projecting iliac blade, via which the girdle attached to the sacral region of the vertebral column. In *Ichthyostega* and *Elginerpeton* as well as in the early anthracosaur *Eoherpeton*, the postiliac process was oriented to lie more or less parallel to the vertebral column, whereas in *Acanthostega*—and indeed most early tetrapods—it lay at an angle of about 30° to the horizontal. What the difference means is unknown, but it might signal differences in the use of the tail and the degree of emphasis in propulsion provided by the tail compared with that by the limbs. In later tetrapods, one or other of the iliac processes may be lost. In temnospondyls, for example, only what appears to be the postiliac process remains, but in others, such as microsaurs and amniotes, the single process seems midway between the two in size and orientation, and its homologies are difficult to draw (Fig. 10.26).

Pubic and ischiadic regions in early tetrapods were hexagonal or rectangular plates, and at the junction of the three regions, the acetabulum forming the articulation point of the femur was situated. In *Ichthyostega* and *Acanthostega,* the anterior margin of the acetabulum was carried forward in a curious extension to the articular surface, whose function is quite obscure. Nothing like it is seen in most later tetrapods, in which the acetabulum has settled into its more usual oval shape. It presumably means there was some locomotory function in the earliest tetrapods that was unlike that of more typical tetrapods, but it is difficult to investigate what that might have been.

It is a strange fact that in many Carboniferous forms, particularly the early ones, the pubis apparently remained unossified throughout life. The ilium with its processes and dorsal contribution to the acetabulum, and the ischium with its almost triangular or oval shape are conspicuous features in animals such as *Silvanerpeton* (Fig. 8.12) and *Eucritta* (Fig. 8.14), but the pubis is missing. This may even have been the case in *Ichthyostega* because one specimen shows an elongated anterior portion, complete with a couple of large obturator foramina, which are absent in most other specimens. The reason for such lack of ossification is obscure, especially given that in the fish girdle, the pubic region of the bone seems to be the most easily homologized with the corresponding region of the tetrapod girdle.

In other Carboniferous forms, such as *Crassigyrinus,* the pelvic girdle was composed of bones that apparently were joined only by cartilage in life. The ischiadic plate was a small, triangular bone that shows no obvious joint surfaces to meet the other pelvic bones, but uniquely, it does show clear scars for a muscle insertion on the external surface.

During the evolution of more terrestrial tetrapods, the platelike form of the pelvic girdle of early tetrapods was modified, particularly in early amniotes, so that each bone was transformed into a slender strut. The girdle took on the characteristic triradiate shape of many later amniotes (Fig. 10.26).

Sacrum

The sacral region by which the pelvic girdle is joined to the vertebral column follows a sequence that probably corresponds closely with increasing terrestriality (Fig. 10.27). The sacrum consists of the vertebra or vertebrae that carry the sacral rib or ribs attaching to the girdle. In

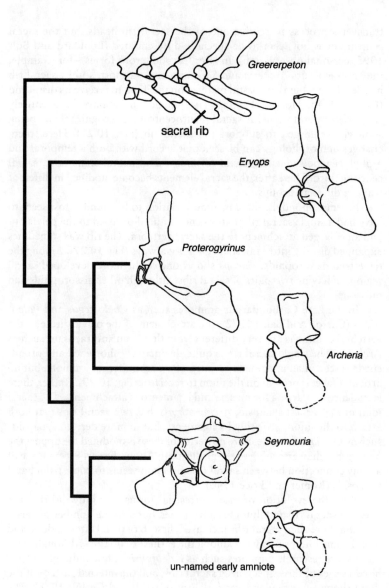

Figure 10.27. *Sacral construction in early tetrapods; the early tetrapod* Greererpeton, *the temnospondyl* Eryops, *the anthracosaurs* Proterogyrinus *and* Archeria, *the seymouriamorph* Seymouria, *and a Carboniferous amniote. Anterior is to the left, except in* Seymouria, *which is shown in anterior view. Not to scale.*

Acanthostega, there appear to be no conspicuous modifications to any vertebra that clearly indicate that it was sacral. There is a sacral rib that shows some broadening of the distal end, but it is not otherwise much modified from the presacral ribs, and the vertebra that bore it is not recognizably different from those fore and aft of it. The sacral rib of *Ichthyostega* has not been described. In *Acanthostega,* the ilium shows no joint surfaces for attachment of the sacral rib, although there is some scarring that might indicate its attachment by soft tissue. There are certainly no articular facets on the end of the sacral rib.

Like *Acanthostega, Greererpeton* has a recognizable sacral rib, but the sacral neural arch is not distinguishable from the rest (Godfrey 1989a). In *Whatcheeria,* the sacral rib is noticeably broader than the pre- or post-sacrals but is shorter than most of the trunk ribs, many of which are greatly expanded at the ends. The sacral neural spine appears to have a broadened

Gaining Ground • 323

transverse process to receive one of the sacral rib heads, but the sacral centrum does not seem to be specially differentiated (Lombard and Bolt 1995; personal observation). In later Carboniferous forms—for example, the anthracosaurs *Archeria* and *Silvanerpeton*—a short, stubby sacral rib has developed, but there is still no facet on the ilium to receive its distal end (Clack 1994b; Holmes 1989b). The junction must have been entirely supported by soft tissue. A sacral pleurocentrum is recognizable by being somewhat enlarged to support the sacral rib (Fig. 10.27). Here, then, changes in morphology can be seen that accord with both a temporal and a phylogenetic sequence: sacral rib first, sacral neural spine next, sacral centrum third. Thereafter, the sacral elements become modified in different ways in different groups.

In temnospondyls, there is some variation to be found. Most seem to have had a single sacral rib that was not sutured or fused to the pelvis but had an enlarged attachment to the sacral vertebra. The rib was sometimes expanded distally into a fan shape, but not always (Fig. 10.27). Among the terrestrial dissorophids, *Cacops* shows one of the more developed sacral regions in having two pairs of sacral ribs, although other dissorophids had only one.

In the Late Carboniferous amniotes such as *Coelostegus* and *Paleothyris* (Carroll and Baird 1972), the attachment of the pelvic girdle to the vertebral column is not very different from that in an anthracosaur such as *Archeria*. The single sacral rib is quite distinctively shortened and broadened, as seen in an unnamed specimen of a Carboniferous amniote, but no articular facet is present on the ilium to receive it (Fig. 10.27). Rather, there is evidence of scarring for muscle and ligamentous attachments. The sacral joint of the earliest synapsid, *Archaeothyris,* has two sacral ribs that both attach to the ilium, probably by ligaments, but in more derived synapsids such as *Ophiacodon,* the more posterior of these is modified to support the first, rather than attaching to the girdle itself. About the earliest evidence of a bony connection between sacral rib and ilium seems to come from basal synapsids (Romer and Price 1940).

With the evolution of more terrestrial forms, not only did the link between sacral rib and ilium become firmer, but also the number of vertebrae that bore such sacral ribs increased, first to two and later to three and more. Some microsaurs were among the earliest to do this. Although most microsaurs resemble amniotes such as *Paleothyris,* a few tuditanomorphs have a series of three specialized sacral ribs, and the internal surface of the ilium shows an articular facet for them (Carroll and Gaskill 1978). The stoutly built terrestrial forms such as *Seymouria* (White 1939) and *Diadectes* (Case 1907) each had two vertebrae modified as sacrals. The first, more anterior of these was heavily built and broadened, and the associated ribs had greatly expanded ends that lined the inner surface of the ilium to a large degree. The second sacral was much less modified, but still the distal end of the rib contacted the ilium. In the young stages of the aquatic larval forms of seymouriamorphs such as *Discosauriscus,* the sacral region was present only as a single unmodified vertebra, and the ilium alone was ossified in the pelvic girdle (Klembara 2000). However, as the animals grew, the remaining components of the girdle ossified, and two sacral ribs developed. The sacral vertebrae themselves are not so conspicuous, although they had extra articulations for the large sacral rib heads.

Once more, adaptation of the sacral region for terrestriality (that is to say, weight-bearing and transmission of limb propulsion to the vertebral column) can be seen to have been a gradual process, with the earliest tetrapods having little in the way of a firm connection between the girdle and

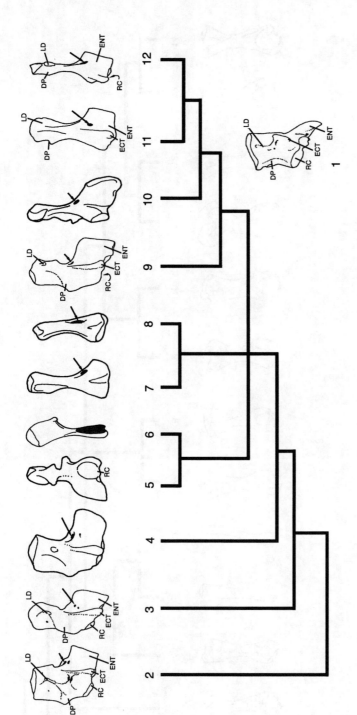

Figure 10.28. Tetrapod humeri.
(1) Outgroup Eusthenopteron.
(2) Acanthostega.
(3) Tulerpeton.
(4) Greererpeton.
(5) Eryops.
(6) Eoscopus (shaded area not preserved).
(7) Pelodosotis (tuditanomorph microsaur).
(8) Microbrachis.
(9) Eoherpeton.
(10) Seymouria.
(11) Casineria.
(12) Westlothiana.
Left humeri, not all in the same aspect, and not to scale. Straight arrow points to entepicondylar foramen.
DP = deltopectoral crest;
ECT = ectepicondyle;
ENT = entepicondyle;
LD = latissimus dorsi process;
RC = radial condyle.
Adapted from Coates (1996).

Gaining Ground • 325

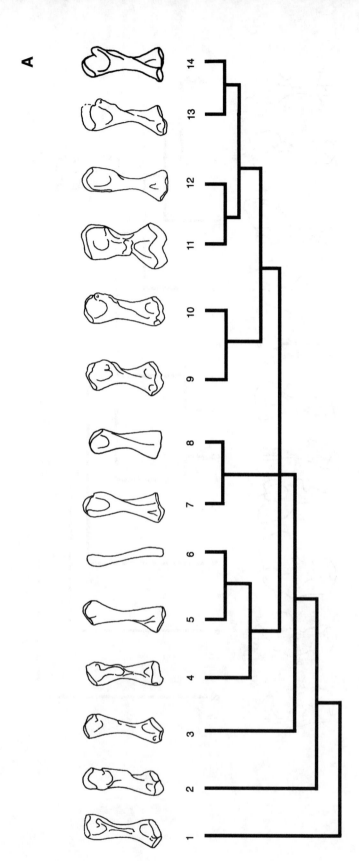

Figure 10.29. Tetrapod femora. (A) Cladogram showing 1, Acanthostega; 2, Tulerpeton; 3, Greererpeton; 4, Eryops; 5, Tersomius; 6, Platyrhinops; 7, Saxonerpeton *(tuditanomorph microsaur)*; 8, Microbrachis; 9, Proterogyrinus; 10, Archeria; 11, Seymouria; 12, Discosauriscus; 13, Westlothiana; 14, Captorhinus. *Not all in the same aspect and not to scale.*

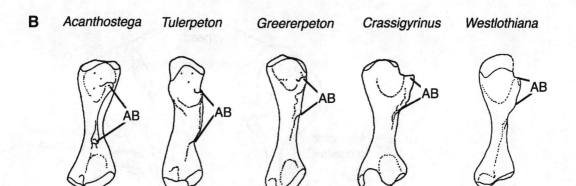

(B) Tetrapod femora aligned to show the reduction in the adductor blade. Adapted from Coates (1996).

the vertebral column. An increase in the size of the sacral elements, the extent of their contact, and the firmness of the complex can be shown to have occurred several times in different groups.

Limb Bones and Joints

Early in tetrapod history, the limb elements were rather different in form from their equivalents in later forms. They tended to be bulky and flattened rather than gracile and elongate. The humerus of *Acanthostega*, for example, is shaped like a broad L. If the limbs of early tetrapods are ranged on a cladogram, a number of significant directions of change are seen, especially to the humerus and femur. The humerus in early tetrapods has an L shape, with large epicondyles for muscles to attach and move the forelimb (Fig. 10.28). The deltopectoral crest is the point where the adductor muscles attach to move the whole limb into the body. These are placed at about the midpoint of the bone, but there is no recognizable shaft to the humerus. Similarly, on the femur of *Acanthostega*, there is a large adductor blade that extends more than halfway along the inside surface of the bone (Figs. 5.24, 10.29A, B). During limb evolution, the deltopectoral crest and the adductor blade become reduced in size and move toward the proximal (body) end of the bone. This seems to be connected with increasing efficiency of the muscles, whose insertion points come to lie closer and closer to the body. This process goes hand in hand with the elongation of the bone and the production of a shaft to give the bone its more familiar shape. The whole system seems to be responding to changing locomotory needs, probably connected with the shift from swimming or paddling to truly terrestrial walking (Coates 1996). Figures 10.28 and 10.29 show humeri and femora on a cladogram and show that slender, elongate long bones arose independently in several terrestrial lineages, including temnospondyls, microsaurs, and amniotes. The picture of limb evolution parallels closely the pattern seen in jaw evolution. Among early tetrapods, the limbs sequentially acquire a suite of characters, including those to the muscle insertions mentioned above, and the elongation of the limb bones themselves. Thereafter, as tetrapods diversify, there is an explosion of convergent evolution as separate groups become more fully terrestrial.

The humeri of *Ichthyostega* and *Acanthostega* retain a series of foramina that are also found in those of *Eusthenopteron* and *Panderichthys*. These are presumed to have been for the passage of blood vessels to and through the bones of the humerus. During the evolution of the tetrapod humerus, most of these were eliminated (Fig. 10.28). All except the en-

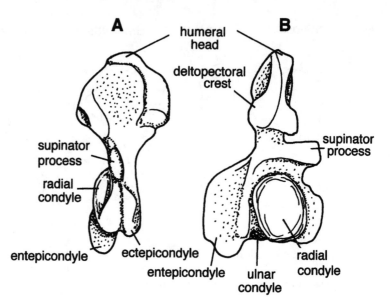

Figure 10.30. Humerus of Eryops showing supinator process and the twist caused by head and condyles being set at 90° to each other. (A) Anterior view of left humerus; (B) ventral view of left humerus.

tepicondylar foramen had disappeared in the humeri of most Carboniferous forms, one of the few exceptions being *Crassigyrinus*. It retains two or three of them, making it apparently very primitive in this respect (Panchen 1985). Given the phylogenetic status of the animal and its reacquisition of secondarily primitive features of the palate, those of the humerus could likewise represent reversals to the more primitive condition. The question remains, however, as to what brought about the loss of the foramina in most other tetrapods. One possibility is that as the humerus slimmed down from its primitive L shape, the blood vessels were simply left to travel in the soft tissue rather than in the bone. However, the loss of the foramina from even the L-shaped humeri of embolomeres argues against this. Another possibility might be that having the blood vessels pass through the bones restricted the amount of blood that could flow through them. While the limb muscles were relatively small, this was not much of a problem, but as the muscles increased in bulk, having the blood vessels enclosed in foramina might have been restrictive. Releasing them into the soft tissue might have allowed a greater volume of blood to pass to more active limbs and bulkier muscles (Lebedev and Coates 1995).

Another conspicuous change to the humerus is evident. In very early tetrapods, the strap-shaped articulatory surface on the head is set at almost the same angle as that formed between the radial and ulnar condyles. In later tetrapods, the angle increases until it is about 90°. The humerus is shaped at that point like two triangles set at right angles to one another and is called tetrahedral. The effect of this twist is to bring the radius and ulna down into the power stroke more easily and sooner than by a humerus that is not so twisted. The tetrahedral humerus is characteristic of many Paleozoic tetrapods that are indisputably terrestrial (e.g., *Eryops*, Fig. 10.30, and *Seymouria*, Fig. 10.29, number 11). It is also found in tandem with changes to the glenoid, which is altered from being a relatively flat surface to one with a distinct twisting curve. The head of the humerus can move along this curved surface, thus exaggerating the effects produced by the torsion of the humeral shaft. It has been suggested that the twisted form of glenoid constrains the path through which the humerus can travel to those

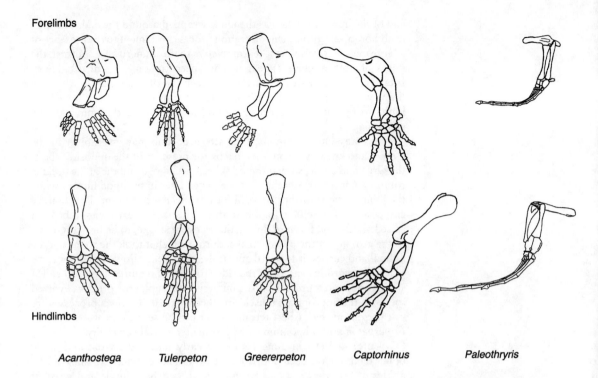

Figure 10.31. *The evolution of limb joints in early tetrapods. Left limbs, anterior to the left.* Acanthostega, Tulerpeton, *and* Greererpeton *are shown with limbs extended,* Captorhinus *and* Paleothryis *with limbs bent.* Captorhinus *based on information from Fox and Bowman (1966).*

movements that allow most efficient walking. The idea is that the bone shape governs the movement rather than having it controlled by the brain, allowing the brain more freedom to perform other tasks (Jenkins 1971).

Along with modifications to the humerus and femur, the evolution of early tetrapods involves the development of elbow and knee joints (Fig. 10.31). Among conventional tetrapods, it has been said that although the elbow and ankle joints are always rotatory in form, the wrist and knee by contrast are always simple hinges (Romer 1962). This is true for most fossil as well as modern tetrapods and can be tested by trying to turn the hand over without using the elbow joint. The radius is so called because it allows for rotatory movement at the elbow. The wrist, however, has no power to twist, as does the ankle. When it comes to very early tetrapods, however, the picture is not so simple.

It has been suggested that the beginnings of the characteristics of tetrapod elbow and knee joints can be seen in the fins of the lobe-finned fish *Ectosteorhachis* (Rackoff 1980), although this fish is rather a derived member of the tetrapod stem lineage and its adaptations may therefore represent an instance of convergent evolution rather than an ancestral condition for tetrapods. *Acanthostega* shows no very clearly defined elbow and knee joints in the sense that neither joint was as mobile as it became in later tetrapods. In *Acanthostega,* the radial and ulnar articulations lay at the extreme end of the humerus (Figs. 5.18, 5.19). This might have allowed the forearm to be raised or lowered with respect to the upper arm, or it may have allowed some capacity for swinging the forearm toward or away from the body, but it was certainly not rotatory. In *Ichthyostega,* the elbow was strictly a hinge and may even have been locked into a right angle bend, judging from the position of the radial condyle with respect to the ulnar in some specimens (Figs. 5.12, 5.14). The radial condyle lay beneath the distal end of the humerus, whereas the ulnar wrapped around the end. The ulna itself bore a very prominent olecranon process, testament to the presence of a large muscle to raise the forearm relative to the humerus. By contrast, the knees of both animals appear to have been to some extent capable of rotation, with a degree of overlap between the tibia and fibula when the limbs were at rest (Coates 1996). The rotation would have allowed the limbs to operate in a form of rowing motion while the animals were swimming.

In many later tetrapods, the pattern of limb maneuverability at the elbows and knees was counteracted by the presence of the opposite type at the wrist and ankle, so that the wrist became a hinge and the ankle became rotatory. Clearly these refinements evolved later in tetrapod history when the joints became further adapted for terrestrial locomotion. The rotatory component of the elbow joint first arose when the forearm had to be both raised and swung forward during the recovery stroke, to be brought down to the ground for the power stroke, during which it took the weight of the animal and carried it forward and at the same time allowed the foot to be placed forward instead of to the side. To do this requires a motion called supination, and a process on the humerus, the supinator process, evolved to which the appropriate muscle attached (Fig. 10.30). Such processes are not present in the earliest tetrapods but can be seen in creatures such as *Eoherpeton* and other animals with some terrestrial capability.

Unfortunately, in almost all very early tetrapods, wrist and ankle bones tend to have been poorly ossified—and hence poorly fossilized. The ankles of *Ichthyostega* and *Acanthostega,* and both ankle and wrist of *Tulerpeton* (Fig. 5.26) constitute conspicuous exceptions to this observation. The ankles of *Ichthyostega* and *Acanthostega* are consequently known to be rather different from those of tetrapods from the mid–Late Carboniferous, with fewer bones and no obvious lines of flexibility that would have allowed the foot to be placed flat on the ground for bearing weight (Bjerring 1985, 1988; Coates 1996). Even in *Tulerpeton,* it is not clear that the joints would have been as flexible as those of later, more terrestrial tetrapods (Fig. 10.31) (Lebedev and Coates 1995). By contrast, the earliest known terrestrial tetrapod, *Casineria,* has few of its wrist bones preserved, although to judge from its digits and humerus, the wrist must have been weight bearing (Fig. 7.3).

Not only the bones of the wrists and ankles in early tetrapods remain elusive as fossils. In many animals, the ends of the limb bones themselves remained poorly ossified. In fully aquatic animals, this is no great surprise, and the skeletons of modern aquatic animals often remain wholly or partly cartilaginous. However, this also seems to be the case in what appear to be terrestrially adapted early tetrapods. An example of this is seen in the development—or lack of it—of the olecranon process. Exceptionally well developed in *Ichthyostega,* a well-ossified olecranon is not seen again in early tetrapods until it appears in early amniotes and some microsaurs. In the otherwise terrestrially adapted *Casineria,* the olecranon is not conspicuously ossified (Fig. 7.3). If nothing else, this makes one wonder whether the function of the olecranon in *Ichthyostega* was comparable with that in later tetrapods or not.

Because of circumstances such as these, the transition from the earliest and presumed non-weight-bearing joints to those which were more fully terrestrial is still poorly understood. Once these features are more closely reflected in the bony skeleton, it becomes possible to say more about terrestrial adaptation of limbs and vertebrae, and a review by Sumida (1997) takes this subject forward.

Conclusions

Although a full understanding of the origin of tetrapods is still a long way off, recent discoveries in paleoecology, paleobotany, invertebrate paleontology, developmental biology, and physiology, as well as invaluable finds of early fossil tetrapods and their relatives, have led to a greatly improved picture of the events and processes involved.

It seems clear that the tetrapods did not invade the land but rather acquired adaptations to terrestriality very slowly, sometimes in a variety of ways. It seems that the colonization of the land was first achieved by plants, then by invertebrates. Only after that could vertebrates follow. Changes to the vegetation involving the evolution of complex ecology were influenced by climatic and tectonic events in the early to mid-Paleozoic. Diversifying plant ecology facilitated the adaptation of invertebrates to life on land, which in turn may have been a critical factor in the gaining of the land by vertebrates. In many circumstances today, this sequence of colonization is still echoed when there is newly exposed land to be exploited. If there is a simple explanation for the evolutionary saga of tetrapod origins, it may be that the time was right.

Looking at the fossils of early tetrapods and their relatives shows how changes to the skull, jaws, and dentition reflect adaptions to terrestrial feeding. They are also inextricably linked to changes to the ear and hearing, and to the neck, vertebral column, and breathing. Changes to the limbs and to locomotory patterns may have been among those that occurred last in the suite of adaptations seen in the transition from fish to tetrapod. Several of these kinds of changes show fairly slow and gradual change to begin with and can be tracked sequentially from the tetrapodomorph fish through the Devonian and early Carboniferous genera. However, when tetrapods are picked up again in the fossil record of the later Early Carboniferous, these features show an explosive radiation with numerous parallel developments to jaws, occiputs, ear regions, vertebrae, ribs, sacrum, pelvic and pectoral girdles, limb proportions, and joint anatomy in response to the demands of terrestriality. They have left their differing legacies for modern forms in the differences between modern amphibians and amniotes, and among the diverse amniote groups. Although changes to soft tissues and anatomy can only be inferred from modern forms, these may not always provide useful models for understanding early tetrapods.

By the closing stages of the Carboniferous, however, it seems as though the foundations were laid for truly terrestrial living among vertebrates, with the development of the amniote egg and the evolution of herbivory well established, as well as adaptations for terrestrial locomotion. A huge diversity of forms had evolved, some of which are like nothing that exists today, and some of which have left no living descendants. Tetrapods appear to have reached some kind of peak in the Late Carboniferous and Early Permian, achieving more or less global distribution. Many groups are represented in far-flung parts of the world.

At the end of the Permian, something catastrophic for all animal life on Earth occurred which resulted in the almost complete extinction of most lineages. Some fortunate survivors of most vertebrate groups were left, although it is not clear that these were the ones with particularly advantageous adaptations. Whatever the truth of this, they were left to repopulate the Earth, building on foundations laid by their Devonian and Carboniferous forebears.

References

Ahlberg, P. E. 1991a. A re-examination of sarcopterygian interrelationships, with special reference to the Porolepiformes. *Zoological Journal of the Linnean Society of London* 103: 241–287.

———. 1991b. Tetrapod or near tetrapod fossils from the Upper Devonian of Scotland. *Nature* 354: 298–301.

———. 1992. Coelacanth fins and evolution. *Nature* 358: 459.

———. 1995. *Elginerpeton pancheni* and the earliest tetrapod clade. *Nature* 373: 420–425.

———. 1998. Postcranial stem tetrapod remains from the Devonian of Scat Craig, Morayshire, Scotland. *Zoological Journal of the Linnean Society* 122: 99–141.

Ahlberg, P. E., and J. A. Clack. 1998. Lower jaws, lower tetrapods—A review based on the Devonian genus *Acanthostega*. *Transactions of the Royal Society of Edinburgh: Earth Sciences* 89: 11–46.

Ahlberg, P. E., and Z. Johanson. 1997. Second tristichopterid (Sarcopterygii, Osteolepiformes) from the Upper Devonian of Canowindra, New South Wales, Australia, and phylogeny of the Tristichopteridae. *Journal of Vertebrate Palaeontology* 17: 653–673.

———. 1998. Osteolepiforms and the ancestry of tetrapods. *Nature* 395: 792–794.

Ahlberg, P. E., and A. R. Milner. 1994. The origin and early diversification of tetrapods. *Nature* 368: 507–514.

Ahlberg, P. E., and N. H. Trewin. 1995. The postcranial skeleton of the Middle Devonian lungfish *Dipterus valenciennesi*. *Transactions of the Royal Society of Edinburgh, Earth Sciences* 85: 159–175.

Ahlberg, P. E., J. A. Clack, and E. Luksevics. 1996. Rapid braincase evolution between *Panderichthys* and the earliest tetrapods. *Nature* 381: 61–64.

Ahlberg, P. E., E. Luksevics, and O. A. Lebedev. 1994. The first tetrapod finds from the Devonian (Upper Famennian) of Latvia. *Philosophical Transactions of the Royal Society of London Series B* 343: 303–328.

Ahlberg, P. E., E. Luksevics, and E. Mark-Kurik. 2000. A near-tetrapod from the Baltic Middle Devonian. *Palaeontology* 43: 533–548.

Alekseyev, A. A., O. A. Lebedev, I. S. Barskov, L. I. Kononova, and V. A. Chizhova. 1994. On the stratigraphic position of the Famennian and Tournaisian fossil vertebrate beds in Andreyevka, Tula Region, Central Russia. *Proceedings of the Geologists Association* 105: 41–52.

Algeo, T. J., and S. E. Scheckler. 1998. Terrestrial–marine teleconnections in the Devonian: Links between the evolution of land plants, weathering processes and marine anoxic events. *Philosophical Transactions of the Royal Society of London Series B* 353: 113–128.

Allin, E. F., and J. A. Hopson. 1992. Evolution of the auditory system in Synapsida (mammal-like reptiles) as seen in the fossil record. In D. B. Webster, A. N. Popper, and R. R. Fay (eds.), *The Evolutionary Biology of Hearing*, pp. 587–614. New York: Springer-Verlag.

Andrews, S. M., and R. L. Carroll. 1991. The order Adelospondyli. *Transactions of the Royal Society of Edinburgh, Earth Sciences* 82: 239–275.

Andrews, S. M., M. A. E. Browne, A. L. Panchen, and S. P. Wood. 1977. Discovery of amphibians in the Namurian (Upper Carboniferous) from Fife. *Nature* 265: 529–532.

Atz, J. W. 1976. *Latimeria* babies are born, not hatched. *Underwater Naturalist* 9: 4–7.

Barker, D., C. C. Hunt, and A. K. McIntyre. 1974. *Handbook of Sensory Physiology*. Vol. 3, Part 2, *Muscle Receptors*. New York: Springer-Verlag.

Barrell, J. 1916. Influence of Silurian–Devonian climates on the rise of air-breathing vertebrates. *Bulletin of the Geological Society of America* 27: 387–436.

Barrett, R. K., and H. Underwood. 1991. Retinally perceived light can entrain the pineal melatonin rhythm in Japanese quail. *Brain Research* 563: 87–93.

Beaumont, E. I. 1977. Cranial morphology of the Loxommatidae (Amphibia: Labyrinthodontia). *Philosophical Transactions of the Royal Society of London Series B* 280: 29–101.

Beaumont, E. I., and T. R. Smithson. 1998. The cranial morphology and relationships of the aberrant Carboniferous amphibian *Spathicephalus mirus* Watson. *Zoological Journal of the Linnean Society* 122: 187–209.

Bell, C. M., and M. J. Boyd. 1986. A tetrapod trackway from the Carboniferous of northern Chile. *Palaeontology* 29: 519–526.

Bemis, W. E., W. W. Burggren, and N. E. Kemp (eds.). 1987. *The Biology and Evolution of Lungfishes*. New York: Alan R. Liss.

Bendix-Almgreen, S. E., J. A. Clack, and H. Olsen. 1990. Upper Devonian tetrapod palaeoecology in the light of new discoveries in East Greenland. *Terra Nova* 2: 131–137.

Benton, M. J. 2000. Stems, nodes, crown clades, and rank-free lists: Is Linnaeus dead? *Biological Reviews* 75: 633–648.

Berman, D. S. 2000. Origin and early evolution of the amniote occiput. *Journal of Paleontology* 74: 938–956.

Berman, D. S., S. S. Sumida, and R. E. Lombard. 1992. Reinterpretation of the temporal and occipital regions in *Diadectes* and the relationships of the diadectomorphs. *Journal of Paleontology* 66: 481–499.

Bernacsek, G. M., and R. L. Carroll. 1981. Semicircular canal size in fossil fishes and amphibians. *Canadian Journal of Earth Sciences* 18: 150–156.

Berner, R. A. 1993. Paleozoic atmospheric CO_2: Importance of solar radiation and plant evolution. *Science* 261: 68–70.

Bjerring, H.-C. 1985. Facts and thoughts on piscine phylogeny. In R. E. Foreman, A. Gorbman, J. M. Dodd, and R. Olsson (eds.), *Evolutionary Biology of Primitive Fishes*, pp. 31–57. New York: Plenum.

———. 1986. Electric tetrapods? In Z. Rocek (ed.), *Studies in Herpetology,* pp. 29–36. Prague: Charles University.

———. 1987. Notes on some annexa oculi in the osteolepiform freshwater fish *Eusthenopteron foordi* from the Upper Devonian Escuminac Formation of Miguasha, eastern Canada. *Acta Zoologica* 68: 173–178.

———. 1988. Armar och ben i utvecklingshistorisk belysning. *Flora Och Fauna* 83: 58–74.

Bolt, J. R. 1969. Lissamphibian origins: Possible protolissamphibian from the Lower Permian of Oklahoma. *Science* 166: 888–891.

———. 1979. *Amphibamus grandiceps* as a juvenile dissorophid: Evidence and implications. In M. H. Nitecki (ed.), *Mazon Creek Fossils,* pp. 529–563. New York: Academic Press.

———. 1991. Lissamphibian origins. In H.-P. Schultze and L. Trueb (eds.), *Origin of the Higher Groups of Tetrapods,* pp. 194–222. Ithaca, N.Y.: Cornell University Press.

Bolt, J. R., and R. E. Lombard. 1985. Evolution of the amphibian tympanic ear and the origin of frogs. *Biological Journal of the Linnean Society* 24: 83–99.

———. 2000. Palaeobiology of *Whatcheeria deltae.* In H. Heatwole and R. L. Carroll (eds.), *Amphibian Biology,* Vol. 4, *Palaeontology,* pp. 1044–1052. Chipping Norton, New South Wales, Australia: Surrey Beatty.

Bolt, J. R., R. M. McKay, B. J. Witzke, and M. P. McAdams. 1988. A new Lower Carboniferous tetrapod locality in Iowa. *Nature* 333: 768–770.

Borgen, U. 1983. Homologizations of skull roofing bones between tetrapods and osteolepiform fishes. *Palaeontology* 26: 735–753.

Boy, J. A. 1974. Die larven der rhachitomen amphibien (Amphibia: Temnospondyli; Karbon-Trias). *Paläontologische Zeitschrift* 48: 236–268.

Boy, J. A., and H.-D. Sues. 2000. Branchiosaurs: Larvae, metamorphosis and heterochrony in temnospondyls and seymouriamorphs. In H. Heatwole and R. L. Carroll (eds.), *Amphibian Biology,* Vol. 4, *Palaeontology,* pp. 1150–1197. Chipping Norton, New South Wales, Australia: Surrey Beatty.

Brainerd, E. L., J. S. Ditelberg, and D. M. Bramble. 1993. Lung ventilation in salamanders and the evolution of vertebrate air-breathing mechanisms. *Biological Journal of the Linnean Society* 49: 16–183.

Briggs, D., and P. Crowther. 1990. *Palaeobiology.* Oxford: Blackwell Scientific.

Burke, A. C., C. G. Nelson, B. A. Morgan, and C. Tabin. 1995. *Hox* genes and the evolution of vertebrate axial morphology. *Development* 121: 333–346.

Bystrow, A. P. 1938. *Dvinosaurus* als neotenische form der stegocephalan. *Acta Zoologica* 19: 209–295.

———. 1944. *Kotlassia prima* Amalitzky. *Bulletin of the Geological Society of America* 55: 379–416.

———. 1947. Hydrophilous and xerophilous labyrinthodonts. *Acta Zoologica* 28: 137–164.

Campbell, K. S. W., and R. E. Barwick. 1987. Palaeozoic lungfishes—A review. In W. E. Bemis, W. W. Burggren, and N. E. Kemp (eds.), *The Biology and Evolution of Lungfishes,* pp. 93–131. New York: Alan R. Liss.

Campbell, K. S. W., and M. W. Bell. 1977. A primitive amphibian from the late Devonian of New South Wales. *Alcheringa* 1: 369–381.

Capdevila, J., and C. J. I. Belmonte. 2000. Perspectives on the evolutionary origin of limbs. *Journal of Experimental Zoology* 288: 287–303.

Caplan, M. L., and R. M. Bustin. 1999. Devonian–Carboniferous Hangenberg mass extinction event, widespread organic-rich mudrock and anoxia: Causes and consequences. *Palaeogeography, Palaeoclimatology, and Palaeoecology* 148: 187–207.

Carlström, D. 1963. A crystallographic study of vertebrate otoliths. *Biological Bulletin* 125: 124–138.

Carr, C. E. 1992. Evolution of the central auditory system in reptiles and birds. In D. B. Webster, R. R. Fay, and A. N. Popper (eds.), *The Evolutionary Biology of Hearing*, pp. 511–544. New York: Springer-Verlag.

Carroll, R. L. 1964. The early evolution of dissorophid amphibians. *Bulletin of the Museum of Comparative Zoology, Harvard* 131: 161–250.

———. 1969. A new family of Carboniferous amphibians. *Palaeontology* 12: 537–548.

———. 1970. The ancestry of reptiles. *Philosophical Transactions of the Royal Society of London Series B* 257: 267–308.

———. 1980. The hyomandibula as a supporting element in the skull of primitive tetrapods. In A. L. Panchen (ed.), *The Terrestrial Environment and the Origin of Land Vertebrates*, pp. 293–317. New York: Academic Press.

———. 1987. *Vertebrate Paleontology and Evolution*. New York: W. H. Freeman.

———. 1995. Problems of the phylogenetic analysis of Paleozoic choanates. In M. Arsenault, H. Lelièvre, and P. Janvier (eds.), *Studies on Early Vertebrates. Bulletin du Muséum National d'Histoire Naturelle, Paris* 17: 389–445.

———. 2000a. Lepospondyls. In H. Heatwole and R. L. Carroll (eds.), *Amphibian Biology*, Vol. 4, *Palaeontology*, pp. 1198–1269. Chipping Norton, New South Wales, Australia: Surrey Beatty.

———. 2000b. *Eocaecilia* and the origin of caecilians. In H. Heatwole and R. L. Carroll (eds.), *Amphibian Biology*, Vol. 4, *Palaeontology*, pp. 1402–1411. Chipping Norton, New South Wales, Australia: Surrey Beatty.

Carroll, R. L., and D. Baird. 1972. Carboniferous stem reptiles of the family Romeriidae. *Bulletin of the Museum of Comparative Zoology, Harvard* 143: 321–364.

Carroll, R. L., and P. Gaskill. 1978. The order Microsauria. *Memoirs of the American Philosophical Society* 126: 1–211.

Carroll, R. L., and R. Holmes. 1980. The skull and jaw musculature as guides to the ancestry of salamanders. *Zoological Journal of the Linnean Society* 68: 1–40.

Carroll, R. L., K. A. Bossy, A. C. Milner, S. M. Andrews, and C. F. Wellstead. 1998. *Handbuch der Paläoherpetologie*, Vol. 1, *Lepospondyli*. Munich: Pfeil.

Carroll, R. L., A. Kuntz, and K. Albright. 1999. Vertebral development and amphibian evolution. *Evolution and Development* 1: 36–48.

Case, E. C. 1907. A restoration of *Diadectes*. *Journal of Geology* 15: 556–559.

Chang, M.-M. 1995. *Diabolepis* and its bearing on the relationships between porolepiforms and dipnoans. In M. Arsenault, H. Lelièvre, and P. Janvier (eds.), *Studies on Early Vertebrates. Bulletin du Muséum National d'Histoire Naturelle, Paris* 17: 235–268.

Chang, M.-M., and M. M. Smith. 1992. Is *Youngolepis* a porolepiform? *Journal of Vertebrate Palaeontology* 12: 294–312.

Chidiac, Y. 1996. Paleoenvironmental interpretation of the Escuminac Formation based on geochemical evidence. In H.-P. Schultze and R. Cloutier (eds.), *Devonian Fishes and Plants of Miguasha, Quebec, Canada*, pp. 47–53. Munich: Friedrich Pfeil.

Clack, J. A. 1983. The stapes of the Coal Measures embolomere *Pholiderpeton scutigerum* Huxley (Amphibia: Anthracosauria) and otic evolution in early tetrapods. *Zoological Journal of the Linnean Society* 79: 121–148.

———. 1987a. *Pholiderpeton scutigerum* Huxley, an amphibian from the Yorkshire Coal Measures. *Philosophical Transactions of the Royal Society of London Series B* 318: 1–107.

———. 1987b. Two new specimens of *Anthracosaurus* (Amphibia: Anthracosauria) from the Northumberland Coal Measures. *Palaeontology* 30: 15–26.

———. 1988. Pioneers of the land in East Greenland. *Geology Today* 4(6): 407–409.

———. 1989. Discovery of the earliest-known tetrapod stapes. *Nature* 342: 425–427.

———. 1992. The stapes of *Acanthostega gunnari* and the role of the stapes in early tetrapods. In D. B. Webster, A. N. Popper, and R. R. Fay (eds.), *The Evolutionary Biology of Hearing*, pp. 405–420. New York: Springer-Verlag.

———. 1993. Homologies in the fossil record—The middle ear as a test case. *Acta Biotheoretica* 41: 391–410.

———. 1994a. *Acanthostega gunnari*, a Devonian tetrapod from Greenland; the snout, palate and ventral parts of the braincase, with a discussion of their significance. Meddelelser om Grønland. *Geoscience* 31: 1–24.

———. 1994b. *Silvanerpeton miripedes*, a new anthracosauroid from the Viséan of East Kirkton, West Lothian, Scotland. *Transactions of the Royal Society of Edinburgh, Earth Sciences* 84: 369–376.

———. 1996. Otoliths in fossil coelacanths. *Journal of Vertebrate Paleontology* 16: 168–171.

———. 1997a. Devonian tetrapod trackways and trackmakers; a review of the fossils and footprints. *Palaeogeography, Palaeoclimatology, and Palaeoecology* 130: 227–250.

———. 1997b. The evolution of tetrapod ears and the fossil record. *Brain, Behavior and Evolution* 50: 198–212.

———. 1998a. A new Lower Carboniferous tetrapod with a mélange of crown group characters. *Nature* 394: 66–69.

———. 1998b. The Scottish Carboniferous tetrapod *Crassigyrinus scoticus* (Lydekker)—Cranial anatomy and relationships. *Transactions of the Royal Society of Edinburgh, Earth Sciences* 88: 127–142.

———. 1998c. The neurocranium of *Acanthostega gunnari* and the evolution of the otic region in tetrapods. *Zoological Journal of the Linnean Society* 122: 61–97.

———. 2001a (in press). *Eucritta melanolimnetes* from the Early Carboniferous of Scotland, a stem tetrapod showing a mosaic of characteristics. *Transactions of the Royal Society of Edinburgh, Earth Sciences*.

———. 2001b. The otoccipital region—Origin, ontogeny and the fish–tetrapod transition. In P. E. Ahlberg, *Major Events in Early Vertebrate Evolution*, 392–505. London: Systematics Association Symposium.

———. 2002 (in press). The dermal skull roof of *Acanthostega,* an early tetrapod from the Late Devonian. *Transactions of the Royal Society of Edinburgh, Earth Sciences.*

Clack, J. A., and P. E. Ahlberg. 1998. A reinterpretation of the braincase of the Devonian tetrapod *Ichthyostega stensioei. Journal of Vertebrate Paleontology* 18(Suppl. 3): 34A.

Clack, J. A., and R. L. Carroll. 2000. Early Carboniferous tetrapods. In H. Heatwole and R. L. Carroll (eds.), *Amphibian Biology,* Vol. 4, *Palaeontology,* pp. 1030–1043. Chipping Norton, New South Wales, Australia: Surrey Beatty.

Clack, J. A., and M. I. Coates. 1995. *Acanthostega*—A primitive aquatic tetrapod? In M. Arsenault, H. Lelièvre, and P. Janvier (eds.), *Studies on Early Vertebrates. Bulletin du Muséum National d'Histoire Naturelle, Paris* 17: 359–373.

Clack, J. A., and S. M. Finney. 1997. An articulated tetrapod specimen from the Tournaisian of western Scotland. *Journal of Vertebrate Paleontology* 3(Suppl. 17): 38A.

Clack, J. A., and S. L. Neininger. 2000. Fossils from the Celsius Bjerg Group, Upper Devonian sequence, East Greenland: Significance and sedimentological distribution. In P. F. Friend and B. Williams (eds.), *New Perspectives on the Old Red Sandstone,* Special Publication 180: 557–566. London: Geological Society Symposium Volume.

Cloutier, R. 1996. The primitive actinistian *Miguashaia bureaui* Schultze (Sarcopterygii). In H.-P. Schultze and R. Cloutier (eds.), *Devonian Fishes and Plants of Miguasha, Quebec, Canada,* pp. 227–247. Munich: Friedrich Pfeil.

Cloutier, R., and P. E. Ahlberg. 1997. Morphology, characters and the interrelationships of basal sarcopterygians. In M. L. J. Stiassny and L. Parenti (eds.), *Interrelationships of Fishes II,* pp. 445–479. London: Academic Press.

Coates, M. I. 1996. The Devonian tetrapod *Acanthostega gunnari* Jarvik: Postcranial anatomy, basal tetrapod relationships and patterns of skeletal evolution. *Transactions of the Royal Society of Edinburgh, Earth Sciences* 87: 363–421.

Coates, M. I., and J. A. Clack. 1990. Polydactyly in the earliest known tetrapod limbs. *Nature* 347: 66–69.

———. 1991. Fish-like gills and breathing in the earliest known tetrapod. *Nature* 352: 234–236.

———. 1995. Romer's Gap—Tetrapod origins and terrestriality. In M. Arsenault, H. Lelièvre, and P. Janvier (eds.), *Studies on Early Vertebrates. Bulletin du Muséum National d'Histoire Naturelle, Paris* 17: 373–388.

Coates, M. I., and M. J. Cohn. 1998. Fins, limbs and tails: Outgrowths and axial patterning in vertebrate evolution. *BioEssays* 20: 371–381.

———. 1999. Vertebrate axial and appendicular patterning: The early development of paired appendages. *American Zoologist* 39: 676–685.

Cope, E. D. 1892. On the phylogeny of the vertebrata. *Proceedings of the American Philosophical Society* 30: 278–281.

Cote, S., R. L. Carroll, R. Cloutier, and L. Bar-Sagil. In press. Vertebral development in the choanate fish *Eusthenopteron foordi* and the polarity of vertebral development. *Journal of Vertebrate Paleontology.*

Couly, G. F., P. M. Coltey, and N. M. Le Douarin. 1993. The triple origin of skull in higher vertebrates: A study in quail-chick chimeras. *Development* 117: 409–429.

Cruickshank, A. R. I., and B. W. Skews. 1980. The functional significance of nectridean tabular horns (Amphibia: Lepospondyli). *Proceedings of the Royal Society of London B* 209: 513–537.

Daeschler, E. B. 2000. Early tetrapod jaws from the Late Devonian of Pennsylvania, USA. *Journal of Paleontology* 74: 301–308.

Daeschler, E. B., and N. Shubin. 1995. Tetrapod origins. *Paleobiology* 21: 404–409.

———. 1997. Fish with fingers? *Nature* 391: 133.

Daeschler, E. B., N. H. Shubin, K. S. Thomson, and W. W. Amaral. 1994. A Devonian tetrapod from North America. *Science* 265: 639–642.

Daly, E. 1994. The Amphibamidae (Amphibia: Temnospondyli), with a description of a new genus from the Upper Permian of Kansas. *University of Kansas Museum of Natural History Miscellaneous Publications* 85: 1–59.

Darwin, C. 1859. *On the Origin of Species by Means of Natural Selection*. London: John Murray.

Dawson, J. W. 1863. *Air Breathers of the Coal Period*. Montreal: Dawson Brothers.

Denison, R. H. 1941. The soft anatomy of *Bothriolepis*. *Journal of Paleontology* 15: 553–561.

de Queiroz, K., and J. Gauthier. 1992. Phylogenetic taxonomy. *Annual Review of Ecology and Systematics* 23: 449–480.

———. 1994. Towards a phylogenetic system of nomenclature. *Trends in Ecology and Evolution* 9: 27–31.

Dilkes, D. W. 1993. Biology and evolution of the nasal region in trematopid amphibians. *Palaeontology* 36: 839–853.

DiMichele, W. A., and R. W. Hook. 1992. Palaeozoic terrestrial ecosystems. In A. K. Behrensmeyer, J. D. Damuth, W. A. Potts, H.-D. Sues, and S. L. Wing (eds.), *Terrestrial Ecosystems through Time*, pp. 206–325. Chicago: University of Chicago Press.

Dudley, R. 1998. Atmospheric oxygen, giant Paleozoic insects and the evolution of aerial locomotor performance. *Journal of Experimental Biology* 210: 1043–1050.

Duellman, W. E., and L. Trueb. 1986. *Biology of Amphibians*. New York: McGraw-Hill.

Eaton, T. H. 1951. Origin of tetrapod limbs. *American Midland Naturalist* 46: 245–251.

Edwards, J. L. 1989. Two perspectives on the evolution of the tetrapod limb. *American Zoologist* 29: 235–254.

Eisthen, H. L. 1992. Phylogeny of the vomeronasal system and of receptor cell types in the olfactory and vomeronasal epithelia of vertebrates. *Microscopy Research and Technique* 23: 1–21.

———. 1998. Evolution of vertebrate olfactory systems. *Brain, Behaviour and Evolution* 50: 222–233.

Erdmann, M. V., R. L. Caldwell, and M. K. Moosa. 1998. An Indonesian "king of the sea." *Nature* 395: 335.

Erwin, D. H. 1995. *The Great Paleozoic Crisis—Life and Death in the Permian*. New York: Columbia University Press.

Erwin, D. H, S. A. Bowring, and J. Yugan. In press. The end-Permian mass extinctions: A review. Geological Society of America Special Volume.

Evans, S. E., and Haubold, H. 1987. A review of the Upper Permian genera *Coelurosauravus*, *Weigeltisaurus* and *Gracilisaurus* (Reptilia: Diapsida). *Zoological Journal of the Linnean Society of London* 90: 205–237.

Ewer, D. W. 1955. Tetrapod limb. *Science* 122: 467.

Falcon-Lang, H. J. 1999. The Early Carboniferous (Asbian–Brigantian) seasonal tropical climate of northern Britain. *Palaios* 14: 116–126.

———. 2001. Fire ecology of the Carboniferous tropical zone. *Palaeogeography, Palaeoclimatology, and Palaeoecology* 164: 373–396.
Feder, M. E., and W. W. Burggren. 1985. Skin breathing in vertebrates. *Scientific American* 253: 106–118.
Forey, P. L. 1981. The coelacanth *Rhabdoderma* in the Carboniferous of the British Isles. *Palaeontology* 24: 203–229.
———. 1998. *The History of the Coelacanth Fishes*. London: Chapman and Hall/Natural History Museum.
Fox, R. C., and M. C. Bowman. 1966. Osteology and relationships of *Captorhinus aguti* (Cope) (Reptilia: Captorhinomorpha). *University of Kansas Paleontological Contributions, Vertebrates* 11: 1–79.
Fraser, N. 1991. The true turtles' story. *Nature* 349: 278–279.
Fricke, H., H. Reinicke, H. Hofer, and W. Nachtigall. 1987. Locomotion of the coelacanth, *Latimeria chalumnae*, in its natural environment. *Nature* 329: 331–333.
Fricke, H., J. Scauer, K. Lissman, L. Kasang, and R. Plante. 1991. Coelacanth *Latimeria chalumnae* aggregates in caves: First conclusions and their resting habitat and social behaviour. *Environmental Biology of Fishes* 30: 281–285.
Fritzsch, B. 1987. The inner ear of the coelacanth fish *Latimeria* has tetrapod affinities. *Nature* 327: 331–333.
———. 1989. Diversity and regression in the amphibian lateral line and electrosensory system. In S. Coombs, P. Görner, and H. Münz (eds.), *The Mechanosensory Lateral Line: Neurobiology and Evolution,* pp. 99–114. New York, Springer-Verlag.
———. 1992. The water-to-land transition: Evolution of the tetrapod basilar papilla, middle ear and auditory nuclei. In D. B. Webster, A. N. Popper, and R. R. Fay (eds.), *The Evolutionary Biology of Hearing,* pp. 351–375. New York: Springer-Verlag.
Gaffney, E. S. 1980. Phylogenetic relationships of the major groups of amniotes. In A. L. Panchen (ed.), *The Terrestrial Environment and the Origin of Land Vertebrates,* pp. 593–610. London: Academic.
———. 1990. The comparative osteology of the Triassic turtle *Proganochelys. Bulletin of the American Museum of Natural History* 194: 1–263.
Gans, C. 1970. Respiration in early tetrapods—The frog is a red herring. *Evolution* 24: 723–734.
Gauldie, R. W., and R. L. Radtke. 1990. Using the physical dimensions of the semicircular canal as a probe to evaluate inner ear function in fishes. *Comparative Biochemistry and Physiology* 96A: 199–203.
Gegenbaur, C. 1878. *Elements of Comparative Anatomy*. London: Macmillan.
Giffin, E. B. 1995. Postcranial paleoneurology of the Diapsida. *Journal of Zoology, London* 235: 389–410.
Gilbert, S. F. 2000. *Developmental Biology.* 6th edition. Sunderland, Mass.: Sinauer Associates.
Gilbertson, T. A., P. Avenet, and S. C. Kinnamon. 1992. Proton currents through amilonide-sensitive Na-channels in hamster taste cells—Role in acid transduction. *Journal of General Physiology* 100: 803–824.
Godfrey, S. J. 1989a. The postcranial skeletal anatomy of the Carboniferous tetrapod *Greererpeton burkemorani* Romer, 1969. *Philosophical Transactions of the Royal Society of London Series B* 323: 75–133.
———. 1989b. Ontogenetic changes in the skull of the Carboniferous tetrapod *Greererpeton burkemorani* Romer, 1969. *Philosophical Transactions of the Royal Society of London Series B* 323: 135–153.

Godfrey, S. J., A. R. Fiorillo, and R. L. Carroll. 1987. A newly discovered skull of the temnospondyl amphibian *Dendrerpeton acadianum* Owen. *Canadian Journal of Earth Sciences* 24: 796–805.

Goin, C. J., and O. B. Goin. 1956. Further comments on the origin of the tetrapods. *Evolution* 10: 440–441.

Gould, S. J. 1977. *Ontogeny and Phylogeny.* Cambridge, Mass.: Harvard University Press.

———. 1980. *The Panda's Thumb.* New York: W. W. Norton.

Graham, J. B., N. Aguilar, R. Dudley, and C. Gans. 1997. The Late Paleozoic atmosphere and the ecological and evolutionary physiology of tetrapods. In S. Sumida and K. L. M. Martin (eds.), *Amniote Origins: Completing the Transition to Land,* pp. 141–167. San Diego: Academic Press.

Grande, L., and W. E. Bemis. 1998. A comprehensive phylogenetic study of amiid fishes (Amiidae) based on comparative skeletal anatomy. *Society of Vertebrate Palaeontology Special Publications* 1–690.

Gregory, W. K. 1915. Present status of the problem of the origin of the Tetrapoda, with special reference to the skull and paired limbs. *Annals of the New York Academy of Sciences* 26: 317–383.

———. 1935. Further observations on the pectoral girdle and fin of *Sauripterus taylori* Hall, a crossopterygian fish from the Upper Devonian of Pennsylvania, with special reference to the origin of the pentadactylate extremities of Tetrapoda. *Proceedings of the American Philosophical Society* 75: 673–690.

Gregory, W. K., and H. C. Raven. 1941. Origin of paired fins and limbs. *Annals of the New York Academy of Science* 42: 273–360.

Gregory, W. K., R. W. Miner, and G. K. Noble. 1923. The carpus of *Eryops.* *Bulletin of the American Museum of Natural History* 48: 279–288.

Günther, C. A. L. G. 1871. Description of *Ceratodus,* a genus of ganoid fishes, recently discovered in rivers of Queensland, Australia. *Transactions of the Royal Society of London* 161: 377–379.

Heaton, M. J. 1979. The cranial anatomy of primitive captorhinid reptiles from the Pennsylvanian and Permian of Oklahoma and Texas. *Bulletin of the Oklahoma Geological Survey* 127: 1–84.

Hinchliffe, R., E. I. Vorobyeva, and J. Géraudie. 2001. Is there a developmental bauplann underlying limb evolution? Evidence from a teleost fish and from urodele and anuran amphibians. In P. E. Ahlberg (ed.), *Major Events in Early Vertebrate Evolution,* pp. 377–391. London: Systematics Association Symposium.

Holder, N. 1983. Developmental constraints and the evolution of vertebrate digit patterns. *Theoretical Biology* 104: 451–471.

Holland, P. W. H., and J. Garcia-Fernàndez. 1996. *Hox* genes and chordate evolution. *Developmental Biology* 173: 382–395.

Holmes, R. 1980. *Proterogyrinus scheeli* and the early evolution of the labyrinthodont pectoral limb. In A. L. Panchen (ed.), *The Terrestrial Environment and the Origin of Land Vertebrates,* pp. 351–376. London: Academic.

———. 1984. The Carboniferous amphibian *Proterogyrinus scheeli* Romer, and the early evolution of tetrapods. *Philosophical Transactions of the Royal Society of London Series B* 306: 431–524.

———. 1989a. The skull and axial skeleton of the Lower Permian anthracosauroid amphibian *Archeria crassidisca* Cope. *Palaeontographica* 207: 161–206.

———. 1989b. Functional interpretations of the vertebral structure in Paleozoic amphibians. *Historical Biology* 2: 111–124.

———. 2000. Palaeozoic temnospondyls. In H. Heatwole and R. L. Carroll (eds.), *Amphibian Biology*, Vol. 4, *Palaeontology*, pp. 1081–1120. Chipping Norton, New South Wales, Australia: Surrey Beatty.

Holmes, R. B., and R. L. Carroll. 1977. A temnospondyl amphibian from the Mississippian of Scotland. *Bulletin of the Museum of Comparative Zoology, Harvard* 147: 489–511.

Holmes, R. B., R. L. Carroll, and S. Godfrey. 1998. The first articulated skeleton of *Dendrerpeton acadianum* (Temnospondyli: Dendrerpetontidae) from the Lower Pennsylvanian locality of Joggins, Nova Scotia and a review of its relationships. *Journal of Vertebrate Paleontology* 18: 64–79.

Holmgren, N. 1933. On the origin of the tetrapod limb. *Acta Zoologica* 14: 185–295.

Hook, R. W. 1983. *Colosteus scutellatus* (Newberry) a primitive temnospondyl amphibian from the Middle Pennsylvanian of Linton, Ohio. *American Museum Novitates* 2770: 1–41.

Hook, R. W., and D. Baird. 1986. The Diamond Coal Mine of Linton, Ohio, and its Pennsylvanian-age vertebrates. *Journal of Vertebrate Paleontology* 6: 174–190.

Hook, R. W., and J. C. Ferm. 1988. Paleoenvironmental controls on vertebrate-bearing abandoned channels in the Upper Carboniferous. *Palaeogeography, Palaeoclimatology, and Palaeoecology* 63: 159–181.

Hotton, N., E. C. Olson, and R. Beerbower. 1997. The amniote transition and the discovery of herbivory. In S. Sumida and K. L. M. Martin (eds.), *Amniote Origins: Completing the Transition to Land*, pp. 207–264. San Diego: Academic Press.

Huxley, T. H. 1861. Preliminary essay upon the systematic arrangement of the fishes of the Devonian epoch. Memoirs of the Geological Survey of the United Kingdom. Figures and descriptions illustrative of British organic remains. *Scientific Memoirs* 2: 421–460.

———. 1882. *A Manual of Anatomy of the Vertebrated Animals*. New York: D. Appleton.

Inger, R. F. 1957. Ecological aspects of the origins of tetrapods. *Evolution* 11: 373–376.

Janis, C. M., and C. Farmer. 1999. Proposed habitats of early tetrapods: Gills, kidneys and the water–land transition. *Zoological Journal of the Linnean Society* 126: 117–126.

Janis, C., and J. Keller. 2001. Modes of ventilation in early tetrapods: Costal aspiration as a key feature of amniotes. *Acta Polonica* 42: 137–170.

Janvier, P. 1996. *Early Vertebrates*. Oxford: Clarendon Press.

Jarvik, E. 1942. On the structure of the snout of crossopterygians and lower gnathostomes in general. *Zoologiska Bidrag fran Uppsala* 21: 235–675.

———. 1944. On the dermal bones, sensory canals and pit-lines of the skull of *Eusthenopteron foordi* Whiteaves, with some remarks on *E. säve-söderberghi* Jarvik. *Kunglinga svenska Vetenskaps Akademiens Handlingar* 21(3): 1–48.

———. 1952. On the fish-like tail in the ichthyostegid stegocephalians. *Meddelelser om Grønland* 114(12): 1–90.

———. 1980. *Basic Structure and Evolution of Vertebrates*. Vols. 1 and 2. New York: Academic Press.

———. 1996. The Devonian tetrapod *Ichthyostega*. *Fossils and Strata* 40: 1–206.

Jenkins, F., Jr. 1971. The postcranial skeleton of African cynodonts. *Bulletin of the Peabody Museum of Natural History* 36: 1–201.

Jenkins, F. A., and D. M. Walsh. 1993. An Early Jurassic caecilian with legs. *Nature* 365: 246–250.

Jeram, A. J. 1994. Scorpions from the Viséan of East Kirkton, West Lothian, Scotland, with a revision of the infraorder Mesoscorpiona. *Transactions of the Royal Society of Edinburgh, Earth Sciences* 84: 283–299.

Jeram, A. J., and P. A. Selden. 1994. Eurypterids from the Viséan of East Kirkton, West Lothian, Scotland. *Transactions of the Royal Society of Edinburgh, Earth Sciences* 84: 301–308.

Jeram, A. J., P. A. Selden, and D. Edwards. 1990. Land animals in the Silurian: Arachnids and myriapods from Shropshire, England. *Science* 250: 658–661.

Johanson, Z., and P. E. Ahlberg. 1997. A new tristichopterid (Osteolepiformes: Sarcopterygii) from the Mandagery Sandstone (Late Devonian, Famennian) near Canowindra, NSW, Australia. *Transactions of the Royal Society of Edinburgh, Earth Sciences* 88: 39–68.

———. 1998. A complete primitive rhizodont from Australia. *Nature* 394: 569–573.

Jones, S. 1999. *Almost Like a Whale—The Origin of Species Updated.* London: Doubleday.

Jones, G. M., and K. E. Spells. 1963. A theoretical and comparative study of the functional dependence of the semicircular canal upon its physical dimensions. *Proceedings of the Royal Society of London* 157: 403–419.

Joss, J. M. P., and G. H. Joss. 1995. Breeding Australian lungfish in captivity. In F. W. Goetz and P. Thomas (eds.), *Proceedings of the Fifth International Symposium on the Reproductive Physiology of Fish,* p. 121.

Jow, L. Y., S. F. Chew, C. B. Lim, P. M. Anderson, and Y. K. Ip. 1999. The marble goby *Oxyeleotris marmoratus* activates hepatic glutamine synthetase and detoxifies ammonia to glutamine during air exposure. *Journal of Experimental Biology* 202: 237–245.

Kardong, K. V. 1998. *Vertebrates—Comparative Anatomy, Function and Evolution.* New York: McGraw-Hill.

Kenrick, P., and P. R. Crane. 1997. The origin and early evolution of plants on land. *Nature* 389: 33–39.

Kinnamon, S. C., and S. D. Roper. 1987. Voltage dependent ionic currents in dissociated mudpuppy taste cells. *Annals of the New York Academy of Science* 510: 413–416.

Klembara, J. 1992. The first record of pit-lines and foraminal pits in tetrapods and the problem of the skull roof bones homology between tetrapods and fishes. *Geologica Carpathica* 42: 249–252.

———. 1994. The sutural pattern of skull-roof bones in Lower Permian *Discosauriscus austriacus* from Moravia. *Lethaia* 27: 85–95.

———. 1995. The external gills and ornamentation of skull roof bones of the Lower Permian tetrapod *Discosauriscus* (Kuhn 1933) with remarks to its ontogeny. *Paläontologische Zeitschrift* 69: 265–281.

———. 1997. The cranial anatomy of *Discosauriscus* Kuhn, a seymouriamorph tetrapod from the Lower Permian of the Boskovice Furrow (Czech Republic). *Philosophical Transactions of the Royal Society of London Series B* 352: 257–302.

———. 2000. The postcranial skeleton of *Discosauriscus* Kuhn, a seymouriamorph tetrapod from the Lower Permian of the Boskovice

Furrow (Czech Republic). *Transactions of the Royal Society of Edinburgh, Earth Sciences* 90: 287–316.

Kurss, V. 1992. Depositional environment and burial conditions of fish remains in Baltic Middle Devonian. In E. Mark-Kurik (ed.), *Fossil Fishes as Living Animals*, pp. 251–260. Tallinn: Academy of Sciences of Estonia.

Laurin, M. 1996. A redescription of the cranial anatomy of *Seymouria baylorensis*, the best known seymouriamorph (Vertebrata: Seymouriamorpha). *Paleobios* 17: 1–16.

———. 2000. Seymouriamorphs. In H. Heatwole and R. L. Carroll (eds.), *Amphibian Biology*, Vol. 4, *Palaeontology*, pp. 1064–1080. Chipping Norton, New South Wales, Australia: Surrey Beatty.

Laurin, M., and R. R. Reisz. 1995. A re-evaluation of early amniote phylogeny. *Zoological Journal of the Linnean Society* 113: 165–223.

———. 1997. A new perspective on tetrapod phylogeny. In S. Sumida and K. L. M. Martin (eds.), *Amniote Origins: Completing the Transition to Land*, pp. 9–59. San Diego: Academic Press.

Lebedev, O. A. 1984. The first find of a Devonian tetrapod vertebrate in the USSR. *Doklady Akademii Nauk SSSR. Palaeontology* 278: 1470–1473 (in Russian).

———. 1985. The first tetrapods: Searchings and findings. *Priroda* 11: 26–36 (in Russian).

Lebedev, O. A., and J. A. Clack. 1993. New material of Devonian tetrapods from the Tula region, Russia. *Palaeontology* 36: 721–734.

Lebedev, O. A., and M. I. Coates. 1995. The postcranial skeleton of the Devonian tetrapod *Tulerpeton curtum* Lebedev. *Zoological Journal of the Linnean Society* 114: 307–348.

Lee, M. S. Y. 1995. Historical burden and the interrelationships of "parareptiles." *Biological Reviews* 70: 459–547.

———. 1997. Pareiasaur phylogeny and the origin of turtles. *Zoological Journal of the Linnean Society* 120: 197–280.

Lee, M. S. Y., and P. S. Spencer. 1997. Crown clades, key characters and taxonomic stability: When is an amniote not an amniote? In S. Sumida and K. L. M. Martin (eds.), *Amniote Origins: Completing the Transition to Land*, pp. 61–84. San Diego: Academic Press.

Lombard, R. E., and J. R. Bolt. 1979. Evolution of the tetrapod ear: An analysis and reinterpretation. *Biological Journal of the Linnean Society* 11: 19–76.

———. 1995. A new primitive tetrapod, *Whatcheeria deltae*, from the Lower Carboniferous of Iowa. *Palaeontology* 38: 471–494.

Long, J. A. 1993. Cranial ribs in Devonian lungfish and the origin of dipnoan air-breathing. *Memoirs of the Association of Australasian Palaeontologists* 15: 199–209.

———. 1995. *The Rise of Fishes*. Sydney, Australia: University of New South Wales Press.

Lull, R. S. 1918. *The Pulse of Life: The Evolution of the Earth*. New Haven, Conn.: Yale University Press.

Lund, R., and W. L. Lund. 1985. Coelacanths from the Bear Gulch Limestone (Namurian) of Montana and the evolution of the Coelacanthiformes. *Bulletin of the Carnegie Museum of Natural History* 25: 1–74.

Magid, A. M. A. 1966. Breathing and function of the spiracles in *Polypterus senegalus*. *Animal Behaviour* 14: 530–533.

———. 1967. Respiration of air by the primitive fish *Polypterus senegalus*. *Nature* 215: 1096–1097.

Marshall, J. E. A., T. R. Astin, and J. A. Clack. 1999. The East Greenland tetrapods are Devonian in age. *Geology* 27: 637–640.

McGhee, G. R. J. 1989. The Frasnian–Famennian extinction event. In S. J. Donovan (ed.), *Mass Extinctions: Processes and Evidence*, pp. 133–151. London: Bellhaven Press.

McKenzie, D. J., G. Piraccini, A. Felskie, P. Romano, P. Bronzi, and C. L. Bolis. 1999. Effects of plasma total ammonia content and pH on urea excretion in Nile tilapia. *Physiological and Biochemical Zoology* 72: 116–125.

McNamara, K., and P. Selden. 1993. Strangers on the shore. *New Scientist* 139: 23–27.

Milner, A. C. 1980a. A review of the Nectridea (Amphibia). In A. L. Panchen (ed.), *The Terrestrial Environment and the Origin of Land Vertebrates*, pp. 377–405. London: Academic.

———. 1994. The aïstopod from the Viséan of East Kirkton, West Lothian, Scotland. *Transactions of the Royal Society of Edinburgh, Earth Sciences* 84: 363–368.

Milner, A. C., and W. Lindsay. 1998. Postcranial remains of *Baphetes* and their bearing on the relationships of the Baphetidae (= Loxommatidae). *Zoological Journal of the Linnean Society* 122: 211–235.

Milner, A. R. 1980b. The tetrapod assemblage from Nýřany, Czechoslovakia. In A. L. Panchen (ed.), *The Terrestrial Environment and the Origin of Land Vertebrates*, pp. 439–496. London: Academic.

———. 1982. Small temnospondyl amphibians from the Middle Pennsylvanian of Illinois. *Palaeontology* 25: 634–664.

———. 1987. The Westphalian tetrapod fauna: Some aspects of its geography and ecology. *Journal of the Geological Society* 144: 495–506.

———. 1988. The relationships and origin of the living amphibians. In M. J. Benton (ed.), *The Phylogeny and Classification of the Tetrapods*, pp. 59–102. Oxford: Clarendon Press.

———. 1990. The radiations of temnospondyl amphibians. In P. D. Taylor and G. P. Larwood (eds.), *Major Evolutionary Radiations*, pp. 321–349. Oxford: Clarendon Press.

———. 1993a. Biogeography of Palaeozoic tetrapods. In J. A. Long (ed.), *Palaeozoic Vertebrate Biostratigraphy and Biogeography*, pp. 324–353. London: Belhaven Press.

———. 1993b. Palaeozoic relatives of lissamphibians. Amphibian relationships. Phylogenetic analysis of morphology and molecules. *Herpetological Monographs* 7: 8–27.

———. 1996. Systematics of the genus *Eryops* (Amphibia: Temnospondyli) and its possible biostratigraphical significance. *Journal of Vertebrate Palaeontology* 16: 53A.

Milner, A. R., and S. E. K. Sequeira. 1994. The temnospondyl amphibians from the Viséan of East Kirkton, West Lothian, Scotland. *Transactions of the Royal Society of Edinburgh, Earth Sciences* 84: 331–362.

Modesto, S. P. 1992. Did herbivory foster early amniote diversification? *Journal of Vertebrate Palaeontology* 12(3): 44A.

———. 1999. Observations on the structure of the Early Permian reptile *Stereosternum tumidum* Cope. *Paleontologica Africana* 35: 7–19.

Moulton, J. M. 1974. A description of the vertebral column of *Eryops* based on the notes and drawings of A. S. Romer. *Breviora* 428: 1–44.

Musick, J. A., M. N. Bruton, and E. K. Balon (eds.). 1991. *The Biology of Latimeria chalumnae and Evolution of Coelacanths*. Dordecht, the Netherlands: Kluwer.

Northcutt, R. G. 1986. Lungfish neural characters and their bearing on sarcopterygian phylogeny. In W. E. Bemis, W. W. Burggren, and N. E. Kemp (eds.), *The Biology and Evolution of Lungfishes,* pp. 277–297. New York: Alan R. Liss.

Novikov, I. V., M. A. Shishkin, and V. K. Golubev. 2000. Permian and Triassic anthracosaurs from Eastern Europe. In M. J. Benton, M. A. Shishkin, D. M. Unwin, and E. N. Kurochkin (eds.), *The Age of Dinosaurs in Russia and Mongolia,* pp. 60–70. Cambridge: Cambridge University Press.

Oelofsen, B. W. 1987. *Mesosaurus tenuidens* and *Stereosternum tumidum* from the Permian of Gondwana of both southern Africa and South America. *South African Journal of Science* 83: 370–371.

Olsen, H. 1993. Sedimentary basin analysis of the continental Devonian basin in North-East Greenland. *Bulletin of the Grønlands Geologiske Undersøgelse* 168: 1–80.

———. 1994. Orbital forcing on continental deposition systems—Lacustrine and fluvial cyclicity in the Devonian of East Greenland. *Special Publications of the International Association of Sedimentology* 19: 429–438.

Olsen, H., and P.-H. Larsen. 1993. Lithostratigraphy of the continental Devonian sediments in North-East Greenland. *Bulletin of the Grønlands Geologiske Undersøgelse* 165: 1–108.

Olson, E. C. 1971. A skeleton of *Lysorophus tricarinatus* (Amphibia: Lepospondyli) from the Hennessey Formation (Permian) of Oklahoma. *Journal of Paleontology* 45: 443–449.

Orton, G. L. 1954. Original adaptive significance of the tetrapod limb. *Science* 120: 1042–1043.

Owen, R. 1841. Description of the *Lepidosiren annectens. Transactions of the Linnean Society of London* 18: 327–361.

Panchen, A. L. 1959. A new armoured amphibian from the Upper Permian of East Africa. *Philosophical Transactions of the Royal Society of London Series B* 242: 207–281.

———. 1966. The axial skeleton of the labyrinthodont *Eogyrinus attheyi. Journal of Zoology* 150: 199–222.

———. 1967. The homologies of the labyrinthodont centrum. *Evolution* 21: 24–33.

———. 1972a. The skull and skeleton of *Eogyrinus attheyi* Watson (Amphibia: Labyrinthodontia). *Philosophical Transactions of the Royal Society of London Series B* 263: 279–326.

———. 1972b. The interrelationships of the earliest tetrapods. In K. A. Joysey and T. S. Kemp (eds.), *Studies in Vertebrate Evolution,* pp. 65–87. Edinburgh: Oliver and Boyd.

———. 1977. On *Anthracosaurus russelli* Huxley (Amphibia: Labyrinthodontia) and the family Anthracosauridae. *Philosophical Transactions of the Royal Society of London Series B* 279: 447–512.

———. 1985. On the amphibian *Crassigyrinus scoticus* Watson from the Carboniferous of Scotland. *Philosophical Transactions of the Royal Society of London Series B* 309: 461–568.

Panchen, A. L., and T. R. Smithson. 1987. Character diagnosis, fossils, and the origin of tetrapods. *Biological Reviews* 62: 341–438.

———. 1988. The relationships of the earliest tetrapods. In M. J. Benton (ed.), *The Phylogeny and Classification of the Tetrapods,* 1: 1–32. Oxford: Clarendon Press.

———. 1990. The pelvic girdle and hind limb of *Crassigyrinus scoticus* (Lydekker) from the Scottish Carboniferous and the origin of the tetrapod pelvic skeleton. *Transactions of the Royal Society of Edinburgh, Earth Sciences* 81: 31–44.

Parrington, F. R. 1956. The patterns of dermal bones in primitive vertebrates. *Proceedings of the Zoological Society of London B* 127: 389–411.

———. 1967a. The identification of dermal bones in the head. *Journal of the Linnean Society (Zoology)* 47: 231–239.

———. 1967b. The vertebrae of early tetrapods. Problèmes Actuels de Paleontologie: Évolution des vertébrés. *Centre National de la Recherchée Scientifique Paris* 163: 269–279.

Parsons, T., and E. Williams. 1963. The relationships of the modern Amphibia: A re-examination. *Quarterly Review of Biology* 38: 26–53.

Paton, R. L., T. R. Smithson, and J. A. Clack. 1999. An amniote-like skeleton from the Early Carboniferous of Scotland. *Nature* 398: 508–513.

Patterson, C. 1993. Naming names. *Nature* 366: 518.

Platt, C. J. 1994. Hair cells in the lagenar otolith organ of the coelacanth are unlike those in amphibians. *Journal of Morphology* 220: 381.

Platt, C. J., and A. N. Popper. 1996. Sensory hair cell arrays in lungfish inner ears suggest retention of the primitive patterns for bony fishes. In *Abstracts*, Vol. 22, p. 1819. Washington, D.C.: Society for Neuroscience.

Popper, A. N., and C. J. Platt. 1993. Inner ear and lateral line. In D. H. Evans (ed.), *The Physiology of Fishes*, pp. 99–135. Boca Raton, Fla.: CRC Press.

Presley, R. 1984. Lizards, mammals and the primitive tetrapod tympanic membrane. *Symposium of the Zoological Society of London* 52: 127–152.

Pridmore, P. A. 1995. Submerged walking in the epaulette shark *Hemiscyllium ocellatum* (Hemiscyllidae) and its implications for locomotion in rhipidistian fishes and early tetrapods. *ZACS Zoology* 98: 278–297.

Rackoff, J. 1980. The origin of the tetrapod limb and the ancestry of tetrapods. In A. L. Panchen (ed.), *The Terrestrial Environment and the Origin of Land Vertebrates*, pp. 255–292. London: Academic.

Rage, J.-C., and Z. Rocek. 1989. Redescription of *Triadobatrachus massinoti* (Piveteau 1936), an anuran amphibian from the Early Jurassic. *Palaeontographica* 206: 1–16.

Randall, D. J., W. W. Burggren, A. P. Farrell, and M. S. Haswell. 1981. *The Evolution of Air-Breathing in Fishes*. Cambridge: Cambridge University Press.

Raup, D. M, and S. M. Stanley. 1978. *Principles of Paleontology*. 2nd edition. New York: W. H. Freeman.

Reisz, R. R. 1972. Pelycosaurian reptiles from the Middle Pennsylvanian of North America. *Bulletin of the Museum of Comparative Zoology, Harvard* 144: 27–62.

———. 1977. *Petrolacosaurus*, the oldest known diapsid reptile. *Science* 196: 1091–1093.

———. 1986. *Pelycosauria*, Vol. 17A, *Handbuch der Paläoherpetologie*. Stuttgart: Fischer.

Reisz, R. R., and M. Laurin. 1991. *Owenetta* and the origin of turtles. *Nature* 349: 324–326.

Reisz, R. R., D. S. Berman, and D. Scott. 1992. The cranial anatomy and relationships of *Secondontosaurus*, an unusual mammal-like reptile

(Synapsida: Sphenacodontidae) from the early Permian of Texas. *Zoological Journal of the Linnean Society* 104: 127–184.

Retzius, G. 1881. *Das Gehörorgan der Wirbelthiere. Morphologisch-Histologische Studien. I. Das Gehörorgan der Fische und Amphibien.* Stockholm: Centraldruckerei.

Rieppel, O., and M. DeBraga. 1996. Turtles as diapsid reptiles. *Nature* 384: 453–455.

Rinehart, L. F., and S. G. Lucas. In press. A statistical analysis of a growth series of the Permian lepospondyl *Diplocaulus magnicornis* showing two-stage ontogeny. *Journal of Vertebrate Paleontology.*

Rogers, D. A. 1990. Probable tetrapod tracks rediscovered in the Devonian of Scotland. *Journal of the Geological Society of London* 147: 746–748.

Rolfe, W. D. I. 1980. Early invertebrate terrestrial faunas. In A. L. Panchen (ed.), *The Terrestrial Environment and the Origin of Land Vertebrates,* pp. 117–157. London: Academic.

Rolfe, W. D. I., E. N. K. Clarkson, and A. L. Panchen, eds. 1994. Volcanism and early terrestrial biotas. *Transactions of the Royal Society of Edinburgh, Earth Sciences* 84: 1–464.

Rolfe, W. D. I., G. P. Durant, A. E. Fallick, A. J. Hall, D. J. Large, A. C. Scott, T. R. Smithson, and G. M. Walkden. 1990. An early terrestrial biota preserved by Viséan vulcanicity in Scotland. *Special Papers of the Geological Society of America* 244: 177–188.

Romer, A. S. 1933. *Man and the Vertebrates.* Chicago: Chicago University Press.

———. 1937. The braincase of the Carboniferous crossopterygian *Megalichthys nitidus. Bulletin of the Museum of Comparative Anatomy, Harvard* 82: 1–73.

———. 1945. *Vertebrate Paleontology.* 2nd edition. Chicago: Chicago University Press.

———. 1947. Review of the Labyrinthodontia. *Bulletin of the Museum of Comparative Zoology, Harvard* 99: 3–366.

———. 1958. Tetrapod limbs and early tetrapod life. *Evolution* 12: 365–369.

———. 1962. *The Vertebrate Body.* Philadelphia: W. B. Saunders.

———. 1966. *Vertebrate Paleontology.* Chicago: University of Chicago Press.

———. 1972a. Skin breathing—Primary or secondary? *Respiration Physiology* 14: 138–192.

———. 1972b. A Carboniferous labyrinthodont amphibian with complete dermal armour. *Kirtlandia, Cleveland, Museum of Natural History* 16: 1–8.

Romer, A. S., and T. S. Parsons. 1986. *The Vertebrate Body.* Philadelphia: W. B. Saunders.

Romer, A. S., and L. I. Price. 1940. Review of the Pelycosauria. *Special Papers of the Geological Society of America* 28: 1–538.

Rosen, D. E., P. L. Forey, B. G. Gardiner, and C. Patterson. 1981. Lungfishes, tetrapods, paleontology, and plesiomorphy. *Bulletin of the American Museum of Natural History* 167: 159–276.

Rowe, N. P., and T. P. Jones. 2001. Devonian charcoal. *Palaeogeography, Palaeoclimatology, and Palaeoecology* 164: 355–371.

Ruta, M., A. R. Milner, and M. I. Coates. 2002 (in press). The tetrapod *Caerorhachis bairdi* Holmes and Carroll from the Lower Carboniferous of Scotland. *Transactions of the Royal Society of Edinburgh, Earth Sciences.*

Sargeant, W. A. S. 1988. Fossil vertebrate footprints [front cover]. *Geology Today* 4: 125–130.

Säve-Söderbergh, G. 1932. Preliminary note on Devonian stegocephalians from East Greenland. *Meddelelser om Grønland* 98(3): 1–211.

Sawin, H. J. 1941. The cranial anatomy of *Eryops megacephalus*. *Bulletin of the Museum of Comparative Zoology, Harvard* 88: 407–463.

Sayer, M. D. J., and J. Davenport. 1991. Amphibious fishes: Why do they leave the water? *Reviews in Fish Biology and Fisheries* 1: 159–181.

Schmitz, B., G. Aberg, L. Werdelin, P. Forey, and S.-E. Bendix-Almgreen. 1991. $^{87}Sr/^{86}Sr$, Na, F, Sr and La in skeletal fish debris as a measure of the palaeosalinity of fossil-fish habitats. *Bulletin of the Geological Society of America* 103: 786–794.

Schultze, H.-P. 1984. Juvenile specimens of *Eusthenopteron foordi* Whiteaves, 1881 (Osteolepiformes Rhipidistia, Pisces), from the Late Devonian of Miguasha, Quebec, Canada. *Journal of Vertebrate Palaeontology* 4: 1–16.

———. 1990. A new acanthodian from the Pennsylvanian of Utah, USA, and the distribution of otoliths in gnathostomes. *Journal of Vertebrate Paleontology* 10: 49–58.

———. 1995. Terrestrial biota in coastal marine deposits: Fossil-Lagerstätten in the Pennsylvanian of Kansas, USA. *Palaeogeography, Palaeoclimatology, and Palaeoecology* 119: 255–273.

Schultze, H.-P., and M. Arsenault. 1985. The panderichthyid fish *Elpistostege*: A close relative of tetrapods? *Palaeontology* 28: 293–309.

Schultze, H.-P., and R. Cloutier, eds. 1996. *Devonian Fishes and Plants of Miguasha, Quebec, Canada.* Munich: Friedrich Pfeil.

Schultze, H.-P., and C. G. Maples. 1992. Comparison of the Late Pennsylvanian faunal assemblage of Kinney Brick Company Quarry, New Mexico, with other Late Pennsylvanian Lagerstätten. *Bulletin of the New Mexico Bureau of Mines and Mineral Resources* 138: 231–242.

Schultze, H.-P., C. G. Maples, and C. R. Cunningham. 1994. The Hamilton Konservat-Lagerstätte: Stephanian terrestrial biotas in a marginal-marine setting. *Transactions of the Royal Society of Edinburgh, Earth Sciences* 84: 443–451.

Scott, A. C., R. Brown, J. Gaultier, and B. Meyer-Berthaud. 1994. Fossil plants from the Viséan of East Kirkton, West Lothian, Scotland. *Transactions of the Royal Society of Edinburgh, Earth Sciences* 84: 249–260.

Shabica, C. W., and A. A. Hay. 1997. *Richardson's Guide to the Fossil Fauna of Mazon Creek.* Chicago: Northeastern Illinois University.

Sharman, A. C., and P. W. H. Holland. 1998. Estimation of *Hox* gene cluster number in lampreys. *International Journal of Developmental Biology* 42: 617–620.

Shear, W. A. 1994. Myriapodous arthropods from the Viséan of East Kirkton, West Lothian, Scotland. *Transactions of the Royal Society of Edinburgh, Earth Sciences* 84: 309–316.

Shear, W. A., P. M. Bonamo, J. D. Grierson, W. D. I. Rolfe, E. I. Smith, and R. Norton. 1984. Early land animals in North America: Evidence from Devonian age arthropods from Gilboa, New York. *Science* 224: 492–494.

Shubin, N. H., and P. Alberch. 1986. A morphogenetic approach to the origin and basic organization of the tetrapod limb. *Evolutionary Biology* 20: 319–387.

Shubin, N., C. Tabin, and S. Carroll. 1997. Fossils, genes and the evolution of animal limbs. *Nature* 388: 639–648.

Skulan, J. 2000. Has the importance of the amniote egg been overstated? *Zoological Journal of the Linnean Society* 130: 235–261.

Smith, J. L. B. 1956. *Old Four-Legs, the Story of the Coelacanth.* London: Longman, Green.

Smithson, T. R. 1982. The cranial morphology of *Greererpeton burkemorani* (Amphibia: Temnospondyli). *Zoological Journal of the Linnean Society of London* 76: 29–90.

———. 1985a. Scottish Carboniferous amphibian localities. *Scottish Journal of Geology* 21: 123–142.

———. 1985b. The morphology and relationships of the Carboniferous amphibian *Eoherpeton watsoni* Panchen. *Zoological Journal of the Linnean Society* 85: 317–410.

———. 1986. A new anthracosaur from the Carboniferous of Scotland. *Palaeontology* 29: 603–628.

———. 1994. *Eldeceeon rolfei*, a new reptiliomorph from the Viséan of East Kirkton, West Lothian, Scotland. *Transactions of the Royal Society of Edinburgh, Earth Sciences* 84: 377–382.

———. 2000. Anthracosaurs. In H. Heatwole and R. L. Carroll (eds.), *Amphibian Biology*, Vol. 4, *Palaeontology*, pp. 1053–1063. Chipping Norton, New South Wales, Australia: Surrey Beatty.

Smithson, T. R., R. L. Carroll, A. L. Panchen, and S. M. Andrews, 1994. *Westlothiana lizziae* from the Viséan of East Kirkton, West Lothian, Scotland. *Transactions of the Royal Society of Edinburgh, Earth Sciences* 84: 417–431.

Steiner, H. 1934. Über die embryonale Hand- und Fuss-Skelett-anlage bei den Crocodiliern, sowie über ihre Beziehungen zur Vogel-Flügelanlage und zur ursprünglichan Tetrapoden-Extremität. *Revue Suisse de Zoologie* 41: 383–396.

Stock, D. W., K. D. Moberg, L. R. Maxson, and G. S. Whitt. 1991. A phylogenetic analysis of the 18S ribosomal RNA sequence of the coelacanth *Latimeria chalumnae*. *Environmental Biology of Fishes* 32: 99–117.

Stössel, I. 1995. The discovery of a new Devonian tetrapod trackway in SW Ireland. *Journal of the Geological Society of London* 152: 407–413.

Sues, H.-D., and R. R. Reisz. 1998. Origins and early evolution of herbivory in tetrapods. *Trends in Ecology and Evolution* 13: 141–145.

Sumida, S. S. 1997. Locomotor features of taxa spanning the origin of amniotes. In S. Sumida and K. L. M. Martin (eds.), *Amniote Origins: Completing the Transition to Land*, pp. 353–398. San Diego: Academic Press.

Sumida, S. S., and R. E. Lombard. 1991. The atlas–axis complex in the late Paleozoic genus *Diadectes* and the characteristics of the atlas–axis complex across the amphibian to amniote transition. *Journal of Paleontology* 65: 973–983.

Sumida, S. S., R. E. Lombard, and D. S. Berman. 1992. Morphology of the atlas–axis complex of the late Palaeozoic tetrapod suborders Diadectomorpha and Seymouriamorpha. *Philosophical Transactions of the Royal Society of London Series B* 336(1277): 259–273.

Szarski, H. 1962. The origin of the Amphibia. *Quarterly Review of Biology* 38: 189–241.

Tabin, C. J. 1992. Why we have (only) five fingers per hand: *Hox* genes and the evolution of paired limbs. *Development* 116: 289–296.

Taylor, D. H., and K. Adler. 1978. The pineal body—Site of extraocular perception of celestial clues for orientation in the tiger salamander. *Journal of Comparative Physiology* 124: 357–361.

Thomson, K. S. 1966. The evolution of the tetrapod middle ear in the rhipidistian–amphibian transition. *American Zoologist* 6: 379–397.

———. 1967. Mechanisms of intracranial kinetics in fossil rhipidistians and their relatives. *Zoological Journal of the Linnean Society* 46: 223–253.

———. 1969. The biology of the lobe-finned fishes. *Biological Reviews* 44: 91–154.

———. 1980. The ecology of Devonian lobe-finned fishes. In A. L. Panchen (ed.), *The Terrestrial Environment and the Origin of Land Vertebrates*, pp. 187–222. London: Academic.

———. 1988. *Morphogenesis and Evolution*. Oxford: Oxford University Press.

———. 1991. Where did tetrapods come from? *American Scientist* 79: 488–490.

———. 1993. The origin of the tetrapods. *American Journal of Science* 293A: 33–62.

Thomson, K. S., and K. H. Bossy. 1970. Adaptive trends and relationships in early amphibia. *Forma et Functio* 3: 7–31.

Thomson, K. S., N. S. Shubin, and F. G. Poole. 1998. A problematic early tetrapod from the Mississippian of Nevada. *Journal of Vertebrate Palaeontology* 18: 315–320.

Thorogood, P. V. 1991. The development of the teleost fin and implications for our understanding of tetrapod limb evolution. In J. R. Hinchliffe, J. M. Hurle, and D. Summerbell (eds.), *Developmental Patterning of the Vertebrate Limb*. NATO ASI Series A, Life Sciences, 205: 347–354. New York: Plenum.

Thulborn, T., A. Warren, S. Turner, and T. Hamley. 1996. Early Carboniferous tetrapods in Australia. *Nature* 381: 777–780.

Trueb, L., and R. Cloutier. 1991. A phylogenetic investigation of the inter- and intrarelationships of the Lissamphibia (Amphibia: Temnospondyli). In H.-P. Schultze and L. Trueb (eds.), *Origins of the Higher Groups of Tetrapods*, pp. 223–314. Ithaca, N.Y.: Cornell University Press.

Ultsch, G. R. 1987. The potential role of hypercarbia in the transition from water-breathing to air-breathing in vertebrates. *Evolution* 41: 442–445.

Van Hoepen, E. C. N. 1915. Stegocephalia of Senekal, Orange Free State. *Annals of the Transvaal Museum* 5: 124–149.

Vorobyeva, E. I. 2000. Morphology of the humerus in the rhipidistian Crossopterygii and the origin of tetrapods. *Paleontogical Journal* 34: 632–641.

Vorobyeva, E. I., and H.-P. Schultze. 1991. Description and Systematics of panderichthyid fishes with comments on their relationship to tetrapods. In H.-P. Schultze and L. Trueb (eds.), *Origins of the Higher Groups of Tetrapods*, pp. 68–109. Ithaca, N.Y.: Cornell University Press.

Walls, G. L. 1942. *The Vertebrate Eye*. Bloomfield Hills, Mich.: Cranbrook Institute of Science.

Walsh, P. J. 1997. Evolution and regulation of urea synthesis and urotely in (batrachoidid) fishes. *Annual Review of Physiology* 59: 299–323.

Walter, H. v., and R. Wernerberg. 1988. Über Liegespuren (Cubichnia) aquatischer Tetrapoden (?Diplocauliden, Nectridea) aus den Rotteröder Schichten (Rotliegendes, Thüringer Wald/DDR). *Freiberger Forschungsheft* 419: 96–106.

Warburton, F. E., and N. S. Denman. 1961. Larval competition and the origin of tetrapods. *Evolution* 15: 566.

Warren, A. A., and N. Schroeder. 1995. Changes in the capitosaur skull with growth: An extension of the growth series of *Parotosuchus aliciae* (Amphibia, Temnospondyli) with comments on the otic area of capitosaurs. *Alcheringa* 19: 41–46.

Warren, A. A., T. Rich, and P. Vickers-Rich. 1997. The last labyrinthodonts? *Palaeontographica A* 247: 1–24.

Warren, J. W., and N. A. Wakefield. 1972. Trackways of tetrapod vertebrates from the Upper Devonian of Victoria, Australia. *Nature* 238: 469–470.

Watson, D. M. S. 1913. On the primitive tetrapod limb. *Anatomischen Anzeiger* 44: 24–27.

———. 1919. On *Seymouria,* the most primitive known reptile. *Proceedings of the Zoological Society of London B* 212: 267–301.

———. 1926. Croonian lecture—The evolution and origin of the Amphibia. *Philosophical Transactions of the Royal Society of London Series B* 214: 189–257.

———. 1927. The reproduction of the coelacanth fish *Undina. Proceedings of the Zoological Society of London* 1927: 453–457.

———. 1929. The Carboniferous Amphibia of Scotland. *Palaontologica Hungarica* 1: 219–252.

———. 1940. The origin of frogs. *Transactions of the Royal Society of Edinburgh* 60: 195–231.

Webb, J. F. 1989. Developmental constraints and evolution of the lateral line system in teleosts. In S. Coombs, P. Görner, and H. Münz (eds.), *The Mechanosensory Lateral Line: Neurobiology and Evolution,* pp. 79–98. New York: Springer-Verlag.

Weinberg, S. 1999. *A Fish Caught in Time.* London: Fourth Estate.

Westoll, T. S. 1938. Ancestry of the tetrapods. *Nature* 141: 127–128.

———. 1943a. The origin of tetrapods. *Biological Reviews* 18: 78–98.

———. 1943b. The origin of the primitive tetrapod limb. *Proceedings of the Royal Society of London B* 131: 373–393.

———. 1961. A crucial stage in vertebrate evolution: Fish to land animal. *Proceedings of the Royal Institution* 38: 600–617.

Wever, E. G. 1978. *The Reptile Ear.* Princeton, N.J.: Princeton University Press.

White, T. E. 1939. Osteology of *Seymouria baylorensis* Broili. *Bulletin of the Museum of Comparative Zoology, Harvard* 85: 325–409.

Wolpert, L., R. Beddington, J. Brockes, et al. 1998. *Principles of Development.* Oxford: Oxford University Press.

Wood, S. P., A. L. Panchen, and T. R. Smithson. 1985. A terrestrial fauna from the Scottish Lower Carboniferous. *Nature* 314: 355–356.

Worobyeva, E. I. 1975. Bemerkungen zu *Panderichthys rhombolepis* (Gross) aus Lode in Lettland (Gauja Schichten, Oberdevon). *Neues Jahrbuch Geolische Palaeontol. Monatshefte* 1975: 315–320.

Wright, P. A. 1995. Nitrogen excretion: Three end products, many physiological roles. *Journal of Experimental Biology* 198: 273–281.

Zardoya, R., and A. Meyer. 1996. Evolutionary relationships of the coelacanth. *Proceedings of the National Academy of Sciences USA* 93: 5449–5454.

Zhu, M., and H.-P. Schultze. 1997. The oldest sarcopterygian fish. *Lethaia* 30: 293–304.

———. 2001. Interrelationships of basal osteichthyans. In P. E. Ahlberg (ed.), *Major Events in Early Vertebrate Evolution,* pp. 289–314. London: Systematics Association Symposium.

Zhu, M., X. Yu, and P. Janvier. 1999. A primitive fossil fish sheds light on the origin of bony fishes. *Nature* 397: 607–610.

Zimmer, C. 1998. *At the Water's Edge.* New York: Free Press.

Index

Page numbers in italic type refer to illustrations.

acanthodian, *88*
Acanthostega, 119–129, 282; air gulping, 135–136, 165; ankles, *127, 137,* 330; anterior tectal, 280; aquatic adaptations, 128–129; atlas-axis, 309, *310,* 311; dating of, *109;* gill skeleton, *123, 124;* lateral line system, 121, *166;* limbs, 124–128, 135, 136, *183, 329;* lower jaw, *7, 8, 284;* nostrils, 121, 142–143, 165; ornament, 122, *149;* palate, 121; pelvic girdle, *27,* 126–128, 135, 136, *321;* primitive or derived tetrapod, 134–137, 272, *274;* reconstruction of, *120, 122, 128;* ribs, 126, 306; saltwater tolerance, 97; sediments where found, 109–110; shoulder girdle, *27, 123, 124, 146, 157, 319;* skull, *7, 36, 120, 121, 122, 123, 287;* stapes, *293, 295;* tabular horn, *120, 148, 253;* tail, 128, 136; teeth, *7, 10, 122, 123,* 124; vertebral column and vertebrae, 126, 306, 312, *313;* wrists, *125, 137, 329, 330.* See also *Eusthenopteron-Panderichthys-Acanthostega;* tetrapods
accessory olfactory bulb, 165
accommodation, visual, 162–171, *164*
acetabulum, 29, 42, 322
Acherontiscus caledoniae, 204
actinopterygians, *16, 88.* See also ray-finned fishes
adaptive radiation, 76
adductor blade, *327; Acanthostega,* 126, *127, 128; Tulerpeton,* 130

adelogyrinids, 200, *201;* skull, 201
Adelogyrinus, 201
Adelospondylus watsoni, 201
Africa, 240, *241*
Age of Fishes, 86
Ahlberg, Per, *90, 110*
Aina Dal Formation, Greenland, 107, *108,* 109
air bladders, 101, 145
air-breathing mechanisms: air gulping, 58, 104, 145–146, 316; book lung, 218; gill breathing, 36, *37,* 174–176. See also cutaneous gas exchange; ventilation
air gulping, 145–146, 316; *Acanthostega,* 135–136, 165; advantage of longer snout, 142; in anoxic water, 104; early tetrapods, 285, 316; lungfish, 58
aïstopods, 199, *200, 260, 262, 268*
Allegheny orogeny, 193
Allenypterus montanus, 52
Amazon Basin, 101, 137
amber, 9
Amia, 175
AMNH 6841, *250*
AMNH 7117, *254*
amniotes, 66, 263–273; atlas-axis vertebrae, 265; brachial plexus, 318; definition, 264; groups, 2; inner ear, *168, 169, 170,* 294, *295;* muscle spindles, 171; no lateral line system, 167; notchlessness, 283; palatal teeth, 286; pterygoid flange, 265, 267, 286, *288;* relatedness to lepospondyls, 258; ribs, 317; sacral region, 324; separation from amphibian

353

lineage, 232; skin, 173; skull, 264, 265, 266, 288; supergroups, 264; tabular-parietal contact, 255; ventilation, 317; vertebrae, 311; vision, 162–163, *164*
Amphibamus, 250; larval specimens, *252, 275;* skull, *275*
Amphibamus grandiceps, 250
amphibians (fossil), 67, 224, 246, 269. See also tetrapods
amphibians (modern), 66; brachial plexus, 318, 319; fossil record of, 274–275; inner ear, 169, *170;* major groups, 1, 273; muscle spindles, 171; nitrogen excretion, 178; occipital condyle of, 249; relationship to dissorophids (temnospondyls), 249, 275–277, *276, 277;* relationship to lepospondyls, 258, 297; saltwater tolerance, 97; selective pressures on, 100–101; separation from amniotes, 232; ventilation, 316–317; visual accommodation, 163
amphibious lifestyle of reproduction, 179–180
Amphioxus, 32
amplexus, 136
analogous characters, 13
anaspids, 87
anatomical terms (list), 28
Andreyevka, Russia, 129
Andrias, 204
ankles, 44, 330; *Acanthostega,* 127, 137, 330; amniote, 265; type of maneuverability, 330
anocleithrum, 159; *Acanthostega,* 123, 124, 159; *Tulerpeton,* 130, 159
anterior tectal, 280
anthracosaurs: cladistic analysis of, 274; ear region, 292; interclavicle, 225; kinetic line, 227; Late Carboniferous, 251, 252–255; tabular-parietal bone feature, 226–227, 251–252. See also *Eldeceeon rolfei; Eucritta melanolimnetes; Silvanerpeton miripedes*
Anthracosaurus russelli, 254
antiarchs, 87, *88,* 89
Antler Highlands, Nevada, 204
Antlerpeton clarkii, 204–205
anurans. See frogs
Apateon, 251, 252, 275
Apateon pedestris, 251
Aphaneramma, 247
apical ectodermal ridge (AER), 184
apodans. See caecilians
apoptosis, 187
appendicular skeleton, 28, 29

aquatic environments: anoxic conditions, 85–86, 101; conditions favoring terrestriality, 103–104; cyclothems, 194; euryhaline character, 50, 98; fossils in, 6; marine to freshwater adaptations, 97–99; marine to freshwater transitions, 194, 239; temperature influence on oxygen uptake, 175–176
Arapaima, 137
Archaeocalamites, 88
Archaeopteris, 84, *85,* 86, *88,* 195; growth rings of, 85, 193
Archaeopteryx, 180
Archaeothyris, 265; vertebrae, *306*
archegosaurs, 247
Archegosaurus, 97
Archeria, 253; pelvic girdle, *321;* sacral rib, 324; vertebrae, *306*
Archeria crassidisca, 254
arid climates: estivation adaptation, 100–101, 260; fossilization in, 9; red-beds indicative of, 99
arthodires, 87, *88,* 89
Arthropleura, 235
arthropleurid myriapod, *83*
arthropods: early behavior, 84; during Silurian, 82; true herbivores, 195
articular, 288
Asia, 65, 76, 193, 240
Asteroxylon mackei, 84
astragalus, 265
atlas-axis vertebrae, 309–311, *310;* amniotes, 265; cladistic analysis, *310;* microsaur, 259, 259–260
atmospheric O_2/CO_2: Carboniferous period, 193–194, 239; Devonian period, 78, *79;* effects on metabolism, 173–174; through time, 79
Australia, 211; Canowindra, 61; Forbes, New South Wales, *134;* Genoa River, 92, 93, *190;* Grampian Mts., Victoria, 95; Queensland, 207; related tetrapodomorphs in, 97
Australian lungfish, breeding program, 171. See also *Neoceratodus*
autostylic skulls, 40, 288; example, 58
axial skeleton, 26, 28, 29

Baird, Bill, 90
Balanerpeton, 222–225, 233; ear region, 224, 297; juvenile specimen, 223
Balanerpeton woodi, 222
Baphetes, 242–243; skull of, 71
baphetids, 241–245, 269; *Baphetes, 71,* 242–243; cladistic interrela-

tionships, 272–273, 274; eye socket of, 242, 244–245; *Loxomma*, 242; *Megalocephalus*, 242; septomaxilla, 280; *Spathicephalus*, 243–245
basal articulation, 270, 271, 280, 294, 297; *Captorhinus*, 281; embolomeres, 280, 281; *Eusthenopteron-Panderichthys-Acanthostega*, 151, 152; temnospondyls, 224
basilar papilla, 169
basioccipital, 22, 23, 151, 265, 280, 304
basipteryoid process, 23, 302
basisphenoid and basioccipital region, 302
batfish, 103
Bathgate Hills Volcanic Formation, Scotland, 214. *See also* East Kirkton locality, Scotland
batrachomorphs, 270, 271
Baylor, Texas, 255
Bendix-Almgreen, Svend, 110
bifurcation, 182
binomial names, 11
birds, 14; accommodation in, 163; condylar process, 302–303; vestibular system, 168
blastopore, 29, 30, 31
body fossils, 6, 89, 91
body scales, 29, 44, 45; advantages to loss, 172; early tetrapods, 173; fish-tetrapod transition, 44, 45; *Osteolepis*, 61
body wall muscles, lungs ventilation, 37, 283, 317
bone, dermal. *See* dermal bone
bone, endochondral. *See* endochondral bone
bones: anatomical terms, 28; fossilization of, 6–7
bony fin rays. *See* fin rays
bony vertebrates, 16
book lung, 218
boron analysis, 98
Borough Lee, Scotland, 196
Bothriolepis, 87–89, 88, 95
bottom walking, 103, 136
brachial plexus, 317–319, 318
Brachydectes elongatus, 261
Brachyops, 247
braincase, 21–24, 22, 24; embryonic development, 300–303, 301, 302; *Eusthenopteron*, 21–23, 22, 300, 301; *Eusthenopteron-Panderichthys-Acanthostega*, 151–155; *Greererpeton*, 300, 301; *Ichthyostega*, 113; neural crest cells, 301, 302
branchial arches, 25, 40, 179, 225

branchiosaurs, 251; larval, 252, 275
Britta Dal Formation, Greenland, 107, 108, 109
buccal pumping: mechanism, 36, 37; mixed-air, 176, 177
buoyancy, 49
Burgess Shale, 180
butt joints, 148

Cabonnichthys, 62, 63
Cacops, 247
caecilians, 45, 273
Caerorhachis bairdi, 245–246
Calamites, 195
Calamites carinatus, 235
Calamoichthys, 173
Calamophyton, 85
Callistophyton sp., 235
Canada: Escuminac Bay, 87; Florence, Nova Scotia, 239, 240; Joggins, Nova Scotia, 229, 235–236, 237, 240; Prince of Wales Island, 65
Canowindra, 61, 63
Canowindra, Australia, 61
captorhinids, 264
Captorhinus, 265, 267, 268; basal articulation, 281; forelimbs, 329; hindlimbs, 329; skull, 267, 281; vertebrae, 306
carbon dioxide metabolism, 174
carbon fixation as coal, 194
Carboniferous period, 5; fossils from, 10; paleoenvironment of, 193–194; phases of, 191, 192. *See also* Early Carboniferous; Late Carboniferous
cartilaginous fishes. *See* chondrichthyans
Casineria kiddi, 197–199, 198, 265; cladistic interrelationships, 273, 274; wrist, 330
catfish, 103
Celsius Berg Group, Greenland, 107, 108
Cenozoic era, 4–5
centra, 26, 42, 305–309, 306; *Antlerpeton clarkii*, 205; tetrapod character, 269
cephalic mesoderm, 301
Ceratodus, 57
ceratohyal, 25
cervical ribs, 313–314, 317–319
characters (features), 12
charophytes, 129
cheek region, 21; amniotes, 265; *Eusthenopteron-Panderichthys-Acanthostega*, 143
Cheese Bay, Edinburgh, Scotland, 197
Cheirolepis, 88, 89
China, 65, 76

choana, 24, 143
chondrichthyans, *15*, 17
chorda tympani, *295*
ciliary muscle, 163
circulatory system, 174, *175*, 319
Clack, Jennifer A., *107, 108, 110*
cladistic analysis, 72–74, 134–135; atlas-axis construction, *310*; crown group Tetrapoda, *65*; ear character, *296, 298, 299*; *Eucritta* (baphetid), 271–273, *274*; femora, *326*; humeri, *325*; lobe-finned fishes, *47*; lower jaws, 284; lungfishes-coelacanths-tetrapods, *15*, 72, *73*, *74*; occipital construction, 270, 271, *304*; osteolepiforms, *61*; pelvic girdles, *321*; reestablishing lepospondyls, 271, 273; reptiliomorph and batrachomorph lineage, 270, 271, 272; rib pattern, *315*; sacral construction, *323*; temnospondyls and lissamphibians, *276, 277*; temporal notch, *282*, 283; tetrapodamorpha, *67*; ventilation mechanisms, *177*; vertebrates, *15, 16*
cladistics, 11–15; crown groups, *65*; paraphyletic groups, 73
classification: of fishes (Huxley), *70*; Linnean hierarchy, 11, *12*; phylogenetic, 11–15
clavicle, 27, 41, 319, *320*; *Eusthenopteron-Panderichthys-Acanthostega*, 158
cleithrum, 27, 41, 124, *146*, 319; *Hynerpeton*, 133; reduction in, 158; *Tulerpeton*, 131
climate: Devonian, 78–82, 84, 106–107; Early Permian, 240–241; East Kirkton of the Viséan, 216; modern East Greenland, 105–106; Tournaisian, 210. *See also* arid climates
clinging, weed, 103, 136
club mosses, 83, 84, 195, 217, *235*
CM 34638, 157
CM 44777, *263*
coal deposits: first, 86; formation, 194
Coal Measures of Wigan, Lancashire, England, 196, 199
cochlea, *168*
cockroaches, 238
coelacanths, 47, 49–50, 77; fin usage and structure, 102; relationship to tetrapods, 58, 72–77; reproductive strategies, 179, 180. *See also Latimeria*
Coelostegus: pelvic girdle, *321*
colosteids, *203*, 204, 269, 281; gill rakers, 283; possibly *Acherontiscus caledoniae*, 204

Colosteus, 203
common ancestor, 13, 33, 68, 77
computer analysis, 14; PAUP, 232
condylar process, 302–303
cones and rods, 163
Congo Republic, Virunga National Park, 216
conservatism in Famennian tetrapods, 137
Cooksonia, 82
Cooksonia caledonica, 84
Cope, E. D., 69
cordaites, 236–237
coronoids, 24, 28, 285, 286, 287
cosmine, 61
costal ventilation, 283
Cowdenbeath, Scotland, 196, 205
cranial hinge, 39, 285, 300; coelacanths, 50; *Eusthenopteron-Panderichthys-Acanthostega*, 143–144; *Osteolepis*, 61
cranial rib, lungfish, *57, 58*
Crassigyrinus, 202, 205–207, *206*, *207*, 269, 284, 322; cladistic interrelationships, 271, 272, *274*; entepicondylar foramen, 328; palatal dentition, 286; probable septomaxilla, 280–281; skin-breathing mechanism, 172; vertebrae of, 207, 307–308
Crassigyrinus scoticus, 205–207, *206*, *207*
creationism, 69
Crinodon limnophyes, 260
crossopterygians, 69, 70, 72; redefined, 72–73
crown groups, 66, *67*, 68
Cryptobranchus, 173
Ctenerpeton, 263
Ctenodus, 57
cutaneous gas exchange, 172–173
cyclothems, 194
Cynognathus, 289
Czech Republic, 239, *240*

Danio, 186
Darwin, Charles, 11, 69
Dawson, William, 242
day length, sensing. *See* pineal gland
deep time, 4
deltopectoral crest, 327
Dendrerpeton, 246; pelvic girdle, *321*; shoulder girdle, 158
Densignathus, 132–133
dentary, 24, 54, 288; teeth on, 285
derived characters, 12, 13
dermal bone: contrasted with endochondral, 17; from neural crest cells, 33; reduction of, 161
dermal rays, 45
dermal scales. *See* gastralia

dermal skull roof, 21
descent with modification, 11, 68
desiccation, 172; and development rate, 210
development rate and desiccation, 210
Devonian-Carboniferous boundary, 196
Devonian period, 5; biogeography and climate, 78–82; fossils from, 10; phylogeny of lobe-finned fishes, 48. *See also* Early Devonian period; Late Devonian period
Diabolepis, 65
Diadectes, 268
diapsids, 264; advanced skull, *289*; cladogram of ear character, 299; earliest known, 265; skull, *264*, *289*
digits, 44; *Acanthostega*, *125*, *126*, *127*, *135*, 186; canonical sets, 181; endochondral origin, 160; hypothesis of similarity, 187; *Ichthyostega*, *115*, *117*; limb bud development, 182; needs for, 103, 136; role of *Hox* genes, 185–187; stabilization at five, 188–189; Tournaisian tetrapod, 209; *Tulerpeton*, 129, 189; *Whatcheeria*, 202
Dimetrodon, 266, *267*
dinosaurs, 6; phylogeny of, *14*; skin preservation, 9
Diplacanthus, 88
Diplocaulus, 262
Diploceratosaurus, 262
dipnoans. *See* lungfish
dipnomorphs, 65
Dipterus, 55, *59*
discosauriscids, 159, 257, 324
Discosauriscus, 159, 257, 324
dissorophids, 247, 249, 275–277, *276*, *277*
Dissorophus, 247
diversity: increased during Early Carboniferous, 196, 209, 211, 331; increased with herbivory, 290
DNA analysis, 74
Doleserpeton, 308
dolphin forelimb, *118*, 119
dragonflies, 238
Drosophila, *32*, 33, 184
"drying pool" theory, 99–101
Dvinosaurus, 281, 282
dwarfing, progenetic, 211

ear stones, 168
Early Carboniferous, 191–196, *192*; diversity explosion among tetrapods, 209–211, 331. *See also* Tournaisian stage; Viséan stage

Early Devonian period: climate, 84; flora and fauna, 82–86
Early Permian: climate, 240–241; Gondwanan amniote radiation, 268; tetrapod stability, 240
ears. *See* hearing; stapes; stapes evolution
earth: geological columns, 4–6, *5*; stratigraphy of Carboniferous, *192*
East Greenland: Britta Dal Formation, 107, *108*, 109; Celsius Berg Group, 107, *108*; Devonian climate of, 106–107; modern climate and terrain, 105–106; trackways, 95; water salinity, 98; Wimans Bjerg Formation, 107
East Kirkton Limestone, 212
East Kirkton locality, Scotland: amniote and amphibian lineage already separate, 232; invertebrates of, 217–221, *218*, *220*, *221*; landscape, *233*; no insects, 221; plants of, 216, *217*; sediments, 212–215, *213*, *215*, *216*; Unit 82, 222; vertebrates of, 221–233
East Kirkton Quarry, *213*
ectoderm, 31
ectopterygoid, 267, 286
Ectosteorhachis, 61, 168
Edaphosaurus, 266, *267*
eels, 102, 103
eggs: amniote development, 263–264, 331; lack of fossil record, 264; size and O_2 conservation, 173
elbows: beginning in *Ectosteorhachis*, 329; *Casineria*, 199; type of maneuverability, 330; *Ichthyostega*, 115, 329–330
Eldeceeon rolfei, 227, *228*, 233
electrosensory organs, 245
elephant seal, *118*, 119
Elginerpeton, 91; contemporary trackways, 92, 95; high level of diversity, 137; lower jaw of, *90*, 91; relationship to tetrapods, 96
Elpistostege, *59*, 89; relationships, 63, 64
embolomeres, 251; basal articulation, *280*, 281; pleurocentra of, 307; skull, 202; supraneural spines of, 157, 312; tabular horn, 252, 253
embolomerous centrum, 307
embryonic development, 30–31, *30*, 33; braincase, 300–303, *301*, *302*; limbs, 182–185, *185*
end-Permian extinction event, 268–269
Endeiolepis, 88
endochondral bone, 17, 33, 41, 161

endoderm, 31
endolymph, 169
England: Coastal Northumberland, 242; Lancashire, 196, 199; Newsham, Northumberland, 196, 239
entepicondylar foramen, 328
Eocaecilia, 276
Eoherpeton, 251, 252, 282, 321
epipterygoid, 23, 24
Equisetum, 195
Eryops, 100, 280; body reconstruction, 248; forelimb, 181; humerus, 328; pelvic girdle, 321; ribs, 315, 316; skull, 248; vertebrae, 306, 308, 311
Eryops megacephalus, 248
Escuminac Bay, Canada, 50, 63, 87, 98–99; *Eusthenopteron*, 63; *Miguashaia*, 50; salinity conditions, 98–99; World Heritage Center, 87
Escuminaspis, 88
Esox, 63
estivation: arid climate adaptation, 100–101, 260; fossilized cocoon, 54; lungfish, 52, 54, 100, 178
Estonia, 64
Eucritta, 228–229, 230, 231; cladistic interrelationships, 271–274, 274; otic region, 297; pelvic region, 322; possible baphetid, 242, 245
Eucritta melanolimnetes, 228–229, 230, 231
Euramerica, 234
Europe, 81, 191
euryhaline character, 50, 98
eurypterids, 217, 219, 220
Eusthenopteron, 22, 62, 88, 269; braincase, 21–23, 22, 300, 301; Escuminac Bay, Canada, 63, 89; fin ossification, 308; fin skeleton, 180, 181; lateral line system, 166; metapterygial axis, 181; pectoral fin, 43; possible eyelid, 164; relationship to tetrapods, 63, 72; stapes attachment, 112; vertebrae of, 26; vertebral differentiation, 312, 313. See also *Eusthenopteron-Panderichthys-Acanthostega*
Eusthenopteron foordi, 90
Eusthenopteron-Panderichthys-Acanthostega braincase comparison: basal articulation, 151, 152; basisphenoid and basioccipital region, 151; hyomandibula to stapes change, 152–154, 153, 155; nasal capsule, 151–152
Eusthenopteron-Panderichthys-Acanthostega postcranial comparison: fin to limb transition, 159–161, 160; pelvic girdle, 159; ribs, 157; shoulder girdle, 157–159; vertebral column, 156–157
Eusthenopteron-Panderichthys-Acanthostega sensory systems comparison: lateral line system, 166–171; olfaction, 165; other mechanoreceptors, 171; vision, 162–165, 164
Eusthenopteron-Panderichthys-Acanthostega skull comparison: cheek region, 143; cranial hinge, 143–144; juvenile proportions, 144, 145; nostrils, 142–143; notch, 145; operculogular bones, 145–147, 146; ornament, 148, 149; shortening, 141, 142, 154–155; sutural types, 147, 148
evolution, 68
exoccipitals, 247, 270, 271, 280; changes in, 303–305; cladistic analysis, 304
extinction events: end-Permian event, 268–269, 331; Frasnian-Famennian extinction event, 85–86; Hangenberg event, 86
extrascapulars, 28, 147; lack of in tetrapods, 38
eye appearance, 165
eye orbits, 144, 162, 222, 229, 244, 279, 283
eyelids, 164

Famennian period. See Late Devonian period (Famennian)
family (classification term), 11
features (characters), 12, 13
feeding habits, higher diversity among herbivores, 290
feeding strategies: amniotes, 265; early tetrapods, 285; first sign of terrestrial feeding, 287; and jaw structure, 148; selective pressures, 104; suctioning, 283, 284–285; temnospondyls, 248, 249; through gills, 283
femur: *Acanthostega*, 125, 126, 127, 135, 136; adductor blade, 127–128, 130, 327; cladistic analysis of, 326; *Ichthyostega*, 117, 119
fenestra ovalis. See fenestra vestibuli
fenestra rotunda, 299, 300
fenestra vestibuli, 39, 153, 291, 295, 299–300, 303
ferns. See seed ferns; true ferns
fertilization, internal, 179
fin rays, 29, 43, 102, 128; jointed, 103, 160

fin to limb comparison, 159–161, *160*
fin to limb theories, 99–104; little relationship to terrestriality, 103; objections to, 102, 103
fin webbing, 45, 102, 160
Finney, Sarah, 9, *107*
fins, 45; coelacanths, 50; *Eusthenopteron*, *181*; fin web, 45, 102, 160; *Hox* genes and, 184–189, *185*; lobe-fin, 17, *18, 43*; lungfishes, 58, 185; metapterygial axis, *180, 181*, 182; *Sauripteris*, 186
fires, forest, 86, 195, 215
first axial radial, 43
fish. *See Eusthenopteron;* fish-tetrapod transition
fish-tetrapod transition, 35; body scales, 44, *45;* fins, 45; hinge loss by tetrapod, 39; hyomandibula, 39–40; limbs, 42–44; loss of lateral commissure, 39; neck development, 38–39; pelvic girdle, 42; shoulder girdle, 41–42; skull, 36–41, 71; ventilation, 36–37
Fleurantia, 59, 88, 89
flight, 239
Florence, Nova Scotia, Canada, 239, 240
focal condensation, 182
Forbes, New South Wales, Australia, *134*
forbidden morphologies, 187, *188*
fore fins compared to hind fins, 184
forelimbs, 28, 42–44; *Acanthostega*, 124, *125*, 136, 183, 329; amniotes, 265; *Balanerpeton*, 224; *Captorhinus*, 329; *Casineria kiddi*, *198*, 199; compared to hindlimbs, 184, 187; dolphin, *118*, 119; *Eryops*, *181;* fish-tetrapod comparison, 42–44, *43;* *Greererpeton*, 329; hypothesis of similarity, 187; *Ichthyostega*, 115, *117*; *Paleothyris*, 329; *Tulerpeton*, 129–130, *131*, 329. *See also* digits; elbows; humerus; limb evolution; wrists
fossil preparation, 9–11
fossil record: incompleteness of, 4, 210; "Romer's Gap," 196
fossilization, 6–9; bones, 6–7; preserved tissues, 8
fossils, 6–11; articulated specimens, 10–11; causes of distortion, 8–9; matrix, 9; molding, 10; types, 6
Frasnian, 137–138
Frasnian-Famennian extinction event, 85–86

freshwater. *See* aquatic environments
frogfish, *Sarassum*, 103, 136
frogs, 66, 273; occipital condyle, 249; palatal vacuities, 224; reproductive strategies, 101, 179; skull, 275; ventilation, 176; vestibular system, *168, 170*, 294, *295*, 297. *See also* amphibians (modern)
fruit fly. *See Drosophila*
fusain, 195, 215, 239; first evidence of, 86

"Gang of Four," 72
Garnett, Kansas, 239
garpike, 284
gastralia, 44, *45*, 161, 233–234; skin breathing possible through, 172
gastrocentrous vertebrae, *306*, 307
gastrulation, stages of, 29, *30*, 31, 33
Gauss Halvø, Greenland, 107, *108*
gene expression, 34
genera (genus), definition, 11
genes. *See Hox* genes; nuclear gene analysis
Genoa River, Australia, 92, *93, 190*
geological column, 4–6, *5*
Gephyrostegus, 254, *255*, 282
Gerrothorax, 247
Gilboa, New York, 83, 84
gill bars. *See* branchial arches
gill breathing, 174–176; conversion to tongue musculature, 40. *See also* buccal pumping
gill rakers, 283, 298
gill skeleton: *Acanthostega*, *123*, 124; *Panderichthys*, 146
gills: CO_2 excretion, 174; feeding through, 283; nitrogen excretion, 178–179
Gilmerton, Scotland, 196, 205
girdle rotation, 320
glaciation, 79–80, 81–82; during Carboniferous, 193; possible link to Frasnian-Famennian event, 85
GLAHM 100815, 207–209, *208, 209*
GLAHM V2051, 222
glenoid, 29
glenoid fossa, 124
gliding, 268
Glyptolepis, 65; pectoral fin of, *43*. *See also* porolepiforms
gnathostomes, 16, 17
gobie fish, 103
Gondwana, *80*, 82; during Carboniferous, 193; Devonian tetrapods confirmed, 207; formation of Pangaea, 234; glaciation of, 85; glaciation period, 239
Goniorhynchus, pelvic girdle, *321*
Goologongia, 60, 97

Index • 359

"Grace," 8, 10
grasping hands, 199
Greenland, 81. See also East Greenland
Greererpeton, 203, 204, 282; atlas-axis of, 310, 311; braincase, 300, 301; forelimbs, 329; hindlimbs, 329; lateral line system, 166; parasphenoid of, 302; pelvic girdle, 321; skull, 203, 204; stapes, 204, 293
groups (taxa), 12
growth rings, tree, 85, 193
gurnards, 136
gymnosperms, 195

Hamilton, Kansas, 239
hands: Acanthostega, 125, 126; first grasping, 199
Hangenberg extinction event, 86, 96
harvestmen, 217, 219–221, 220
hearing, 290–291. See also stapes; stapes evolution
heart, three-chambered, 175
hellbender, 173
herbivory, 195, 237–238; amniotes, 266–267; among arthropods, 84; gut conditions for, 267; need for static pressure system, 290
herpetology, 69
Hibbertopterus, 219, 220, 233
Hibbertopterus scouleri, 220
hindlimbs, 28, 42–44; Acanthostega, 126–128, 127; amniotes, 265; Balanerpeton, 224; Captorhinus, 329; compared to forelimbs, 184, 187; Eldeceeon rolfei, 227; evolutionary progression, 329; fish-tetrapod comparison, 42–44; Greererpeton, 329; hypothesis of similarity, 187; Ichthyostega, 114, 116, 117, 119, 183; Paleothyris, 329; Tulerpeton, 131, 329. See also ankles; femur; knees; limb evolution
Hitchin, Becky, 107
Holoptychius, 65, 89, 110
holospondylous vertebrae, 307
homeobox genes, 33
Homo sapiens, 6
homologous characters, 13
horsetails, 84, 85, 88, 195, 236
Horton Bluff, Nova Scotia, Canada, 197
Hox genes, 32–34, 32; limb formation, 184–187; stabilization, 188; vertebral patterning, 317
humerus, 43–44; Acanthostega, 124, 125; cladistic analysis, 325; Crassigyrinus, 328; entepicon-dylar foramen, 328; Eryops, 328; Ichthyostega, 115, 116; shape of, 327; tetrahedral, 328; Tournaisian tetrapod, 208–209; Tulerpeton, 129–130
Huxley, T. H., 69
Hylonomus, 229, 265, 266, 268
Hyner, Pennsylvania, 133
Hynerpeton bassetti, 131–133
hyobranchial skeleton, 25, 37, 40
hyoid arch, 25
hyomandibula, 39–40, 145, 152–154, 153, 155; evolutionary progression, 291–303; function in fishes, 40. See also stapes; stapes evolution
hyostylic skulls, 40
hypothesis of similarity, 187

Iapetus Ocean, 80, 193
Ichthyostega, 111–119, 269; ankle joints, 330; braincase, 113; elbows, 115, 329–330; hindlimbs, 183; jaws, 113, 284; limbs, 115, 116, 117, 119; otic notch, 112–113, 282; palate, 172; reconstruction, 107, 114, 120; ribs, 113–115, 114, 315, 316; sediments where found, 110, 111; skull, 111, 112, 113, 114; stapes, 112; tail, 115–116; teeth, 113; vertebrae, 115, 306; wrists, 115
Ichthyostega stensioei, 111
ichthyostegalians, 111
ilium, 42, 247, 322
Indonesian, 49
inner ear, 168, 169, 170, 294, 295
insects: buzzing, 300; flight, 239; Late Carboniferous, 238, 239; none found at East Kirkton, 221; winged, 195, 221, 235
interclavicle, 27, 319; Acanthostega, 157; Dendrerpeton, 247; Silvanerpeton, 225; Tulerpeton, 131
interdigitating joints, 148
intermedium, 125
interpterygoid vacuities, 297
intertemporal, 251; lost by amniotes, 255; lost in nectrideans, 262; Loxomma, 242; seymouriamorphs, 255
intracranial joint. See cranial hinge
invagination process, 29, 30
Ireland, Valentia Slate in, 93
ischiadic region, 42, 322

Jacobson's organ, 165
jaw closure, 287, 288

jaw evolution, parallels limb evolution, 327
jawed vertebrates. *See* gnathostomes
jawless fishes. *See* anaspids; osteostrachans
jaws, of *Panderichthys*, 285–286
jaws, lower, 22, 24; *Acanthostega*, 7, 8, 284; character of early tetrapods, 147, 284, 286; cladistic analysis, 284; *Densignathus*, 132; Famennian vs. Frasnian tetrapods, 137–138; *Hynerpeton*, 132–133; *Ichthyostega*, 113, 284; *Livoniana*, 64, 65; *Metaxygnathus*, 132, 133–134; *Obruchevichthys*, 91; *Panderichthys*, 284; temnospondyls, 284, 286; *Ventastega*, 130
Joggins, Nova Scotia, Canada, 229, 235–236, 237, 240
Jørgenson, Birger, 110
juvenile characters. *See* pedomorphic processes
juvenile to adult skull transformations, 144, 145

Kejser Franz Joseph, Greenland, 108
Keraterpeton, 262
keratin, 173
kidney, urea processing, 178–179
kinetic inertial jaw mechanism, 288
kinetic line, 227
knees, 44, 330; beginning in *Ectosteorhachis*, 329; type of maneuverability, 330. *See also* tibia-fibula articulation

labyrinthodont ear, 292–293
labyrinthodonts, 257–258, 269; phylogenetic interrelationships, 269–270
lacrimal glands, 163
lagenar pouch, 168, 168
Late Carboniferous: arthropods, 237–238, 239; atmospheric conditions, 239; diversity explosion, 234, 247–248; plants, 235–237, 235, 236; tetrapod localities, 239–240
Late Devonian period (Famennian), 104; first evidence of fusain, 86; flora and fauna, 86–89; Frasnian-Famennian extinction event, 85–86; Hangenberg extinction event, 86; isolated as tetrapod origin, 96; tetrapod origin, 109
lateral commissure, 39, 153
lateral line grooves and pineal foramen, 142

lateral line system, 44, 166, 166–167; *Acanthostega*, 121; *Eusthenopteron-Panderichthys-Acanthostega* comparison, 166–171; none for *Balanerpeton*, 225; rhizodonts, 60
lateral otic fissure, 22, 23, 300
lateral sequence walk, 189
Latimeria, 47, 49–50, 51, 77; inner ear, 169, 170; nostrils, 49; pectoral fin of, 43. *See also* coelacanths
Latimeria chalumnae, 49–50, 51
Latvia, 64, 91; marginal marine locality, 99; Venta River region, 130
Laugia groenlandica, 53
Laurussia, 80; during Carboniferous, 193, 195; probable tetrapod origin, 96–97
lepidodendroids, 86, 235–236
Lepidodendron sp., 235
Lepidosiren, 52, 58; distribution map, 53; estivation ability, 100; nostrils, 69. *See also* lungfish
lepidotrichia, 29, 43; neural crest origin in fishes, 160
Lepisosteus, 175
lepospondyls, 258, 269, 277; cladistic analysis, 271, 273; holospondylous vertebrae, 307; notchlessness, 283; skull modifications, 279
Lethiscus, 199, 200; skull, 200
Lethiscus stocki, 200
limb evolution: parallels jaw evolution, 327; progression, 329; selection pressures, 99–104, 136–137
limb formation: embryonic development, 31, 182–185, 185; Hox genes, 184–187
limbless tetrapods: *Adelogyrinus*, 200, 201; aïstopods, 260; *Lethiscus*, 199, 200
limbs. *See* forelimbs; hindlimbs; limb evolution
Linnean classification, 11, 12
Linton, Ohio, 10, 211, 239, 240
Lissamphibia, 179, 249. *See also* amphibians (modern)
Livingston Development Corporation, 227
Livoniana, 59, 64, 65; geographical location, 97; high level of diversity, 137
Livoniana multidentata, 64, 65
lizards: accommodation in, 163; brachial plexus, 318
Loanhead, Scotland, 196

lobe-finned fishes, 46–48, *47;* dipnomorphs, 65–66; fin support structure, 17, *18;* maneuverable fins, 17; pectoral fin of, *43;* phylogeny, *15,* 48, 59; saltwater tolerance, 98; tetrapodomorphs, 59–64, 67; ventilation, 37
locomotion, 42, *189–190;* girdle rotation, *320*
Lode, Latvia, 63
Low Main Seam, Northumberland, England, 196
Loxomma, 196, 242
lungfish, 50, 52–58, 77, *88;* air dependency, 52, 58; circulatory system, 174, *175;* considered as tetrapod ancestor, 58, 72–77; cranial rib of, *57,* 58; estivation, 52, *54,* 100, 178; fins, 58, 184; inner ear, *168,* 169, 295; nitrogen excretion, 178; no visual accommodation, 162; nostrils of, 55; reproduction strategy, 179; skull, 52–58, *55, 56, 57;* soft tissue fossils, 73; tooth plates of, *54, 55, 57;* ventilation, 58, 176. See also *Lepidosiren; Neoceratodus; Protopterus*
lungs, 145; book lung, 218; and CO_2 excretion, 174; swim bladder, 49; temperature influence on oxygen uptake, 175–176
lycopods, 217
lycopsids, 83, 84, *195, 235*
lysorophids, 260, *261*

Macropomoides orientalis, 52
mammals, 2, 66; circulatory system, 174, *175;* muscle spindles, 171; vertebrae, 312; vestibular system, *168;* visual accommodation, 163
Mandageria, 63
manus, 199
marginal tooth-bearing bones, 28, *54,* 285, 286
marine water. See aquatic environments
Mastondonsaurus, 247
maxillary bones, 50, *54, 55,* 279
Mazon Creek, Illinois, 240
mechanoreceptors: *Eusthenopteron-Panderichthys-Acanthostega* comparison, 171
Meckelian bone, 24, 147, 285
Megalichthys, 61
Megalocephalus, 242; lower jaw, *284;* otic region, 297; skull, *243*
Megarachna, 238
mesoderm, 31; cephalic, 301, *302;* somitic, 301, *302*

mesosaurs, 240, *241,* 268
Mesosaurus, 268
Mesozoic era, 4–5
metapterygial axis, 43, *180, 181,* 182
Metaxygnathus, 132, 133–134; dating of, *109*
Metoposaurus, 247
MGUH 6033, *120*
MGUH f.n. 1227, *121*
MGUH f.n. 1258, *123*
MGUH f.n. 1300, *8*
MGUH f.n. 1349, *116*
MGUH VP 6115, *117*
MGUH VP 6158, *114*
Microbrachis, 259
microbrachomorphs, 258, *259*
Micromelerpeton credneri, 251
microsaurs, 258–260; atlas-axis vertebrae, *310,* 311; crushing teeth, 287; East Kirkton individual, 232–233; possible ancestors of caecilians and urodeles, 276; sacral region, 324
Middle Devonian, no osteolepiforms before, 96
middle ear: formation, 288; fossil record of, *290*
Miguashaia, 50, 89
Milankovitch cycles, *81,* 82; coal layers, 194
millipedes. See myriapods
Mississippian (Lower Carboniferous), 191
mitochondrial gene analysis, 75
molecular data analysis, 74–75
Morganucodon, 290
morphogens, 34, 184
mouse, *Hox* genes, *32*
mudskippers, 103
mummification, 9
muscle spindles, 171
muscles, germ layer origin of, 31
Myakka State Park, Florida, *236*
myriapods, 217, 219–221, *220;* size of, 237

nasal capsule: *Eusthenopteron-Panderichthys-Acanthostega,* 151–152; ossification of, 151–152
National Geographic Society, *106*
natural group, 76
Natural History Museum, London, England, 69
natural selection, 11
neck development, 38–39, 303–305, 311; *Acanthostega,* 126; *Eusthenopteron-Panderichthys-Acanthostega,* 147, *148;* resulting skull shortness, 154

nectrideans, *261–263*
Neininger, Sally, *107*
Neoceratodus, 43; body form of, *59;* breeding program, 185; distribution, 50, *53;* fins, 160; gill skeleton compared to *Acanthostega,* 124; pectoral fin of, *43*
nerves: brachial plexus, 317, *318;* cranial nerve X, 24
neural arches, 26, 157, 312; nectrideans, 261, 262. *See also* vertebrae
neural crest cells, *30,* 31; form anterior braincase, 301, *302;* form lepidotrichia, 160; form otic capsule, 154, 301
neural spines: elongated in amniotes, 268; seymouriamorphs, 256
neurulation, *30,* 31
New Synthesis, 76–77
NEWHM:2000.H845, *243*
Newsham, Northumberland, United Kingdom, 239
newts, 66, 249, 273
nitrogen excretion, 178–179
NMS 1992.14, *231*
NMS G1990.72.1, *231*
NMS G1992.21, *220*
NMS G1986.39.1p, 228
NMS.G1993.54.1, *198*
node-based analysis, 67, 68
nomenclature, 11. *See also* classification
North America, 81, 191
nostrils (external): *Acanthostega,* 121, 165; *Crassigyrinus scoticus,* 205, *206; Eusthenopteron-Panderichthys-Acanthostega,* 142–143; *Latimeria,* 49; modifications, 279–281; *Panderichthys,* 165
nostrils (internal), 24; *Eusthenopteron-Panderichthys-Acanthostega,* 143; lungfish, 55
notchlessness, 283
notochord, 26, 31, 39, 42, 303
nuclear gene analysis, 75

Obruchevichthys, 91, *92;* geographical location, 97; high level of diversity, 137; relationship to tetrapods, 96
occipital arch. *See* basioccipital; exoccipitals; supraoccipital
occipital condyle, 304–305; amniotes, 265; *Dendrerpeton,* 247; frogs, 249; *Greererpeton,* 302; microsaur, 259–260
occipital construction, 270, 280; cladistic analysis of, 270, *271, 304. See also* basioccipital; exoccipitals; occipital condyle; otic capsule
odontoid, 311
Oestocephalus, 262
Old Red Sandstone Continent, 107
olecranon process, 115, 125, 199, 330
olfaction, 165
olfactory bulb, accessory, 165
operculogular series, 21, 28; *Eusthenopteron-Panderichthys-Acanthostega,* 145–147, *146;* loss of in *Acanthostega,* 135–136; reduction of, 28, 36–37, 303
Ophiacodon, 289
Ophiderpeton, 232
opilionids. *See* harvestmen
opisthotics, 270
opossum, *290*
orbit. *See* eye orbits
order (classification term), 11
Origin of the Species (Darwin), 69
ornament, 269; *Acanthostega,* 122, 149; *Eusthenopteron-Panderichthys-Acanthostega,* 148, *149;* lacking on tetrapod cleithrum, 158
ossification: *Eusthenopteron,* 308; nasal capsule, 151–152; pedomorphosis, 151–152; pelvic girdle, 320, 322; vertebrae, 308, 309
osteichthyans, 17; represented at Escuminac Bay, 89
osteolepiforms, *61, 88;* as hypothetical tetrapod ancestor, 77; inner ear, 168; jaw changes, 147, 286. *See also* tristichopterids
Osteolepis, 61, 62; skull of, *71*
Osteolepis macrolepidotus, 62
osteostrachans, 87, *88*
ostracods, 220–221
otic capsule, 22, 23, 39, 270, *271,* 280; changes precede those of inner ear, 169; *Eusthenopteron-Panderichthys-Acanthostega* comparison, 151–154
otic notch. *See* temporal notch
otolith, 168
otolithic organs, *168*
ovoviviparous, 49, 179
Owen, Richard, 69
oxygen metabolism, 173–174

Pachygenelus, 290
Palaeogyrinus, skull of, *71*
palatal teeth, *122,* 286
palatal vacuities: *Balanerpeton,* 222, 223–224, 247; convergent in nectrideans, 262; *Dendrerpeton,* 247

palate: *Acanthostega*, 121; *Balanerpeton*, 222, 223–224; *Crassigyrinus scoticus*, 205, 206, 207; fish, 40; *Ichthyostega*, 121, 172; seymouriamorphs, 256, *256*; *Silvanerpeton miripedes*, 227
Paleoherpeton, 293, 294
Paleothyris, 289; forelimbs, *329*; hindlimbs, *329*; sacral region, 324; skull shape, 287
Paleozoic Era, 4–5
Panderichthys, 59, 62, 64, 89; cheek region, 143; geographical location, 97; jaws, *284*, 285–286; nostrils, 165; pectoral fin of, *43*; relationship to tetrapods, 76, 96; tail fin, 160. See also *Eusthenopteron-Panderichthys-Acanthostega*
Panderichthys rhombolepis, 11, 64
Pangaea, *80*, 82; formation of, 234, 240
Pantylus, 259, 306
paralogous gene groups, *32*, 33
paraphyletic groups, 73
parareptiles, 264
parasphenoid, of *Greererpeton*, 302
parasymphysial teeth, 66, 147
pareiasaurs, 264, 268
parietal foramen. See pineal foramen
paroccipital process, 288, 305
Parotosuchus aliciae, 297–298
PAUP analysis, 232
pectoral fin skeleton, *43*
pectoral girdle. See shoulder girdle
pedomorphic processes, 136, 151–152; *Greererpeton*, 203; juvenile to adult skull transformations, *144*, *145*; in lateral line system, 167; otic region, 153–154; pelvic girdle, 320, 322; temnospondyls, 249
Peltobatrachus, 308
pelvic girdle, 25, 27, 28, 29, 42; *Acanthostega*, *126*, 135, 136, *321*; cladistic analysis, *321*; *Eusthenopteron-Panderichthys-Acanthostega*, 159; *Ichthyostega*, 116
Pennsylvanian (Upper Carboniferous), 191
pentadactyly, 182; evolution, 102, 188–189. See also polydactyly
perilymphatic space, 169, *170*
Periophthalmus, 103
Permian period, *5*; end-Permian extinction event, 268–269, 331
Pertica, 83, 84
Pertica quadrifaria, *84*
Petrolacosaurus, 265

Phanerozoic interval, 4
pheromone detection, 165
Pholiderpeton, 253, *253*, 280, 281; anocleithrum, 159; stapes, *293*
phylogenetic classification: analysis methods, 67–69; character chart, *13*; cladistics, 12; terms for, 12, 13
pineal foramen, 142
pineal gland, 169, 171
placoderms: at Escuminac Bay, 87–89, *88*; sediments where found, 110; trackways, *93*, 95
plants: Carboniferous, 193–194; deciduous forms, 85; Early-Middle Devonian, 83–86; Silurian, 82
Platyrhinops lyelli, 250, 275
Pleurdosteus, *88*, 89
polydactyly, 129, 182, 186–187. See also pentadactyly
Polypterus, 145, 175
Polypterus bichir, *44*
population pressure, 100, 101
porolepiforms, 77. See also *Glyptolepis*
postaxial radials, 44
postbranchial lamina, 37, 41, *123*, *124*; *Tulerpeton*, 130
postcranial skeleton, 25, 28
postiliac process, 322
Powichthys, 65
preaxial radials, 43
predation pressures, 101, 104, 210
preopercular bone, 147
pressure relief window (PRW), 299, 300
primitive characters, 12, 13
Prince of Wales Island, Canada, 65
proatlases, 311
procolophonids, 264
progenetic dwarfing, 211
proprioceptors, 171
Proterogyrinus, 253; vertebrae, *306*, *312*, *313*
Protopterus, 52; body form, 58, *59*; distribution map, *53*; estivation ability, 100; skull of, *56*, *57*. See also lungfish
prototetrapods, evolutionary pressures on, 99–104, 136–137
Psarolepis, 76
Psaronius sp., *235*
Pseudobornia ursina, *85*
Psilophyton crennulatum, *84*
pteridosperms, 86, 217, 236
pterygoid flange, 265, *267*, 286, *288*
pterygoideus muscle, *288*
pubic region, 42, 322
Pulmonoscorpius kirktonensis, 218, 219

quadrate, 22, 23, 24, 288; lungfish, 55, 56
Queensland, Australia, 207

radial fins, 17, 18, 45, 186
radius-ulna articulation, 44. *See also* elbows; wrists
ray-finned fishes: dominance at end of Paleozoic, 47; fin support structure, 17, 18; lungs, 175–176; no visual accommodation, 162; reproduction strategy, 179; ventilation, 37
red-bed sediments, 99–101; associated conditions, 101
Red Hill, Hyner, Pennsylvania, 133
Reekie, Bob, 90
reproduction: egg size and O_2 conservation, 173; parental care of young, 267; selective pressures, 101; strategies, 179–180; use of limbs, 136
reptiles, 2, 66, 69; cutaneous gas exchange, 173; definition, 263; scales of, 45
reptiliomorphs, 270, 271, 272; vertebra, 307
respiration, 172–178, 175, 177
retinoic acid, 184
retractor bulbi, 164
Rhabdoderma, 51, 52, 98
Rhabdoderma elegans, 51, 52
rhachitomous centra, 306, 307
Rhacophyton, 85, 86, 88, 195
Rhacophyton ceratangium, 85
rhipidistians, 65; redefined, 72–73; relationship to tetrapods, 72
rhizodonts, 59; skull, 60
Rhynie Chert, Scotland, 83
rib flanges, 314
rib walking, 260
ribs, 28, 313–317; *Acanthostega*, 126, 306; aïstopods, 260; amniotes, 176, 317; cladistic analysis, 315; *Eusthenopteron-Panderichthys-Acanthostega*, 157; *Ichthyostega*, 113, 114, 115, 306, 315, 316; K-shaped, 199, 260; Tournaisian tetrapod, 209; *Whatcheeria*, 202
Ricnodon, pelvic girdle of, 321
RNA analysis, 74
Robinson, Kansas, 239
rods and cones, 163
Romer, A. S., 100, 196, 269
"Romer's Gap," 196
Royal Museum of Scotland, 227, 229
Royal Society of Edinburgh, 214
Russia, 63, 91, 99, 129, 138

sacculus pouch, 168, 169

sacral rib, 322, 323, 324
sacrum, 322–327; cladistic analysis, 323
Sagenodus, 57
salamanders, 249; adductor blade of, 128; ancestry, 76–77; *Andrias*, 204; circulatory system, 174, 175; occipital condyle, 249; palatal vacuities, 224; pineal gland, 169; ventilation, 176, 177
salinity, water: indicator species of, 129
salt gland, 244
Sarassum frogfish, 103, 136
sarcopterygians, 16, 18, 43. *See also* lobe-finned fishes
Sauripteris, 60, 72
Sawdonia ornata, 84
Saxonerpeton, 259
scales, body. *See* body scales
scapula blade, 158
scapulocoracoid, 27, 41, 319; enlargement in *Panderichthys*, 158
scarf joints, 148
Scat Craig, Scotland, 90, 91
Scaumenacia, 88, 89
sclerotic ring, 163, 165
scorpions, 217, 218, 219
Scotland: Auchenreoch Glen near Dumbarton, 208; Bathgate outside Edinburgh, 212, 213, 214; Borough Lee, 196; Cheese Bay, Edinburgh, 197; Cowdenbeath, 196, 205; Gilmerton near Edinburgh, 196, 205; Loanhead, 196; Midland Valley, 211, 213, 240, 242; Rhynie Chert, 83; Scat Craig, 90, 91; Tarbat Ness, 92; Wardie Shales, Edinburgh, 199
scutes. *See* gastralia
seed ferns, 88, 217
segmentation, 182
selective pressures, 99–104
sensory systems, 23; hearing, 167–169, 290–291, 291 (*see also* stapes); lateral line system, 166–167, 166; mechanoreceptors, 171; olfaction, 165; pineal organ, 169–171; vision, 162–165, 164 (*see also* eye orbits)
septomaxilla, 279, 280–281
serpulid worms, 129
Seymour, Texas, 255
Seymouria, 280, 281, 324; humerus, 328; pelvic girdle, 321; vertebrae, 306
Seymouria baylorensis, 255, 256; skeletal model, 257
seymouriamorphs, 255–257, 256

shagreen, 224, 246, 285
shoulder girdle, 27, 28, 29, 41–42,
319, 320; *Acanthostega*, 27, 123,
124, 146, 157, 319; adelogyrinids, 201; *Densignathus*, 133;
Eusthenopteron, 27, 146;
*Eusthenopteron-Panderichthys-
Acanthostega* comparison, 157–
159; fish-tetrapod transition, 41–
42; *Hynerpeton*, 132, 133;
Ichthyostega, 115, 117; *Panderichthys*, 146; of rhizodonts,
60; shift in functions, 159;
Tulerpeton, 130; *Ventastega*, 131
sight. *See* eye orbits; vision
Sigillaria sp., 235
Silurian period, living organisms of, 82
Silvanerpeton, 225–227, 226, 227,
233, 322; notch, 283; sacral rib,
324
Silvanerpeton miripedes, 225–227,
226, 233
simple hinge joints, 137
size: conservatism of Famennian
tetrapods, 137–138; decrease in
amniotes, 199; increase of
rhizodontids, 60; increase of
tristichopterids, 63; not an
indication of maturity, 308
skin: early tetrapod, 173; preservation,
9
skin breathing. *See* cutaneous gas
exchange
skull, 279–281; *Acanthostega*, 7, 36,
120, 121, 122, 123, 287;
adelogyrinids, 201; amniotes,
264, 265, 266; *Amphibamus*,
275; *Baphetes*, 71; *Captorhinus*,
267, 281; coelacanths, 50;
Crassigyrinus scoticus, 205, 206;
Dendrerpeton, 246; diapsid, 264,
289; embolomeres, 202; *Eryops*,
248; *Eucritta melanolimnetes*,
231; *Eusthenopteron*, 36; fish-
tetrapod comparison, 36–41, 71;
frog, 275; *Greererpeton*, 203,
204; *Ichthyostega*, 111, 112, 113,
114; *Lethiscus*, 200; lungfish, 52–
58, 55, 56, 57; *Megalocephalus*,
243; microsaur, 259; *Oestocephalus*, 262; *Osteolepis*, 71;
Palaeogyrinus, 71; *Paleothyris*,
287; *Protopterus*, 56, 57;
Rhabdoderma, 51; rhizodonts,
60; seymouriamorphs, 256;
Silvanerpeton miripedes, 226,
227; *Spathicephalus*, 243;
synapsids, 264, 289; temnospondyls, 247; *Ventastega*, 130–
131, 132; *Whatcheeria*, 202. *See
also Eusthenopteron-
Panderichthys-Acanthostega* skull
comparison
skull architecture: autostylic, 40, 58;
hyostylic, 40; juvenile to adult
transformation, 144, 145;
landmarks for comparison, 142,
300, 301; shortening, 40, 143,
147, 154–155, 279, 281, 288;
snout lengthening, 40, 141, 142,
154–155; tabular-parietal
contact, 226–227, 251–252, 262;
tetrapod characters, 269
skull table, 28
smell, sense of, 165
Smithwoodward Bjerg, Greenland,
108
snout elongation, 28, 40–41, 141,
142, 144, 154–155; *Acan-
thostega*, 121, 122
Soederbergia, 134
soft tissue analysis, 75–77
soft tissue fossilization, 8, 9
Solnhofen Limestone, 180
somites, 30, 31
somitic mesoderm, 302, 303
sonic hedgehog gene, 184, 187
sour tastes, detection of, 165
South America, 240, 241
Spathicephalus, 243–245; otic region,
297; skull, 243
Spathulopteris obovata, 217
species, definition of, 11
sphenophylls, 88
sphenopsids, 85, 88, 236
Sphenopteridium crassum, 217
spiders, 238
spiracular cleft, 145
spiracular notch, 21, 282–283. *See
also* temporal notch
spiracular pouch, 145
stabilization and variation, 188
Stanwoodia, 216
stapes: *Acanthostega*, 293, 295;
amniote, 265, 288; *Balanerpeton*,
224, 297; cladistic analysis of,
296; *Dendrerpeton*, 246;
Greererpeton, 204, 293;
Ichthyostega, 112; *Paleoherpeton*, 294; *Pholiderpeton*,
293; *Seymouria*, 292; seymouriamorphs, 256, 257, 292;
Tournaisian tetrapod, 208
stapes evolution, 39–40, 152–154,
153, 155, 291–303; cladistic
analysis, 296, 298, 299
static pressure jaw mechanism, 287,
288
static pressure system: herbivory, need
for, 290

statolith, 168
stem lineage of Tetrapoda, 65, 67, 68
Stensiö Bjerg, Greenland, 81, 82, *108*, 110
stereospondylous centrum, 307
stretch receptors, 171
stride movement, *320*
stromatolite, 129
strontium ratio analysis, 98
subtemporal fossa, 23
supracleithrals, 38, 41, 147
supraneural spines, 157, 160
supraoccipital, 265
suspensorium, 21, 23
sutural types, 148; *Eusthenopteron-Panderichthys-Acanthostega*, *147*, 148
swim bladder, 49–50, 176
swimming motion, 189
synapsid lineage, 265
synapsids, 264, 267; advanced skull, *289;* skull, *264, 289*

tabular horn, 120, 148, 252, 253
tabular-parietal contact, 226–227, 251, 252, 255, 262
tail: *Acanthostega*, 128, 136; *Eusthenopteron*, 63; *Ichthyostega*, 115–116; movements, 50, 189, 322; postiliac process, 322
Tarbat Ness, Scotland, 92
taxa (classification term), 12
teeth: *Acanthostega*, 7, 10, 122, *123*, 124; adelogyrinids, 201; *Balanerpeton*, 224; characters of early tetrapods, 147, 285; coelacanths, 50; *Crassigyrinus*, 205; crushing, 287; fangs, 130, 285, 286; *Ichthyostega*, 113; labyrinthodonts, 269; *Livoniana*, 64, *65;* lungfish, 54, 55, 57; onychodont, 65, 66; rhizodont, 60; *Ventastega*, 130
teleost fish, *175*
temnospondyls: cladistic analysis of, *274;* closed palate of, 223–224; diversity explosion, 247–248; ear region, 292; formerly within labyrinthodonts, 257; jaws, 284, 286; notch, *281, 283;* occipital formation, 304–305; palatal vacuities, *222*, 223–224, 247; relationship to amphibians, *277;* relationship to amphibians (modern), 249, *276;* ribs, *315*, 316; sacral region, 324; skull outline comparison, *247;* ventilation, 316. *See also* *Balanerpeton; Dendrerpeton*
temporal fenestra, 264, 265

temporal notch, 145, 292, 297; *Balanerpeton*, 224; cladistic analysis, *282, 283; Crassigyrinus scoticus*, 205; *Eusthenopteron-Panderichthys-Acanthostega*, 145; *Ichthyostega*, 112–113; lacking in amniotes, 265; lacking in anthracosaurs, 251; lacking in *Dvinosaurus, 281;* lacking in *Eoherpeton*, 251; lacking in *Greererpeton*, 204; seymouriamorphs, *256, 257;* through tetrapod history, *282, 283*
Terrestrial Ecosystems through Time, 79
terrestriality: adaptations of sacral region, 324, 327; effects of continental movement on, 80–81; little relationship to limb development, 102–103; plant activity and implications of, 85–86; selective pressures, 99–104, 136–137; vertebrae character, 305, 308, *320*, 329–330
tetrapodomorphs, 59–64, 65, 67
tetrapods, 66–77; body fossils of, 91, 95; diversity explosion, 209–211; early classification, 269–270; first evidence of, 81, 89, 96; geographic origins, 96–97; phylogenetic analysis, 15–19, 67–69, 72–77, 270–277; selective pressures, 99–104, 136–137; trace fossils of, 92–95, *94;* unique characters, 66, 68. *See also Acanthostega*; amphibians (fossil); tetrapodomorphs
tetrapods (modern), 2, 66
Thrinaxodon, 290
tibia-fibula articulation, 44. *See also* ankles; knees
tongue-and-groove joints, 148
tongue musculature, 40
tooth enamel, of rhizodonts, 60
tooth plates, of lungfish, 54, 55, 57
total group analysis, 68
Tournaisian stage: forest cover disappears, 195; unidentified tetrapod fragments, 197
Tournaisian tetrapod (articulated), 202, 207–209, *208, 209;* notch, 283; vertebrae, 306
Tournaisian tetrapod (fragments), 197
trace fossils, 6, 89
trackways, 92–95, *92, 94, 95;* arthropod, 237; interpretation, 96, *190*
transcription factors, 34
transverse abdominal muscle, 176, *177*, 317

trematosaurs, 247
Trematosaurus, 247
trigonotarbids, 82, *83*
trimerophytes, 83, *84*
triploblastic animals, 31
tristichopterids, 59, 60, *61,* 63. See also *Eusthenopteron*
true ferns, *88,* 217, 236
tuditanomorphs, 258, *259*
Tuditanus, 259
tuff, 212, 214, *215*
Tula, Russia, 99, 129
Tulerpeton, 129–130, *131;* forelimbs, 129–130, *131, 329;* hindlimbs, *329;* shoulder girdle, 158; stem lineage based on digits, 188
Tulerpeton curtum, 131
turtles, 173
tympanic ears, 292; first in temnospondyls, 297; frog and amniote compared, 294, *295*
tympanum, *291*
Tyrannosaurus rex, 60

UCLA VP2802, *261*
UMZC GN 243, *53*
UMZC GN 766, *62*
UMZC GN 790, *90*
UMZC T250, *248*
UMZC T955, *251*
UMZC T970, *251*
UMZC T1300, *123*
UMZC T1313, *223*
UMZC T1317, *226*
UMZC T1342, *241*
UMZC T1347, *230*
underclay, 194
Undina, 180
United States: Antler Highlands, Nevada, 204; Catskill Basin, 99; Gilboa, New York, 83, 84; Hyner, Pennsylvania, *133;* Linton, Ohio, 10, 211, 239, *240;* Mazon Creek, Illinois, 237, *240;* Red Hill, Hyner, Pennsylvania, *133;* Robinson, Kansas, 239; Texas, *255;* West Virginia, 203; What Cheer, Iowa, 202
University of Cambridge (UK), *9*
urea processing, 178–179
urodeles, 249, 273. See also salamanders
utriculus pouch, 168

Valentia Slate, Ireland: trackways from, *92, 93*–95, *94, 190*
variation and stabilization, 188
Variscan orogeny, 193
Ventastega, 130–131, *132;* lower jaw, 284
Ventastega curonica, 130–131

ventilation: buccal pumping mechanism, 36–37; costal ventilation, 283; evolution of, 176–178, *177, 283,* 319; mixed-air buccal pumping, 176, *177;* role of ribs, 316–317
ventral cranial fissure, 300
vertebrae: *Acanthostega,* 126, 306, 312, 313; adelogyrinids, 201; amniotes, 266, 268; *Caerorhachis bairdi,* 246; *Crassigyrinus,* 207, 307–308; *Eryops, 306;* gastrocentrous, 307; *Ichthyostega,* 115, 306; nectrideans, 261–262, *262;* ossification of, 308, *309;* *Proterogyrinus,* 253; rhachitomous, 305–306, 307; seymouriamorphs, *256;* terms, 26; Tournaisian tetrapod, 306; *Whatcheeria,* 306. See also centra; vertebral column
vertebral column: *Acanthostega,* 126; *Casineria kiddi,* 197, *198;* *Eusthenopteron-Panderichthys-Acanthostega* comparison, 156–157; regional differentiation of, 26, 311–312
vertebrates, *15;* broad phylogeny of, *15, 16;* neural crest cell behavior, 31, 33
vestibular fontanelle, 23, 153
Victoria, Australia, 95
Virunga National Park, Congo Republic, 216
Viséan stage, 191, *192,* 195; grasping hand, 199
vision: binocular, 279; *Eusthenopteron-Panderichthys-Acanthostega,* 162–165, *164*
vulcanism, effects of, 214–215

walking, 101, 102; girdle rotation, 320. See also trackways
walking trot, 189
Wallace, A. R., 11
Wardie Shales, Edinburgh, Scotland, 199
water habitats. See aquatic environments
Watson, D. M. S., 49, 275
weed clinging, 103, 136
West Lothian District Council, 229
West Virginia, United States, 203
Westlothiana, 308
Westlothiana lizziae, 229, *231,* 232, *232*
What Cheer, Iowa, 202
Whatcheeria, 147, 202; cladistic interrelationships, 272, *274;* notch, 283

Whatcheeria deltae, 202, 203
Wimans Bjerg, Greenland, 107, *108*
Wood, Stan, 212, 225, 229
World Heritage Center, Escuminac Bay, Canada, 87
wrists, 44, 330; *Acanthostega,* 125, 137, 329; *Ichthyostega,* 115; type of maneuverability, 330

Xenopus, 188

Yellowstone National Park, 216
Ymer Ø, *106*
Youngolepis, 65
YPM 794, *250*

zone of polarizing activity (ZPA), 184, 187
zosterophylls, 83, *84*
zygapophyses, 26, 42, *312*

JENNIFER A. CLACK is Reader in Vertebrate Palaeontology and Senior Assistant Curator, University Museum of Zoology, Cambridge, and author of numerous papers on Devonian and Carboniferous life. A shorter version of *Gaining Ground* was published in Japanese in 2000.